DOMINANCE BY DESIGN

DOMINANCE
BY DESIGN

Technological Imperatives and
America's Civilizing Mission

MICHAEL ADAS

The Belknap Press of
Harvard University Press
Cambridge, Massachusetts, and London, England
2006

Printed in the United States of America

Library of Congress Cataloging-in-Publication Data

Adas, Michael, 1943–
Dominance by design : technological imperatives and
America's civilizing mission / Michael Adas.
p. cm.
Includes bibliographical references and index.
ISBN 0-674-01867-2 (alk. paper)
1. United States—Foreign relations—20th century.
2. United States—Territorial expansion.
3. Technology—United States—History—20th century.
4. Technological innovations—United States—History.
5. Messianism, Political—United States.   I. Title.
E744.A243 2005
327.73′009′04—dc22   2005048240

For my grandfather, Lawrence B. Rivard

engineer and inventor

and

my grandson, Malcolm Shaw Wilson

who always wants to know the purpose of things

CONTENTS

DOMINANCE BY DESIGN

A woodblock impression by Hiroshige, one of Japan's most popular nineteenth-century artists, of the black-hulled steamships and miniature train that were the centerpieces of Admiral Matthew Perry's expedition to Japan in 1843–1844. Library of Congress.

The Japanese are remarkable for their inordinate curiosity and, in the display of so many of the inventions of our ingenious countrymen, they had ample means of gratifying this propensity.

—Matthew C. Perry, 1854

INTRODUCTION: A TRAIN FOR THE SHOGUN

Commodore Matthew Calbraith Perry had overseen the preparations for the naval expedition with an attention to detail that would be well appreciated by his reluctant Japanese hosts. He had read the fifty or sixty books that were available on the little-known islands and reclusive people that many influential Americans had become convinced must be drawn—preferably by persuasion, by force if necessary—into the rapidly expanding global system. While his agents gathered reading material in the

bookstores of New York and London in the early 1850s, Perry interviewed merchants and studied reports by American and European
sea captains who had preceded him on missions to the islands. Incensed by the curt rebuffs and humiliations these earlier envoys had
endured, Perry was determined to put together an embassy so imposing that the Japanese ruler would not dare refuse the summons he
carried in a letter from the president of the United States, Millard
Fillmore, which exhorted the shogun to act vigorously to open his
kingdom to international commercial exchange and political intercourse.

An accomplished diplomat, who had successfully negotiated with
leaders from states as diverse as Naples, Liberia, and Mexico, Perry
was an obvious choice to head the expedition to Japan. He decided
long before he set out from Norfolk, Virginia, in late November
1852, to present himself as a somewhat inscrutable but decidedly dignified plenipotentiary who would tolerate no slights to his country
and would persevere until its demands were met. His diplomatic experience and his personal inquiries had convinced him that he must
sail into the Uraga Channel that led to the Japanese capital at Edo
with a powerful fleet of warships that included the most modern
steam frigates. Though the sizeable flotilla Perry had been promised
had been reduced to four (only two of which were steam-driven) by
the time the expedition reached Japan on July 8, 1853, he choreo-
graphed a military display for his first venture ashore to meet with
the emissaries of the shogun, Tokugawa Ieyoshi. On the morning of
July 14, more than a hundred well-armed U.S. marines marched up
the wharf near the village of Kurihama and snapped to attention in
two long lines along the path to a hastily constructed reception hall
where Japanese dignitaries awaited the landing party. A hundred
sailors and two bands playing "Hail, Columbia!" led Perry and his
fellow officers between the smartly attired columns to present President Fillmore's letter to the warrior-aristocrats representing the "emperor" of Japan. The letter and Perry's credentials were inscribed in

vellum bound in blue velvet, and the seal for each document was encased in a box of fine gold. The documents were carried ahead of the commodore by two boys, and he was flanked on each side by tall, muscular "Negro" guards, with two sailors of "stalwart proportions" following behind.[1]

Perry's admission in his official account of the expedition that the splendid procession "was but for effect" suggests that he may have been quite aware of the special resonance his grand show would have for the Japanese officials, who were drawn from an elite samurai culture that revered elaborate and well-performed rituals. Samuel Wells Williams, an American missionary serving in China who joined the expedition as an interpreter, had the distinct impression that "the Japanese were sorry to see the show end so soon."[2] Perry had carefully calibrated the impressions he intended to evoke through successive revelations of technological wonders. These marvels would be dramatically displayed by both the mighty ships that transported the embassy and the presents that had been meticulously chosen for the shogun and his emissaries, many of whom were powerful *daimyos* or lords of the patchwork of feudal fiefdoms *(han)* that made up the Tokugawa realm. Perry's tireless efforts to persuade often skeptical congressmen and Washington bureaucrats to provide him with an impressive fleet of the most advanced ships were driven by his conviction that a clear demonstration of U.S. naval superiority was essential if significant concessions were to be extracted from the Japanese. To this end, he seized every opportunity to invite Tokugawa notables to tour his frigates and to examine the steam engines below decks and the formidable Paixhans cannon above.[3]

Despite these successes, Perry was discomfited by the fact that he had been compelled, by reports of attempted landings by Russian rivals, to hurry to Japan with fewer than half the ships that had been assigned to the expedition. That concern may in part explain his abrupt departure from the islands soon after the ceremonies at

Kurihama. Fearful that the limited supplies his truncated fleet carried would leave him dependent on the Japanese for vital provisions, Perry gave orders to sail out of the Uraga Channel on the morning of July 17, after informing the Japanese commissioners that he would return within months to receive the shogun's reply to the president's letter. The expedition then made its way back to China, where Perry planned to rendezvous with the ships that had not arrived in time for the initial voyage.

A good part of the layover time in China was devoted to inspecting and refurbishing the array of presents that Perry had chosen to dazzle the Japanese monarch and his most exalted officials. Perry's researches may have made him cognizant of the importance of the exchange of gifts in Japanese culture. Certainly, from the earliest stages of planning for the expedition, he had envisioned carefully selected gifts as the centerpiece of an elaborate effort to awe the Japanese with the superiority of American inventions and manufactures. Such a demonstration, he believed, would make it difficult for them to resist entering into commercial exchange with a nation that was so enterprising and ingenious, so clearly more materially advanced than their own.

When Perry's East India squadron—more than twice the size of the first flotilla, with a third steamship and supply ships added—returned to Edo Bay on a blustery winter day in mid-February 1854, the full complement of gifts was ready for presentation to the Tokugawa court. It included expensive perfumes for the "empress" and whiskey and fine cloth for the shogun's councilors. But the gifts that most delighted the Japanese were the sophisticated machines, weapons, and gadgets that Perry had solicited from American firms. A quarter-size steam engine, complete with tracks, tender, and passenger coach, was the greatest source of fascination for the islanders, whether they were nobles or commoners (though only the former were invited to ride atop the miniature cars). The shogun's commissioners were also enthralled by the two telegraph sets, Colt revolvers, clocks, and copper boats. Lower-level officials and local peasants witnessed demonstrations of farm machines, including a straw cutter, a garden cultivator, and a rice huller, which American observers reported the Japanese found far more efficient than the oversized mortars and pestles they had used to process their staple crop for millennia.[4]

The crowd-pleasing spectacles as American technicians demonstrated the utility of Perry's well-chosen machines rivaled the arduous negotiations and even the signing of the Treaty of Kanagawa on March 31, 1854, as the main attractions for both the Japanese and their visitors. As Perry intended, the technological wonders provided the mix of incentives, including an implicit threat of force, that he deemed essential for achieving the "opening" of Japan, which was cautiously—and only partially—ratified by the terms of the treaty. He had been able to negotiate access to the small ports of Shimoda and Hakodate for U.S. vessels in need of provisions, refitting, or coaling, and had extracted pledges of assistance for American castaways from shipwrecks along the coasts of Japan. The shogun's commissioners also agreed to allow the United States to establish a consulate at Shimoda, and to grant America any privileges subsequently negotiated with other Western nations. And though there were no provisions for facilitating commercial exchanges, Perry rightly concluded that those would soon follow. By the time his East India squadron sailed for the United States in late June, Perry was confident that the expedition had been a resounding success, an assessment that has subsequently been confirmed by most historians—at least those on the American side of the Pacific.

On his expedition to Japan, Matthew Perry made use of displays of American technological prowess to secure a treaty that would prove a turning point in the development of both nations—and in global history. In defiance of skeptics throughout Europe and even many at home, Perry not only negotiated an end to Japan's centuries of seclusion but dramatically launched the U.S. quest to be included in the exclusive circle of world powers. The embassy to Japan was both the culmination of a process of national development that had its origins in the English settlement of North America and the beginning of a new phase in that historical trajectory. And yet, despite Perry's own emphasis on the importance of advanced technology and displays of technical aptitude to the success of the venture, these aspects of the expedition have been typically treated by historians as "side shows," as they are dubbed in one of its standard histories.[5] This peripheralization of the technological dimensions of both nation-building within the United States and its subsequent emergence as a global power has been the norm rather than the exception in mainstream American historiography. The marginalization of technological factors in the writing of American history is by no means exceptional. But it is remarkable for a people "whose most notable and character-forming achievement for almost three centuries," in the words of Thomas Parke Hughes, "has been to transform a wilderness into a building site," and a nation whose citizens have long seen technological innovation and technical aptitude as the most readily verifiable measures of their superiority over both the "Old World" of their ancestors and the "exotic" nonwestern cultures they encountered in North America and elsewhere. The neglect of technological pursuits in the history of a nation in which they have arguably been the "most characteristic activity" has often resulted in skewed or very partial understandings of the American experience.[6]

In this book I focus on the motivations technological imperatives supplied and the means they provided for Anglo-American and U.S. expansion. I explore the effects of technological development on the ways in which Americans perceived and depicted, interacted with and sought to influence—and very often dominate—nonwestern peoples and cultures. The Perry embassy encapsulates many of the themes and cultural currents I will discuss in the chapters that follow. Support for the expedition, for example, was directly linked to technological innovations that were transforming the U.S. economy and society in the mid-nineteenth century as well as making possible a more fully integrated global system. Merchants and entrepreneurs, who were among the earliest and most determined advocates of diplomatic relations and trade with Japan, viewed the islands as a potential market for both manufactured goods and agricultural produce. America's industrialization, including the steady mechanization of farming, sparked a search for market outlets to absorb the increased productivity of new machines and modes of organization. Like the Philippines at the end of the century, Japan was often viewed as a way station to China with its prodigious population to be clothed and fed. But those who had visited or read about the islands were also hopeful that the energetic and hard-working Japanese themselves would prove to be avid consumers. This prospect accounts for the ease with which Perry was able to persuade leading merchants and manufacturers to donate their wares—from fancy garments and Webster's dictionary to the Norris Brothers' miniature train and Colt revolvers—as presents for the "emperor" and his courtiers.[7]

By the early 1850s, major advances in transportation and communications technologies were binding together what had become a transcontinental nation, as well as providing much improved access to the markets and resources of the Pacific basin and Latin America. Railroad lines formed the backbones of networks of commerce that spread overland across the farmlands and on to the mining centers of the West, making it economically feasible to export the produce grown and raw materials extracted from these once remote areas. Swift clipper ships not only linked the Atlantic seaboard with recently acquired territories on the Pacific coast but greatly shortened traveling time across the oceans. These sleek sailing vessels were increasingly rivaled by steamships both on ocean routes and along the great rivers that opened the interior of North America to the rest of the world. During the 1840s and 1850s the number of American whaling vessels in the north Pacific peaked, and by 1846 more than 75 percent of U.S. ships working the waters in the vicinity of Japan were whalers, in pursuit of oil for lamps and lubricants as well as bones for fashionable corsets. In 1855, the year following Perry's second visit to the islands, shipbuilding in the United States reached its nineteenth-century apex, and in the next decade the tonnage exchanged in domestic and foreign commerce achieved heights that would not be matched again until America entered the First World War.[8] As these innovation-driven enterprises transformed American society and increased its involvement in overseas trade, eminent politicians—including Senator Daniel Webster and Presidents Fillmore and Pierce—joined merchant and missionary lobbyists as strong proponents of American efforts to "open" Japan.

As Matthew Perry's highly successful career as a naval officer and his approach to the Japan expedition suggest, America's emergence as a global power has been consistently driven by a sense of can-do confidence, a faith in scientific and technical solutions, and a missionary certitude that the United States was destined to serve as a model for the rest of humanity.[9] Each of these cultural predispositions has been grounded in foundational American myths and ideals that coalesced in the first decades of settlement and have persisted with varying degrees of intensity through centuries of cross-cultural exchanges until the present day.

Perry himself personified many of the dominant values, favored pursuits, and prevailing attitudes of a society that had long been enthralled by discovery and invention. He shared the abiding American commitment to enlisting science and technology in the grand project of building a prosperous and powerful nation. In his naval career Perry championed steam-powered warships (often to the displeasure of senior officers) so enthusiastically that he came to be known in the military establishment as "the father of the steam navy."[10] He advocated the founding of a training school for naval engineers, and he was the first to command a steam frigate, which he used as a lab to test newly invented (by a Frenchman) Paixhans cannon, explosive shells, and innovative designs for gun carriages. Long before the Civil War, he was involved in the planning of the first U.S. ironclad warship. And his insistence that scientists ought to be retained by the navy and the other services ensured that naturalists would be prominent members of his embassy, and that surveyors and mapmakers would work long hours whenever his ships visited a new locale. Perry was very much the product of a nation that abounded in tinkerers, a people fascinated with gadgets, inventions, and the power (and pleasure) that machines could bestow.

Technological pursuits and material achievement have strongly influenced the construction of American identity, on both an individual and a collective basis. Perry's inventiveness, technical aptitude, and such personal attributes as self-discipline and industriousness go far to account for his enduring stature as a national icon. For centuries these qualities have been associated with heroic American figures—whether intrepid frontiersmen or rags-to-riches entrepreneurs—and the national ideals they have been seen to embody. The sense of the larger purposes of the Japan expedition that Perry shared with many of its advocates was informed by an age-old teleology that celebrated American expansion as part of a grand cosmic design. Depending upon the prophetic voice in question, the domination of nonwestern peoples into whose homelands Americans expanded was justified by various missions, from the need to render productive the earthly bounty granted to God's favored people to the call to complete the millennia-long westward progression of the locus of civilization from China to the ambitious young republic in North America. As Perry and other members of his expedition made clear, they regarded America's undeniable technological superiority over Japan as a compelling gauge of its higher level of civilized development. America's advancement in this regard provided rationales for those who backed an embassy to compel the Japanese to alter the historical trajectory of their long-secluded country and to choose the United States as their tutor in the ascent to civilization.[11] Similar assumptions would inform the policies pursued by the American engineers who sought to "uplift" the Filipinos, the medical missionaries who viewed applied science as a means of "awakening" China from its "medieval torpor," and the modernization theorists in the cold war era who were convinced that the technological edge of the United States would prove decisive in its competition with communist rivals to offer a model for development for the nations of the "Third World."

To avoid perpetuating national stereotypes or essentializing the diverse population of a multiethnic nation, it is important to point out that a substantial majority of contemporary Americans, much like their forebears, know little about how their appliances work or what is under the hood of their ubiquitous automobiles. Nevertheless, familiar tools and basic machines are critical determinants of the patterns of their everyday lives. These devices physically manifest the technical advancement and material prosperity that Americans have long regarded as definitive aspects of an exceptional national experience. The fascination, for example, on the part of Perry and many of his contemporaries with steam engines and new designs for naval vessels was mirrored a century later by a much more broadly based American fondness for big and showy cars. And this attachment to the automobile has been seen as expressive of presumed national characteristics as diverse as a love of freedom ("on the road" in one's own private vehicle), an ambition to improve social status, and even a preoccupation with owning and refurbishing highly visible machines.[12]

The sense of superiority that Perry and his entourage linked to American advances in technology encompassed more than just tools, machines, and weaponry—and that broader conception of technology will be employed throughout this book. The discipline, precision, and teamwork, for example, that were exhibited by the sailors' handling of the black-hulled ships of the American flotilla and the marines' well-rehearsed parades exemplified skills and organizational aptitudes that Perry and his fellow officers associated with advanced industrial societies. Although American observers admitted that the Japanese also valued some of these qualities, and such related virtues as cleanliness and orderly town planning,[13] members of the embassy believed that they had been far more fully cultivated in the United States. The Americans were therefore confident that they had much to teach the islanders.

Long before the United States possessed the power or even the motivation for its leaders to contemplate projecting its influence overseas, similar perceptions and conclusions played vital roles in Anglo-American expansionism. For centuries Anglo-American settlers had dismissed the indigenous peoples of North America as primitives, who lacked not only the farm implements but also the cultivation techniques and work habits to make effective use of the abundant resources of the vast lands they inhabited. In the decades after Perry's embassy to Japan, Americans often judged and ranked societies from the Philippines and China to the Caribbean and the Middle East according to their demonstration of technical proficiency and related personal qualities such as efficiency, foresight, a strong work ethic, punctuality, and self-reliance.

When, as was often the case, nonwestern peoples were found wanting in these respects, the dissemination of these presumed virtues through missionary schools, employment on public works projects, and recruitment into American-led police and military units became—along with the transfer of technology—a major component of U.S. civilizing offensives. Advocates of U.S. expansion have rather consistently assumed that the adoption of American technologies (and material culture more broadly) would also entail the incorporation of American values, ways of thinking, and modes of organizing everything from factory workers to political systems. And Americans who have embraced these civilizing projects have repeatedly proclaimed that these transfers would vastly enhance the capacity of recipient societies to exploit natural resources, produce material goods, raise living standards, expand market profitability, and secure sociopolitical stability.

The unwelcome but incessant surveying operations conducted in Japanese coastal waters by members of Perry's expedition exhibited many of the culturally ingrained organizational proclivities and personal aptitudes that Americans had come to regard as key sources of their nation's prosperity and power. And the determination of Perry and his officers to continue these "scientific investigations" despite the obvious hostility of both the shogun's officials and the general populace[14] underscores the importance of broadening the concept of technology to take into account the complex interrelationships between scientific and technological pursuits in American society. These intricate linkages have confounded attempts to draw clear distinctions between scientific pursuits, which are seen to be focused on amassing knowledge about and seeking understanding of the cosmos and the mundane world, and technology, which some have sought to differentiate from science by its emphasis on solving practical problems and exerting control over nature. But from the early nineteenth century onward (and in some cases a good deal earlier), the boundaries between science and technology became increasingly difficult to maintain with regard to Great Britain, the United States, and other industrial nations. By the last decades of the century, in societies in transition to a new phase of industrialization, particularly the United States and Germany, scientific research was becoming an essential component of corporate conglomerates and (more slowly) government agencies, whose directors were intent on rationalizing resource extraction, production, and marketing. A long-standing American valorization of applied, as opposed to purely theoretical, scientific inquiry gave great impetus to these trends, which particularly in relation to the history of U.S. interventions overseas have rendered distinctions between the roles of these "mirror-image twins" arbitrary and unnecessary.[15]

The diversity of the social groups to which supporters of the Perry expedition belonged and the mixture of motivations expressed by the demands set forth in President Fillmore's letter to the "emperor" of Japan suggest the need to adopt a multifaceted, nonreductionist approach to the analysis of the patterns of American expansionism. Consequently, I have sought to be attentive to other major motivations that have interacted with technological imperatives in shaping American frontier advances or interventions overseas. Alternative, but often complementary, rationales for expansionism have ranged from religious proselytism and a determination to build democratic nations to great-power rivalries and a desire to extend market capitalism. And the blend of these motives with technological factors has varied considerably. Nonetheless, in each phase of expansion technological imperatives have strongly influenced America's encounters with the peoples and cultures of the non-western world.

From their first encounters with the Japanese, American officers and sailors were impressed by the seemingly insatiable curiosity of the shogun's emissaries, and especially their intense interest in the design and equipment of the ships of the East India squadron. Perry and his entourage were delighted when the Japanese admired their ships' powerful artillery and navigational instruments. "Nothing," recorded Edward McCauley, the young acting master on the sloop-of-war *Powhatan*, "has astonished the natives so much as our impudence—Paixhan guns—Electric Telegraph—Steam—and firearms, all called for their admiration."[16] Like many of his compatriots, McCauley was struck by the intellectual vigor of the Japanese officials and their knowledge of U.S. history. He and other members of the expedition were also pleased to discover that the Japanese were well aware of the advanced state of America's technological endowment. At the same time, the Americans were taken aback by the readiness of the Japanese emissaries to admit that their own country was far behind the Western powers in manufacturing, weaponry, and capacity to exploit natural resources.

McCauley was surprised that the shogun's envoys who toured the *Powhatan* understood the principles that made its steam engine run, and he thought they were clever to bring along artists to make detailed sketches of the engine and numerous other parts of the ship. He and others also commented approvingly on the emissaries' eagerness to operate and understand the ship's magnetic telegraph machine. James Morrow, a scientist who had been assigned to set up and demonstrate the mechanical contrivances that were presented as gifts, singled out the alacrity with which a Japanese carpenter learned to use unfamiliar tools when invited to assist the Americans in assembling several agricultural implements. Morrow found the Japanese workman "almost as handy and intelligent as the excellent machinist, Dozier, . . . who was working with me."[17] Morrow and the other officers would have been even more impressed had they been able to foresee that several years later a delegation of Japanese notables, paying a return visit to America, would present to the president of the United States an *improved* version of the Sharps rifle that Perry had given to the shogun's officials.[18]

The determination of the Japanese to improve on the guns and machines introduced by Western visitors was instilled by their recognition that the successful adoption of applied science and industrial technologies would be crucial to the very survival of their nation. A new generation of samurai leaders, emerging from the intensified social and political turmoil in the shogunate following the American interventions, quickly grasped that scientific investigation and technological innovation were the mainsprings of the power and wealth of the industrialized West. The obvious military superiority of the American fleet to any force the Japanese could conceivably assemble against it left the shogun's emissaries little choice but to accede to the intruders' demands. And the strategy devised by Japan's leaders in the next decades was premised on the assumption that further concessions would be necessary until the country was sufficiently industrialized to break free from the informal domination that had been imposed by the Western imperialist powers.[19] As they soon realized, this process of national strengthening would also entail far-reaching changes in everything from Japanese political and educational institutions to patterns of gender and generational interaction.

Despite the fact that Perry and some of his advisors, especially the missionary and "expert" on China Samuel Wells Williams, had acquired some knowledge of Japan, the leaders of the expedition were largely oblivious to the deep sense of humiliation their forced entry evoked at all levels of Japanese society and particularly among the warrior aristocracy. Like Perry, most contemporary social commentators and the informed public in the United States regarded the "opening" of Japan as one of the great international triumphs of the early republic. For those who had supported the expedition, its success was confirmed over time by the fact that it had served as a catalyst for sweeping transformations within Japanese society. Like Perry and his fellow officers, all too often the agents of U.S. expansionism have evinced scant concern for the anxiety or outright opposition of nonwestern peoples who have become the objects of America's civilizing projects. And American expansionists have often reflected far too little on the possible long-term consequences of the changes their intrusions have set in motion.

Although Perry, McCauley, and others who wrote accounts of the expedition considered many aspects of Japanese society to be backward, and some utterly barbarous, they saw the enthusiasm of the Japanese for American technology and material culture as an indication that the islanders had the potential to make good use of the new tools and machines to which they had been so abruptly introduced.[20] The cultural standards that informed the American envoys' positive estimate of Japanese capabilities contrasted sharply with the contemporary, often viciously racist, denigration of "Oriental" migrants who had begun to make their way across the Pacific to Hawaii and California. This obvious discrepancy in American perceptions of the Japanese had much to do, of course, with marked differences in the social circumstances in which impressions were formed and stereotypes constructed. But the technological measures so evident in the responses of members of Perry's embassy to the Japanese very often had a decisive impact on American assessments of other nonwestern peoples as well.

On the one hand, evidence of significant achievement in material endeavors emboldened agents of American expansion to modify racist strictures that were pervasive in the United States itself at the time. They surmounted racist assumptions by privileging more mutable cultural characteristics, such as a people's past material accomplishments or its presumed capacity to acquire Western knowledge and operate "modern" machinery. The presumption that the obstacles impeding the fulfillment of America's civilizing mission were culturally not racially grounded gave great impetus to ambitious schemes for the development of the recently colonized Philippine islands launched in the early 1900s by Progressive proconsuls and engineers. From the mid-nineteenth century until well into the 1930s, this willingness to override racist skepticism buttressed the conviction of numerous missionaries and philanthropists that "decadent" China could be rejuvenated with sufficient infusions of Western education, science, and technology. And after the Second World War the explicit rejection of the notion of innate, racially embedded differences between human groups emerged as a fundamental presupposition of technology-driven modernization theories. For decades thereafter these highly ethnocentric, but often explicitly anti-racist, paradigms dominated policymaking with regard to American assistance to the developing nations of Africa, Asia, and Latin America.

As these disparate examples suggest, sites as remote in time and space as Japan in the 1850s and Iraq in the early twenty-first century provide the basis for exploring the diverse ways in which technological imperatives have shaped the nature of cross-cultural perceptions, the ideology of America's civilizing mission, and the policies and programs actually pursued. Each of the case studies in the chapters that follow recounts a major episode in U.S. expansionism and the interactions between American and foreign peoples and cultures. Each case example also illustrates broader patterns of the interplay among technological changes in American society, their transfer to foreign cultures, and the responses of nonwestern peoples to American interventions. Through a comparison of the case examples, I seek to account for significant differences in the nature and impact of technological imperatives in each phase of expansionism, and to analyze their cumulative effects on patterns of cross-cultural interaction that recur from one phase to the next. The decades of English colonization in North America in the mid-1600s, when a mix of religious proselytization and major differences in material culture between Anglo-Americans and indigenous Indian societies shaped the initial conception of America's civilizing mission, are the focus of Chapter 1. Chapter 2 traces the persistence and transformation of these themes in the nineteenth century, when industrialization and conquest of the western frontiers provided the resource base and ideological impetus for American expansion across the Pacific to China and Japan. The multifaceted impact of technology on America's rise to global power and a growing emphasis on nation building as a rationale for U.S. interventionism in the decades on either side of 1900 are illustrated by the case studies in Chapters 3 and 4 on the colonization of the Philippines, interventions in the Caribbean, and continuing efforts to shape political and social development in China and Japan.

The second half of the book is devoted to the half-century after the Second World War, when the United States became first a superpower and then, after the collapse of the Soviet Union, the global hegemon. The centrality of technological achievement and material prosperity in shaping the rivalry between the United States and the Soviet Union for influence in the emerging nations is explored in Chapter 5 on the deployment of modernization theory as an antidote to Marxist socialism and converging approaches to development assistance in the postcolonial world. America's failed nation-building efforts and military interventions in Vietnam, which exemplified the technocratic hubris and false assumptions of its anti-communist crusade in the cold war era, are the focus of Chapter 6. The crisis and conflict in the Persian Gulf in 1990–1991 provide the basis for the analysis in Chapter 7 of the roles of technological innovation and post-Vietnam military reforms in the emergence of the United States as the sole global hyperpower. I argue in Chapter 7 and the Epilogue that the failure to follow through on post-Vietnam efforts to overhaul America's development assistance programs and the "blowback" from erratic military interventions in the Middle

East have become key sources of U.S. vulnerability in the era of intense globalization that has ushered in the twenty-first century.

In contrast to American confidence that the Japanese or Filipinos could with proper tutelage master Western scientific thinking and technologies, in numerous other instances of interaction between agents of U.S. expansionism and non-European societies this potential was strenuously questioned or flatly denied. The widely shared opinion, for example, of Perry and his contemporaries (European as well as American) that China's once splendid technological endowment had been allowed to fall into disrepair was repeatedly linked to the conservative nature of the "Middle Kingdom's" scholar-gentry elite and the decadence of Chinese civilization. Just over a hundred years after Perry's visits to Japan, Charles Gillispie, a prominent American historian of science, articulated the fears of many of his fellow citizens that "the instruments of power created by the West [would] come into the hands of men not of the West, formed in cultures and religions which leave them quite devoid of the western sense of some ultimate responsibility to man in history." Ignoring the fact that the United States was then (and remains) the only country that had used atomic weapons against another human society, Gillispie was alarmed by the prospect that Communist China or Egypt might soon develop nuclear devices because he believed they would not be restrained by the Christian tradition or Western "history and values." His categorical assertion that none of the "Oriental civilizations" had "graduated beyond technique or thaumaturgy [magic] to curiosity about things in general" provided a scientific imprimatur for speculation in the 1950s and 1960s by prominent foreign policy specialists such as W. W. Rostow regarding what they viewed as "pre-rational" or "pre-Newtonian" responses by peoples in postcolonial societies to America's civilizing mission—at that point dressed up in modernization jargon.[21]

The very direct ways in which perceptions of technological backwardness have shaped assumptions of cultural and, by the early 1800s, racial inferiority have a long lineage that can be traced back from these musings in the early atomic age to the first decades of English colonization in North America. Whether they were peoples of the eastern forests or the western plains, the indigenous societies of North America were almost invariably judged to be technologically primitive, materially impoverished, and backward in their modes of social organization. Even in the earliest phase of interaction between Anglo-Americans and Indians in the seventeenth century, many of those who wrote about Indian peoples or formulated policies for them assumed, at least implicitly, that these collective and individual deficiencies were not only obvious markers of inferiority but so deeply socially embedded that they were in effect innate. Though seldom articulated in explicitly racist terminology, these views tacitly called into question cultural explanations for Amerindian differences from Europeans, which were favored by many observers, including Thomas Jefferson, at least until the end of the eighteenth century. Over the course of the nineteenth century, inferences of inferiority drawn from evidence of indigenous peoples' material backwardness were tapped to justify the application of increasingly elaborate racial theories to the diverse peoples who were lumped together as Indians.

For his first foray ashore in Japan in 1853, Matthew Perry included two well-armed African-American guards in his personal entourage. Although he never explained why he accorded them this place of prominence, Perry was apparently quite deliberate in his decision to present them to the Japanese as trusted and worthy Americans[22] even though they were denied the rights and privileges of citizenship in the United States. As he must have been aware, his gesture was directly at odds with the fact that Americans of African descent had long been targets of prejudice and discrimination. Very often their denigration and oppression were excused, in part, by claims that they were incapable of rational or scientific thinking and lacking in technical dexterity. The racial theories formulated to explain Amerindian difference (which at the time connoted inferiority), and to question the feasibility of assimilating the remaining Indians into American society, were even more vociferously applied to African Americans. Empirical tests that most contemporaries accepted as scientific proofs of the mental and physical inferiority of "Negroes"—and all other "races"—to "whites" appeared to validate the arguments of both those who defended the continuance of slavery at home and those who championed European and American dominance over most of the rest of humanity overseas. Prominent defenders of slavery in the American South insisted that the innate indolence of the African Americans and their carelessness or incompetence in wielding farm implements, much less in operating sophisticated machines, meant that they needed constant supervision. In the era of abolition, similar arguments were advanced to discourage the recruitment of freed slaves into the industrial labor force, and, if they were fortunate enough to be hired, to restrict or altogether deny their eligibility for skilled positions.[23]

Connections were often drawn among technological imperatives, social engineering projects designed for foreign societies, and the historical experience of African Americans. Booker T. Washington's stress on technical training at the Tuskegee Institute was seen by colonial officials in the Philippines as an apt precedent for the colony's educational programs; the rhetoric and many of the policies advocated by modernization theorists for the uplift of developing nations in the 1950s and 1960s were also enlisted in Lyndon B. Johnson's War on Poverty in America's inner cities.[24] But the influence of technological imperatives on slavery, segregation, or racial discrimination against Americans of African descent is beyond the scope of this book. Even though the African-American minority has been long oppressed and often physically segregated, from at least the second half of the seventeenth century it has been one of the core groups in the ethnic mélange making up American society. Unlike the Amerindians—even after they were forced onto the reservations[25]— African Americans have rarely lived under independent political regimes, and those who did found refuge in runaway slave or mixed Afro-Indian communities struggling to survive on the margins of American society. Despite the fact that enslaved African Americans managed to preserve significant elements of African culture, soon after they were introduced into England's North American colonies they ceased to be regarded by the colonizers as *African* aliens. In contrast, the peoples and societies that I consider in the case studies that follow were regarded as foreign. From the early decades of settlement in North America, African Americans have been a major force in the domestic socioeconomic, political, and cultural history of the United States. The Amerindians, Chinese, Vietnamese, and Arab peoples who are the focus of this book struggled to ward off U.S. domination in each of these spheres or to find ways to accommodate American imports to their societies' distinctive historical trajectories.

The diplomatic, and potentially military, confrontations that Perry's expedition set in motion when it sailed into Edo harbor in 1853 were emblematic of a broader contest between social classes that had long been under way in much of the world. Like the samurai notables who served as the shogun's emissaries, Matthew Perry was the scion of a prominent military family. His father, Christopher, had fought in the American War of Independence as a soldier and a privateer, and Matthew's brother, Oliver Hazard Perry, had commanded the naval force that bested the British in the Battle of Lake Erie during the War of 1812. Matthew himself had joined the navy as a teenager, fought as a midshipman in the 1812 conflict, and made a distinguished career as a naval officer. But in contrast to the daimyos and samurai he negotiated with, all of whom served the interests of Japan's feudal aristocracy, Matthew Perry exemplified the values and sought to advance the fortunes of the better-educated, more affluent sectors of the American middle class. Despite his family connections, he was very much a self-made man, whose physical courage, intelligence, and social skills had carried him to the upper levels of the naval hierarchy and positioned him to be appointed commander of the embassy to Japan. It is not surprising then that the main supporters of Perry's expedition were prominent middle-class merchants, manufacturers, and expansion-minded politicians.

Perry's samurai counterparts shared many of his values, including the premium he placed on self-control and maintaining personal dignity, and they would surely have admired his military accomplishments. But their responses to him and to his nation's demands were dictated by their determination to defend a dying feudal order. Their opposition to trade concessions and reluctance to open Japan to Western influences were premised on their conviction that warding off the foreigners was essential for the preservation of the rigid social divisions, hierarchical institutions, and isolation that had secured dominance for the families of the military aristocracy for centuries. The shogun's emissaries negotiated with little concern for the needs of the rising merchant classes, which had grown prosperous but not yet politically influential, as Edo, Osaka, and other cities flourished in the Tokugawa era.[26] Partly because they monopolized political power, members of the samurai elite designated by the shogun were the only mediators who mattered in American-Japanese cross-cultural exchange at the time of the Perry mission. But their position as brokers also depended on the exclusive access to information about the West enjoyed by the physicians, naturalists, and other scholars who were patronized by powerful daimyos and drawn mainly from the lower military aristocracy. Since the mid-seventeenth century, when the Tokugawa shogun Ieyasu had forcibly put an end to most foreign contacts, the Dutch trading post on Deshima island near Nagasaki had been Japan's only entrepôt for books, maps, and other artifacts from the West. It provided the most capacious window that Japan's rulers left open to the outside world, but only a small, highly educated circle of scholars from or connected to aristocratic families were permitted to peer through it.[27]

Agents from a wide range of social levels—from trappers, cowboys, and itinerant traders to engineers, missionaries, and railway magnates—have been involved in American expansion in different phases. But like Matthew Perry, most of those who conceived, financed, and directed the enterprises that made possible the republic's rise from a continent-spanning nation to global hegemon have been drawn from the middle class. While individuals of patrician lineage, such as Thomas Jefferson and Theodore Roosevelt, have served at different junctures as visionaries or the driving force behind expansionist projects, the extension of American dominance, whether through outright colonization, commercial enterprise, development assistance, or military campaigns, has largely been a middle-class project. As encounters between members of Perry's embassy and the Japanese warrior aristocrats illustrate, in the early phases of U.S. expansion, its largely middle-class agents on the frontiers and overseas mainly had to negotiate with, appease, or make war against upper-class elites: Latin American oligarchies, the Confucian-educated scholar-gentry in China, the privileged *ilustrados* in the Philippines. From the turn of the nineteenth century, however, the decline or transmutation of these elites meant that Americans overseas increasingly sought collaboration with Western-educated—and often Christianized—bureaucrats, professionals, and political leaders who were in effect their middle-class counterparts. In the twentieth century, America's civilizing projects throughout the colonized and, after 1945, the developing world came to depend on alliances with such indigenous groups, from members of the *compradore* social stratum in the Philippines and the leaders of the Guomindang in China to the Western-educated nationalists who spearheaded decolonization movements in Africa and Asia.

The choice of a successful naval commander to head the expedition to Japan, and the fact that he and his officers dealt mainly with samurai during their visits, suggest a persisting dynamic of U.S. interventions in nonwestern societies: an oscillation between constructive projects and violent reprisals. Although the uninvited entry of Perry's East India squadron into Japanese territorial waters did not result in military clashes, both sides assumed it could and prepared for that eventuality. As we have seen, Perry insisted on being provided with a sizeable squadron of warships, and he deployed them in ways intended to remind the Japanese that force was a very real option should they refuse the concessions set forth in the letter from President Fillmore. The members of the expedition were confident that their well-armed "embassy" could overwhelm any military opposition the Japanese might offer. And this assessment appeared to be confirmed by the discovery that the Japanese had only antiquated weapons and appeared to know little about modern rifles, much less artillery. After viewing Edo from his flagship, Perry declared that the whole city could be "destroyed by a few steamers," and he and his officers made it clear that they were prepared to do battle if the Japanese attempted to obstruct their surveying operations or threatened their landing parties. Bayard Taylor, a correspondent for the *New*

York Tribune who had joined the expedition in Hong Kong, exulted in "belonging" to the "great American Navy—that glorious institution which scatters civilization with every broadside and illuminates the dark places of the earth with the light of its rockets and bombshells."[28]

Taylor's confidence in the righteousness of using America's advanced weaponry to force recalcitrant foreigners to cooperate with its civilizing initiatives echoed the sentiments of those who had championed frontier expansion in North America at the expense of Amerindian peoples who refused to give up what Anglo-American settlers deemed their backward, savage ways. A half-century after Perry's visits to Japan, similar rationales were offered for the brutal suppression of Filipino nationalist resistance, which American colonizers argued was essential if U.S. engineers and schoolteachers were to succeed in bringing those benighted islands into the modern age. As the global reach of the United States has expanded, politicians, military strategists, missionaries, and development theorists have justified forcible—at times preemptive—interventions in the name of ambitious projects for the progressive transformation of societies as diverse as China, Haiti, Vietnam, and Iraq. But constructive applications of technology have also been important in diverse areas and time periods. These include contributions America has made to enhance food production and distribution, to expand international communications networks, to facilitate interregional migration, and to extend what has become a global exchange of ideas and material culture to the most remote areas of the planet.

In his fierce pride, self-control, and martial prowess, Matthew Perry was more than just a paragon of American middle-class virtue, he exemplified qualities that were widely admired as manly. Among the more refined circles of the American middle classes, Perry might well have been judged overbearing and lacking in subtlety, but his subordinates revered his somewhat taciturn, highly disciplined, and very direct approach to commanding the East India squadron or negotiating with the Japanese. Though some Japanese artists depicted Perry as a grotesque, hairy demon, the samurai emissaries who reluctantly acceded to most of his demands treated him with a respect that they had denied to earlier Western envoys who in their eyes had lacked the "manly dignity" that Perry possessed in abundance.[29] These earlier emissaries had also lacked the leverage provided by the powerful fleet Perry had so deliberately assembled and the well-armed and -drilled marines he paraded with such effect. For Perry and his officers, and equally for their samurai counterparts, military posturing and the diplomatic maneuvers it was intended to bolster were unquestionably gendered masculine. More than in many other episodes of American expansion, exchanges between members of Perry's expedition and the Japanese were confined largely to males. And indeed in Perry's day (and until the last half of the twentieth century), the scientific breakthroughs and technological innovations that played such critical roles in America's expansionism were devised, fabricated, and deployed almost exclusively by men. The contemporary consensus—shared by men and women in both Japan and America—assigned science and technology, like politics and diplomacy, to the professional, public sphere that was a masculine realm.

The narratives of Perry's expedition are noteworthy for their scant and often dismissive references to women. On at least one occasion when envoys from the two sides were dining together on Perry's flagship, toasts were proffered in honor of their wives. And predictably, a number of the American chroniclers paid a good deal of attention to the physical appearance of Japanese women sighted in boats near the visitors' ships or in the countryside explored by landing parties. The handful of American observers who have left their impressions seem to have considered Japanese women quite attractive, despite their habit of blackening their teeth. But in an echo of the earlier responses of English settlers to Amerindian women in Virginia and Massachusetts, Perry and his compatriots judged Japanese women immodest in dress and behavior. Some described them—as Indian women had been described in the English colonies and on the western frontiers—as being burdened with physically demanding labor that the Americans deemed unsuitable for females. After noting the staggering burdens carried by a number of women he passed while ashore, for example, Edward McCauley exclaimed, "I have not paper enough to tell what the women I saw were doing."[30] This "squaw drudge" trope has recurred over the centuries in American images of widely varying nonwestern societies around the globe. And putting an end to what was perceived as the exploitation or degradation of women has been cited by advocates of American intervention from the early colonization of North America to the twenty-first-century wars against the Taliban in Afghanistan and the Ba'athist regime in Iraq.

Even though the exchanges mediated by technology transfers between Perry's American emissaries and their Japanese hosts were virtually monopolized by men, on a number of occasions Japanese women became recipients of American technical advice. In one telling encounter, James Morrow, the young scientist who had been placed in charge of the farm implements and other machinery, fashioned a simple bamboo hook and convinced an elderly peasant woman that it was far superior to her fingers for putting the warp in her loom. Morrow reported that she was "well pleased," and that a cluster of other old women who "had congregated as usual to see what the foreigner was doing" tested the hook themselves, then "begged that I would make one for them."[31] Though American civilizing initiatives involving technology transfers have usually targeted male technicians and laborers, some have been designed to ease the burden of women's work and others to put indigenous men in charge of tasks that Americans thought women should not perform. Despite their apparent marginalization in contemporary accounts and much of the history that has been written about technology, American expansionism, and U.S. foreign relations more generally, women have played important roles in these endeavors, whether as pioneers, missionaries, schoolteachers, or political activists in the United States or overseas. Enduring images of foreign societies conveyed to those back home by American travelers, missionaries, museum curators, and diplomats have often been highly gendered. And these constructions have in turn informed the civilizing agendas and actual projects that agents of U.S. expansionism have promoted over the centuries, from the American frontier and China to the Philippines and the Middle East.[32]

The two visits of Perry's expedition to Japan provided the first opportunities for extensive interaction between Americans and Japanese. Until then the two peoples had known very little about each other. A number of prominent Americans, including Perry himself, had become concerned about this state of affairs in the previous decade. But most of what they had been able to learn about Japan was gleaned from sporadic reports by sea captains, shipwrecked sailors, and European envoys who had earlier attempted to establish contacts with the reclusive kingdom. The vague impressions the Japanese held of the United States—and before Perry's ships sailed into Edo bay in 1853 only a handful of scholars had paid much attention to the distant republic—had been taken mainly from Dutch books or Dutch interpreters on Deshima. The paucity of knowledge on both sides meant that during the weeks when the East India squadron was in harbor American officers and midshipmen and Japanese samurai officials, peasants, and fisher folk scrutinized each other intensely. In official reports, diaries, sketches, block prints, and tall tales, they recorded all manner of impressions from the physical appearance of the alien others, including their fashions in clothing, to the excellent planning of Japanese towns or the marvels of the intruders' steamships. And observers on each side worked these rumors, images, perceptions, evaluations, and understandings into composite representations of the other's society and culture. Greatly expanded intercourse between the two nations in the following decades resulted in refinements of and elaborations on these representations, which profoundly shaped American-Japanese relations until well into the twentieth century.

Although in the case studies I will give some attention to the responses of Amerindian, Asian, or Middle Eastern peoples to the agents of U.S. expansionism, my discussion centers on the multifaceted effects of American representations of both themselves and nonwestern peoples and cultures. These representations were fashioned from amorphous and ever-shifting combinations of eyewitness comparisons, recurring metaphors, and cross-cultural projections. In each case, we need to go beyond the content of these representations to explore the ways in which stereotypes of Amerindian or overseas peoples and essentialized notions about their societies and cultures affected America's civilizing agendas. Despite significant variations from case to case in the mix of gender, racial or ethnic, class, and cultural markers, these distinctions combined with perceptions of technological proficiency and material accomplishment to shape prejudices that affected everyday social interaction and informed American ideologies of dominance.[33] Although these ideologies were usually articulated by American politicians and members of social and intellectual elites, they encompassed a "lived system of meanings and values" that were believed to embody the nation's virtues and inspire its mission to civilize. Both values and sense of mission have been widely shared for centuries across class levels and the political spectrum. As with all ideologies, formulations of America's civilizing credo have been mutable and contested. But they have been consistently legitimized by an enduring confidence that a superior aptitude for technological endeavors ensures that American designs for the transformation of non-European societies will be realized.[34]

The inchoate American representations of Japan and the Japanese that had begun to be formulated in the early decades of the nineteenth century were deepened and elaborated by the relative flood of information that became available after Perry's expedition and the European embassies that followed it to the islands. These more complex, and sometimes conflicting, Euro-American representations interacted with ever more strident exhortations to the citizenry of the United States to take up the "White Man's Burden" to civilize the Japanese as well as other nonprogressive, hence "backward," peoples. America's emergence in these decades as a global economic and naval power meant that these representations increasingly shaped not just ideology but U.S. diplomatic relations with and active involvement in overseas societies, particularly in East Asia and Latin America. Tracing the linkages between representations, expansionist ideology, technological imperatives, and the policies and civilizing projects that the United States pursued first in North America and later over much of the globe is one of my central concerns in this book. Although these projects were often resisted, subverted, or significantly modified by the peoples whose societies were targeted for civilizing, American interventions have transformed local cultures—in both improving and debilitating ways—and also engendered long-term, often global effects.

In contrast to most later English accounts of the Atlantic coast of North America, which deplore the Indians' nomadic way of life and their failure to make use of the continent's resources, John White's late sixteenth-century drawing of the "Towne of Secota" in the Chesapeake region portrays substantial village dwellings and the cultivation of a variety of crops. From Thomas Hariot, *A Briefe and True Report of the New Found Land of Virginia* (Dover, 1972).

These Indians being strangers to arts and sciences, and being
unacquainted with the inventions that are common to civilized
people, are ravished with admiration at the first view of any
such sight.

—WILLIAM WOOD, 1834

1

"ENGINS" IN THE WILDERNESS

From their first sightings of the rocky northern
coasts of North America, explorers and settlers
projected ambivalent images of the new lands. Some likened New
England or Virginia to the Garden of Eden, a lush paradise, where
sweet smells filled the air. Others found the immense forests dark
and foreboding, the rampant nature surrounding their tiny settlements oppressive. For Thomas Morton, in beauty and "faire endowments" the New England was "farre more excellent" than the old,

comparable only to Canaan, the promised land of the ancient Israelites. But William Bradford dwelt on the "sharp and violent" winters and the "cruel and fierce storms" of his new home, finding it a "hideous and desolate wilderness" akin to that where the Israelites wandered for decades before entering Canaan. The sheer size of the North American continent and the fecundity of its exotic plants and animals gave rise to visions of abundance and unbounded wealth. Ralph Lane assured Richard Hakluyt, an indefatigable promoter of English exploration, that Virginia was "the goodliest and most pleasing territory of the world," with the most fertile soil under heaven. But John Smith complained that, in the environs of Jamestown, "grasse there is little or none, but what groweth in lowe Marishes: for all the Countrey is overgrowne with trees, whose droppings continually turneth their grasse to weedes, by reason of the ranknesse of the ground."[1]

At the turn of the seventeenth century the promotional literature of the chartered and joint stock companies organized in England to establish settlements along North America's Atlantic coast made much of the bountiful natural resources to be found there. Many accounts by those who resided in North America included quite detailed catalogues of plants and animals, commodities that could be grown or extracted, and the variety of English goods that might someday be marketed there.[2] But these sanguine assessments belied the often desperate struggle of the early colonists to survive in their tiny enclaves along the coast. The writers intentionally finessed the fact that the early colonists were dependent on seasonal shipments of provisions from England or on corn and other staples given freely, purchased, or extracted by force from the indigenous peoples.[3] The Indians themselves were portrayed with ambivalence in representations of the newly settled lands. Company servants initially depicted them as friendly, docile, and a potential source of labor and trade goods. But misunderstandings and quarrels, some of which ended in violent confrontations, fed counterimages of treachery, savagery, and sloth that soon came to dominate the colonizers' discourse.

Whether the New World was seen as a paradise populated by pliable natives or a howling wilderness where brutish savages roamed, a consensus soon developed among the English that North America was wasted on its primitive inhabitants. The Indians were thought to lack the tools and skills needed to put the continent's abundant resources to proper use. Subduing these "backward" peoples and transforming their sparsely populated and poorly cultivated lands into towns teeming with settlers and ringed with productive farms became key objectives for those engaged in the colonial enterprise. This assessment of the past of the newly explored land and this corrective vision of its future were shared by company directors and promoters in England and migrant groups as dissimilar as the religious dissenters of New England and the aristocratic adventurers of Jamestown. In each case the colonizers' mission in the wilderness was informed by their determination both to convert the heathen Indians to Christianity and to rescue them from their savage state.

At least in their rhetoric, some early colonizers privileged converting the Indians to Christianity over all the other transformations they hoped to effect in the New World. There is good reason to trust that the Puritans, in particular, sincerely believed that this effort was part of God's purpose in prompting them to embark on perilous journeys and to settle in the wilderness. Spreading the gospel among the heathen was the first of nine reasons offered by John Winthrop in his defense of a proposed plantation in New England against criticisms raised by his Puritan brethren. As their writings and sermons make clear, those associated with the New England expedition were confident that their enterprise was divinely inspired and that they themselves were instruments of God, hence "moral wardens" of the world. These convictions led the Puritans to see events that advanced their efforts to build a new Zion in the wilderness—such as the epidemic that decimated the Indian population in coastal New England in 1616–1617 or their victories in the early wars with local tribes — as providential interventions that sanctified their struggles as part of the divine purpose.[4]

Puritan colonists and their supporters in England were well aware that trade and agriculture would be necessary to sustain the settlement. But uppermost in their designs for the "new Eden" was the opportunity to develop a community of the faithful unsullied by the papist and worldly corruptions of the Church of England and free of the strictures of rival Protestant sects. Winthrop's enduring image of the Puritan plantation as "a city upon the hill" is revealing in this regard. Not only was it to be carved out of the wilderness by the Lord and to serve as a sign that the Puritans, like the ancient Israelites, were God's chosen people, but it was to stand as a model for all humankind. Thus, even before the tiny settlement was founded, the colonizers shared Winthrop's sense that the "eyes of all people" were on the New World society they sought to build, and that the outcome of the Puritan experiment "would be a story and a by-word through the world."[5]

Proselytism among the Indians as part of God's design for humanity was also cited as a major motive by the company officials and settler-adventurers who were struggling to establish plantations farther south along the coast. In a 1609 promotional tract exhorting his countrymen to support the colonization of Virginia, Robert Gray repeatedly alluded to biblical precedents and to the importance of converting the benighted inhabitants of the New World.[6] But in Gray's arguments, religious designs for the future of the continent and its inhabitants were intimately intertwined with civilizing projects of a more mundane character. Like many of his contemporaries, Gray paired Christian conversion with the need to raise the "barbarous and savage" Indians to a level of "civilitie" comparable to that enjoyed by the English. Though the meaning of "civility" or "civil" varied somewhat from one writer to the next, it usually connoted stable communities based on sedentary agriculture, a more intrusive and centralized government than most of the Indian peoples had experienced before the coming of the English, and improved tools and working habits that would allow both migrants and the native peoples to exploit the rich resources of North America more fully.

In the early centuries of colonization, most observers believed it was necessary to render the Indians civil before they could genuinely embrace the Christian faith. But some saw Christianity and civility as synonymous. Thomas Hariot went so far as to suggest that the Indians' intense interest in English technology and material culture could prove critical to their religious conversion. Once they had seen the navigational instruments, guns, and clocks of the English, he reported, the "priests" of the indigenous societies began to doubt their own gods and beliefs, and they were forced to conclude that if truth were to be found, "it was rather to be had from us, whom God so specially loved."[7]

Until well into the eighteenth century, religious affiliation was the foremost marker of identity for the great majority of English settlers and the attribute that above all distinguished them from the indigenous peoples. But from the very first encounters, the pervasive and obvious contrasts between the material culture of the European intruders and that of the Indians profoundly influenced the way each perceived, represented, and interacted with the other. The English were convinced that a great technological gap separated their culture from those of any of the Indian peoples, regardless of the levels of social complexity different peoples had attained. With few exceptions, the explorers and settlers who wrote about early contacts with the inhabitants of North America implicitly assumed or stated outright that the Indians realized that their tools and weapons were no match for those of the colonizers. Depictions of Indians as awestruck by the seeming magic of European mechanical contrivances were standard fare. William Wood reported that the peoples of New England were "strangers to the arts and sciences" and ignorant of "the inventions that are common to a civilized people." Consequently, they were "ravished with admiration" for English ships, which according to Wood they thought to be "walking island[s]" with masts for trees and sails for clouds. Having nothing in their own culture to compare to the English cannon, Wood imagined, the Indians compared them to lightning and thunder. Windmills, he claimed, they esteemed as "the world's wonder," and ordinary plows, which could "tear up more ground in a day than [the Indians'] clamshells could scrape up in a month" they considered to be the work of "the Devil."[8]

A number of chroniclers asserted that the Indians' recognition of the superiority of the colonizers' technology was confirmed by their eagerness to obtain European firearms, axes, or kettles. The fact that the Indians would accept simple artifacts, such as pins, needles, bells, and "such like trash," from the English in exchange for much-needed food, furs, and other marketable commodities heightened the settlers' sense of superiority and confidence that they could dominate the indigenous peoples.[9] Both settlers and company officials recommended demonstrations of technological mastery to impress the Indians with English power and perhaps convince them that the intruders were supernatural beings. Frequently citing Spanish precedents, Gray, Hariot, and others reasoned that such displays would discourage resistance to English settlement and render the Indians amenable to schemes to enlist them as laborers and provisioners for migrant communities that could not support themselves.[10]

Although some English chroniclers stressed physical differences, in the early phases of colonization, cultural rather than racial markers provided the matrix of both settler identity and interaction with the Indians. Numerous accounts contain descriptions of the Indians' comely physiques, agility, and great endurance.[11] But these qualities were no more remarkable or definitive in the representations of the indigenes by the English than the nature of the Indians' housing, their admirable canoes, or their weapons. Until well into the eighteenth century, Indians were only rarely described as having red skin, and throughout the colonial era they were neither widely identified with slavery nor seen as a distinct racial type.[12] The iron-age technology of the English was a more pervasive measure of their difference from and sense of superiority to the stone-age peoples they encountered. As the intruders were well aware, tools and weapons rather than religion were the chief source of whatever influence they were able to exert over the tribes living in the vicinity of their small and vulnerable settlements.

Technology and disease were the two most important sources of English power vis-à-vis the Indians in the early decades of settlement.[13] Though the colonizers possessed greater immunity to many of the pathogens they unwittingly carried to the New World, they had no more control over them than the Indian peoples whom the new diseases devastated and demoralized. But the newcomers' metal implements and weapons could be deployed both to enhance the settlers' power in various kinds of encounters with the Indians and in their efforts to transform the wilderness into blessed Zions and prosperous communities within an expanding English empire.

Like many of his contemporaries, John Winthrop discerned both historical and supernatural meaning in the transoceanic migrations and the effects of European settlement on the indigenous peoples. And like so many of the writers who promoted exploration and settlement, the future governor of the Massachusetts Bay Colony had not yet set foot in the new lands. But that did not deter him from asserting that before the coming of the Europeans the "whole continent, as fruitful and convenient [as it might be] for the use of man," had lain in "waste without any improvement." In Winthrop's view, the Indians' failure to render the land productive violated both God's purposes and the inexorable propensity of human societies to advance through their steadily increasing domination of nature. He concluded that the Indians' contravention of biblical injunctions for "the sons of men" to "replenish and subdue" the earthly realm that God had created to ensure that they would "increase and multiply" compelled migration from England's overcrowded and overworked lands to the New World. He was confident that English settlement and cultivation would render the lands so productive that the "natives" would "find benefit . . . by our neighborhood and learn of us to improve part [of their lands] to more use than before they could do the whole." Thus the dispossession of Amerindian peoples and the transformation of their lands were sanctioned by both God's commands and what Winthrop and his contemporaries perceived as the natural, hence inevitable, course of human history.[14]

The apparently boundless wilderness that confronted the immigrants on their arrival in North America was charged with manifold and often disturbing associations. For the Pilgrims and Puritans, it was a wasteland that defied God's designs for a bountiful earth. It was also a place of evil, where the Devil and his minions held sway. In the wilderness, the self-discipline and social control so prized by the Puritan brethren were likely to grow lax or dissolve, creating opportunities for a multitude of sins. Runaways and misfit settlers, such as the notorious Thomas Morton, who spurned the closely surveilled world of the Puritan settlements to live in the woods, were seen to personify the corruption, debauchery, and anarchy that the sectarians took to be synonymous with life in the wilderness.[15]

Although most early settlers shared the Puritans' sense that the forest beyond their farms and villages "eluded Christian norms and the established framework of Christian society," it was also a place fraught with more mundane perils. At least since medieval times, Europeans had usually represented heavily forested regions as dark, overgrown, and disordered spaces whose crude human inhabitants were prone to melancholy, isolation, and rough living. Woodlands were seen as places of danger, where wild beasts prowled and bandits and legendary wild men lay in wait for travelers or those who strayed too far from settled areas. These associations are underscored by the derivation of the concept of savagery from the Latin word *silva*, which referred to a wooded area.[16] The English who first colonized North America took "Especial Care [to] Choose a Seat for habitation that Shall not be over burthened with Woods." They stressed the need for constant vigilance to ward off predatory animals lurking in nearby forests. John Winthrop, for example, carried a gun whenever he left his home because he feared he might be attacked by wolves from the woods. But above all, the settlers regarded the wilderness as the home and refuge of indigenous peoples who were deemed, even by sympathetic observers, to exist in a savage or bestial state.[17]

The pervasive influence of this vision of North America as a wilderness where nature reigned unchecked by human laws or artifice is suggested by the fact that it is explicitly acknowledged by Thomas Hobbes as the inspiration for the most famous passage of his *Leviathan*, which he began writing in 1649. For Hobbes, the "solitary, poore, nasty, brutish, and short" lives of the indigenous peoples of realms such as America made it impossible for them to attain the prerequisites of civilized life: "In such condition, there is no place for Industry; because the fruit thereof is uncertain: and consequently no Culture of the Earth; no Navigation, nor use of the commodities that may be imported by Sea; no commodious Building; no Instruments of moving, and removing such things as require much force; no Knowledge of the face of the Earth; no account of Time; no Arts; no Letters; no Society."[18]

Hobbes's reasoning was decidedly circular: the wilderness could produce only savages, and savages could not develop the tools, work habits, or scientific understandings that would allow them to break through the constraints imposed by their subjugation to nature. It is likely that his thinking was informed by settlers' accounts that blamed the savagery of the indigenous inhabitants for what the writers perceived to be their failure to exploit the abundant resources of North America. In this view, the Indians' feeble stone age tools and weapons, their indolence (or at least that of the men), and their deeply flawed social organization made them content to adjust to the forces of nature rather than striving to subdue them in ways that would enhance their wealth and comfort. At times these views anticipated the "easy life in the lush tropics makes lazy natives" trope that would be ubiquitous in the literature of European colonization in the eighteenth and nineteenth centuries. A number of colonial writers wondered whether the temperate climate, fertility, and wealth of resources of coastal North America were responsible for its inhabitants' apparent contentment with what the intruders viewed as primitive and subsistence—though amply provisioned—patterns of livelihood. From untapped minerals and huge tracts of rich but uncultivated soil to stands of fine timber that had never been cut and broad rivers that had known neither ships nor commerce, the Indians had wasted the continent that had been bequeathed to them by the Creator.[19]

The English were confident that they possessed the skills, technology, and work ethic to tame the vast wilderness they equated with North America. Exemplifying this assurance, Robert Gray exulted: "Nature had emptied her selfe in bestowing her richest treasures upon that Countrie; so that if Art and industrie be used, as helpes to Nature, it is likely to produce the happiest attempt that ever was undertaken by the English."[20] Taking advantage of the opportunities the Indians were believed to have squandered to exploit the resources of the New World became a major arguing point for promoters petitioning for royal charters and recruiting prospective settlers. The scantily populated wilderness of North America was seen literally as a heaven-sent solution to the disturbing increase in the numbers of landless laborers and vagabonds in England due to the acceleration of the enclosure movement in the Tudor era. At a time of deepening political crisis that would culminate in the English civil war in the 1640s, those who promoted colonization argued that social tensions could be alleviated if members of the "burthenous, changeable, & unprofitable" underclasses were encouraged to seek employment and land in the New World.[21]

In a broader perspective, colonization opened up promising new enclaves in which the English could extend their sectors of the European mercantile capitalist system, which was expanding on a global scale. Advocates of exploration and settlement maintained that exploitation of the resources in America would contribute significantly to the growth of English commerce, enhancing the prosperity of the mother country. They envisioned a flow of precious metals, forest products, and foodstuffs into England in exchange for English manufactured goods that, they were certain, would find a ready market in the colonies. Though English merchants and the royal bureaucracy that levied taxes on their transactions would clearly be the greatest beneficiaries of this exchange, the tools, weapons, and other goods shipped to the New World were thought to be essential to the daunting task of transforming the wilderness into productive farmlands and prosperous towns.[22]

Although those actually laboring to settle the American wilderness acknowledged these transatlantic dimensions of their enterprise, other objectives were more central to their own sense of mission. Clearing the forests and cultivating them in the English manner would bring about an advance of civil society and the retreat of the forces of savagery. For the English settlers, plowed fields, fences, and cows in the pasture were themselves vital signs of civilization. Cutting trees not only made agriculture possible, it removed obstacles to moral advancement and social improvement.[23] It was essential to the growth of colonial economies based on the production and sale of timber and wood by-products. Though iron products, both imported and increasingly manufactured locally, were also important, the material culture of early Anglo-American society was overwhelmingly oriented to wood. Housing (for humans and domesticated animals), furnishings, storage containers, and many household, handicraft, and farming implements were made primarily of wood. Transport, whether by wagon or boat, depended on vehicles of wood construction. Timber, pitch, wood tar, and resins were key exports from most of the early plantations. And early colonial industries, from shipbuilding to the manufacture of iron goods, relied on charcoal rather than coal firing, and thus contributed heavily to the deforestation of large swaths of the coastal areas.[24]

Clearing the forest to bring the land into cultivation provided a major rationale for colonial expansion both in the early decades of settlement and during the following centuries. Few of those engaged in colonizing the New World seriously challenged the often iterated proposition that if the Indians could not make proper use of the lands and resources, the settlers were justified in occupying them and making them productive.[25] This assumption undergirded the legal precedents that various authors cited to legitimize the occupation of lands deemed to be vacant or the outright dispossession of indigenous groups. Many invoked the concept of *res nullius*, which held that "empty" or "waste" lands belonged to humanity as a whole until they had been cultivated or put to other productive use. These vacant and undeveloped areas *(vacuum domicilium)* then became the property of those who had brought them into production. Those peoples who did not practice agriculture were denied any say in the disposition of their lands. Once these areas had been farmed by the

English, the cultivators had the right to use force to deny the Indians access to them. Some writers argued that lands the indigenous peoples had actually planted were their property and should be left in their possession or should at least be purchased rather than seized by the settlers. But others held that since the Indians had no concept of property (or at least none that the English could discern or would recognize by law), the colonizers could simply appropriate their lands for farming or settlement.[26]

All these claims for the dispossession of the Indians were rooted in the English sense that colonial domination involved advancing in space rather than exerting control over people.[27] But advocates of colonization also believed that there was a critical moral dimension to domesticating the North American wilderness and its inhabitants. Settlers were bringing into production territory that had formerly been wasted, and they were heeding God's injunction for man to turn the earth into his garden. From the first decades of English exploration, some writers explicitly argued that this transformation would greatly benefit the indigenous peoples. Because the wilderness was corrupt and the dwelling place of Satan, subduing it increased the portion of the earthly realm won for Christ and occupied by the righteous. Therefore, explorers and settlers "opening up" new lands, particularly sectarians such as the Puritans, could see themselves as specially chosen agents carrying out a divine plan. Their success as farmers or merchants trading in the bountiful products of the New World was a sign of God's approval and guiding hand as well as a confirmation of their standing among the elect.[28]

The Indians' failure to turn the land's resources to productive use was linked by most settler chroniclers with their failure to advance beyond the state of savagery. Though the fact that they were not

Christians was for most of the English the definitive attribute of the Indians' savagery,[29] it was rarely seen as the cause of their cultural backwardness. In fact, because England's adventurers and merchants were active in West Africa and the Indian Ocean in the early seventeenth century, the chroniclers were well aware that non-Christian peoples had produced great (if religiously misguided) civilizations.[30] Most conceded that since the Indians had not yet been exposed to Christian teachings, they were not to blame for their heathen condition. But their ignorance, indolence, organizational ineptitude, and feeble technology were clearly responsible for their miserable, hand-to-mouth existence. These failings were almost invariably assessed with reference to English standards of resource exploitation, social organization, work patterns, and technological artifice. With rare exceptions, the colonizers assumed that English ways of thinking and doing were both normative and "natural." The ambivalence that is so pronounced in explorers' and travelers' accounts of their first contacts with Indian peoples is usually, and rather quickly, resolved through dismissive or even contemptuous assessments of those peoples' exotic customs and primitive material cultures.

Customs, level of material attainment, and moral disapprobation came together in English responses to the partial or complete nakedness of many of the indigenous peoples they encountered along the Atlantic coast (depending, in the northernmost reaches, on the season of the year). An undercurrent of envy on the part of the English travelers—constricted and sweating in their wool and leather garments and metal breastplates—for the Indians' bodily freedom and apparently innocent sensuality can be detected in a number of the early European accounts of initial contact.[31] But from the time of Columbus's first landfall in the New World, the Indians' state of undress was closely associated with the primitive state of their existence. Columbus's remark that the first Amerindian peoples he encountered "go quite naked as their mothers bore them" follows immediately upon his observation that "these people were very poor in everything." After noting the Indians' "very handsome bodies and very good faces" and brief descriptions of their hairstyles and bodily decorations, he comments at some length on their inability to comprehend the technology employed by his company of sailors and men-at-arms: "They bear no arms, nor know thereof; for I showed them swords and they grasped them by the blade and cut themselves through ignorance; they have no iron. Their darts are a kind of rod without iron, and some have at the end a fish's tooth and others, other things."[32]

In sixteenth- and seventeenth-century English accounts, the paradisiacal resonance of Amerindian nudity gave way to inferences of savagery, sinfulness, and material want. The influence of the writings of Columbus and other Iberian explorers and conquistadors on English advocates of colonization in this era is well established.[33] But the English appear to have outdone the Spanish in transforming what were first seen as Indian virtues into evidence of their technological backwardness and incapacity to comprehend the principles of commercial exchange and material advancement. Early travelers represented gifts of fish or maize, for example, as indicating the Indians' openness and generosity. By the first years of settlement, though, the Indians' willingness to trade fur pelts or essential foods for trinkets, mirrors, and iron pots was widely reported, despite considerable evidence of their aptitude for hard bargaining. Similarly, early travelers' impressions of the Indians' affinity for the lush environment, in which they were seen to move so effortlessly, were soon displaced by disparaging remarks regarding their utter submission to the whims of nature.[34]

As their plantations expanded and became more widely dispersed in the first half of the seventeenth century, English settlers came to see the Amerindian peoples' failure to master their natural environment as the underlying reason for their primitive and precarious modes of existence. This view helped to account for the paucity of Indians in areas the settlers were convinced could support far larger populations. It also resolved the apparent contradiction the English discerned between the insecurity and scarcity in the Indians' day-to-day lives and the abundant resources of the lands they inhabited.[35] Explanations for this presumed failure were usually inserted into discussions of issues that settler chroniclers or pamphleteers in England considered more immediate and pressing. The causes set forth varied depending on the writer, the time period, and the plantation area in question. But taken together they amounted to a potent indictment of Indian culture and character as these were understood and represented by the English invaders. In all cases, censure was a result of decidedly ethnocentric comparisons between English and Indian epistemologies, institutions, material culture, and modes of social organization. As with nudity and the giving of gifts, Indian attributes that were admired in the decades of initial encounters often came to be seen as emblematic of fundamental flaws in the indigenous cultures. These shortcomings were identified as both the cause and consequences of the Indians' savage state. And not surprisingly, they were untiringly elaborated by those promoting English colonization in North America and seeking to justify the dispossession of the Amerindian peoples that was the inevitable outcome of that enterprise.

Even before they encountered the Amerindians, the English had developed deep-seated prejudices against peoples who led a peripatetic existence. In seventeenth-century England, the hostility of aristocrats, bourgeois townsmen, and rural gentry was directed mainly against pastoralists—migratory peoples who based their subsistence on the maintenance of large herds of sheep, cattle, and other domesticated animals. Similar animosity was expressed in dismissive comparisons of the Indians of the New World to "barbarian" peoples, such as the Mongols, Turks, and other nomads, who had periodically threatened Christian Europe from Central and West Asia. English settlers' disparagement of the Indians' itinerant way of life may have also reflected more diffuse anxieties with regard to the menacing wild men of the European forests and mysterious gypsy wanderers. But the most immediate and visceral sources of the English colonizers' hostility toward migratory peoples were their country's earlier confrontations with pastoralists in Ireland—who, like the Indians, were deemed savages—and their fears of the bands of vagabonds who drifted through the towns and countryside of early modern England.[36]

The ever-expanding ambition of state and local officials to surveil and regulate subject populations was increasingly a source of seventeenth-century strictures against peoples on the move.[37] But control was not initially the main issue for English settlers confronting the migratory Indian populations of North America. Rather, they believed that continual movement—or in the rendering of more knowledgeable reporters, seasonal migration—was a key reason for the backward state of the technology and material culture of the Amerindian peoples. In this view, flimsy housing was easier than more substantial structures to move or abandon; cultivating small patches of land in the midst of the forest required only the crudest of tools. English observers also saw a clear link between the migratory habits of the Indian peoples and their political and social systems. With the exception of Powhatan's confederacy in Virginia, the invaders represented these systems as localized and decentralized. Because the English were certain that sedentary agriculture was the best way to use fertile land, settler farmers assumed that their rights to ownership were far superior to the feeble claims of the indigenous inhabitants. They concluded that excessive mobility had left the Indians with little or no sense of private property and utterly lacking in ambition for the long-term improvement of the areas they inhabited.[38]

In many English descriptions of Amerindian societies, prejudice against migratory peoples was expressed through dismissive assessments of the prospects of hunting and gathering cultures. Hunting was seen to be synonymous with savagery, and Indian hunters were likened to the predatory animals that roamed the American wilderness. Some accounts linked the bestial natures of hunting peoples to the keenness of their sensory perceptions, particularly sight and smell, at the expense of their rational faculties.[39] Given the exalted status of hunting in England, this harsh appraisal was perverse at best. In the British Isles, the hunt was considered an art form that not only contributed to a participant's physical well-being and moral development but provided valuable lessons in endeavors as disparate as warfare and natural philosophy.[40] But writers who dismissed the Indians as savages saw little correspondence between the hunt as it was staged in England and hunting as it was practiced by the peoples of North America. In England, hunting was a leisure pastime of the privileged classes with only marginal economic significance. It was, in sum, an anachronistic and inessential adjunct of a society based on agricultural and handicraft production. In the New World, it was a subsistence activity, pursued by virtually all able-bodied males, that yielded significant supplements to the diet of the Indians, and ironically to that of the settlers as well.

One might expect production-minded Englishmen to have valued the Indian version of the hunt over their own. But in fact they considered the Amerindians' dependence on food killed in the wild, rather than raised on settled farms, a key sign of primitivism. In their view it degraded, or at best stultified, Indian societies rather than contributing to their advancement as genuine productive labor would have done.[41] This assumption rested on the colonizers' disparagement of the Indians' agricultural skills. There was a striking and obvious discrepancy between what the settlers actually encountered (and what they often reported) in New England and on the Chesapeake and the dominant image of the indigenous peoples of these areas as hunters and gatherers rather than agriculturists. Colonial observers, from Thomas Hariot in the late sixteenth century to Robert Beverly in 1705, penned quite detailed, and at times laudatory, descriptions of Indian agricultural techniques, crops, and tools. It was widely known both in England and in the colonies that the Indians cultivated a great variety of staple crops from the ubiquitous maize to beans, melons, and potatoes. Early settlers were also aware of their own periodic reliance on foods bartered or given as gifts by neighboring Indian groups. Though often unrecognized as such, this dependence has been enshrined in the annual American commemoration of the Pilgrims' thanksgiving feast. The food at this three-day celebration consisted principally of local game, including wildfowl shot by the settlers and venison supplied by Indian hunters, as well as maize, squash, and other vegetables that the Wampanoags, who lived in the vicinity of the Plymouth settlement, had shown the Pilgrims how to cultivate. Farther south, it was the export of tobacco, an Indian crop, that finally made the Virginia colony economically viable. Some English chroniclers even recognized that Indian cultivation patterns represented ingenious and quite productive adaptations to the varying environmental conditions in the diverse ecosystems on the coast between Maine and the Carolinas.[42]

However much the success of the early English plantations may have hinged on Indian agriculture, as settlement expanded the colonizers compared Indian crop production unfavorably with their own methods of farming. The Indians' mobility was seen to be a major source of their shortcomings as cultivators. Because of their continual migration, they seldom fully cleared the areas where they planted their crops, but were content to burn off the forest undergrowth and insert their seeds in holes poked in the ashes. Prefiguring a standard Western response to shifting cultivation practices by peoples from Southeast Asia to the Amazon basin in the following centuries, English settlers dismissed the Indians' burning of the forests to clear them for planting as ineffectual and destructive.[43] Seasonal wanderings, which included moving entire villages, meant that the Indians found it unnecessary to plough and manure their fields, or to build fences, barns, or elaborate granaries. Because there were no large domesticated animals in North America before the coming of the Europeans, the Indians made no use of animal power or manure in agricultural production. Most English writers who bothered to comment deemed Indian tools—wooden digging sticks and hoes—crude in the extreme, even though it would be decades later before the English themselves made extensive use of plows.[44]

Perhaps no aspect of Indian agricultural practices elicited more disdainful English responses than the fact that in Amerindian societies cultivation was overwhelmingly women's work. In seventeenth-century England, women participated in various aspects of agricultural production, especially those involving the processing of dairy products and the manufacture of clothing. Women also worked in the fields alongside the men in peak seasons, such as harvest time, and women of the laboring classes hired themselves out for field work throughout the growing season. But this mixing of productive roles deviated from the English ideal which held that women's work ought to be confined to the household and the farmyard. Cultivation, particularly that connected to operations involving the plow, was held to be an activity for males. In fact, plowing was strongly associated with manliness among non-aristocratic social groups in rural areas.[45]

The gendering of food production in the Amerindian societies along the Atlantic coast was virtually the exact opposite of the pattern followed in England. Though men helped with clearing and planting in some societies,[46] the cultivation of staples like maize, peas, and squash was women's work. English observers, even though some conceded that Indian women were quite skilled in producing a surprisingly wide range of crops, devalued Indian agriculture both as a female pursuit and often as a supplemental rather than essential source of foodstuffs.[47] Perhaps in part because the English, in the colonies and especially in England itself, were intensely interested in acquiring marketable animal pelts, they represented the Indian peoples primarily as hunters and gatherers. Moreover, depicting the Indians as the skilled farmers they often were might have cast doubt on the settlers' justifications for appropriation of land. Consequently, settler chroniclers tended to understate or ignore the substantial contribution of agriculture to the Indians' subsistence-oriented economies.[48]

The colonizers' devaluation of both the hunting and agricultural modes of production pursued by the indigenous peoples they encountered was central to their representation of the Indians as savages. Largely oblivious to the strenuous labor, specialized skills, and physical endurance that were essential for successful hunting,[49] and usually encountering Indian men only when they were resting in their villages between expeditions into the wilds, English observers derided them as lazy and idle. By contrast, Indian women were depicted as "squaw drudges" who did virtually all the hard work in their communities, from the construction of housing to cooking to the fabrication of clothing and the cultivation of crops.[50] The Indians' patterns of gendering work, as the English perceived them, not only violated the colonizers' ideal of a clear distinction between male and female spheres of activity along public and domestic lines, they prompted moral indignation regarding the ways in which Indian women were misused and degraded.

The Indian men's alleged brutalization and oppression of women was linked to other facets of their savage natures. In contrast to their overworked wives and daughters, Indian men were represented as

incorrigibly indolent. Rather than assume their proper roles as cultivators and builders, they were reported to laze about in their villages—smoking, gossiping, or sleeping when they should have been
engaged in productive labor. Part of the colonizers' disapproval may
have arisen from the fact that most of the Amerindian peoples had
little interest in providing labor for the settlers.[51] Like all primitive
peoples (and children), Indians were said by English writers to have
little sense of the worth of time and hard work. They were seen to be
shiftless, untruthful, improvident, and fatalistically resigned to the
ravages inflicted by natural forces and human rivals.[52] From the
English point of view, all these character flaws were even more reprehensible given that the Amerindian peoples inhabited lands so fertile that it should have been possible for them to enjoy lives of plenty
and prosperity with a minimum expenditure of labor.

Very often the English linked the Indians' deficiencies as subsistence hunters to what were dismissed as the primitive and unmanly
ways in which they engaged in warfare. Despite significant settler casualties in early clashes that suggested otherwise, Indian weapons
were deemed crude and ineffectual. Conversely, the Indians' admiration for English firearms was reported to be profuse. The settlers gave scant notice to the Indians' rather remarkable aptitude for
adopting European firearms and adjusting to English military tactics
and weaponry, which has been well documented by historians. The
English chroniclers also played down the considerable adaptations
that the settlers made in the ways they employed their weapons and
the changes they made in military tactics to counter those of their
Indian adversaries.[53]

What the colonizers did emphasize was, as they derisively termed
it, the Indians' "skulking" way of making war.[54] The Indians' approach to battle was often perceived as involving a cowardly reliance
on ambushes and surprise assaults. Here again the Amerindians' ingenuity in making use of the forest and other forms of natural cover

was usually reported as evidence of unmanly and treacherous temperaments rather than as effective exploitation of the environment. Predictably, advocates and agents of English colonization tended to stress petty, vindictive motives for conflicts between different Indian peoples and perfidious designs for their assaults on settler communities. Though usually depicted as physically strong and fierce in hand-to-hand combat, Indian warriors were seen by English adversaries as undisciplined and prone to flight in the face of determined resistance.

No writer summarized the English view of shortcomings of the Indians' ways of warfare more succinctly than William Wood. He implicitly contrasted Indian approaches to battle with the principles favored in European military theory and practice, which included an emphasis on innovative weaponry, bodily discipline, regular military formations, and coordinated troop movements. He ridiculed "the antic warriors [who] make towards their enemies in a disordered manner, without any soldierlike marching or warlike postures, being deaf to any word of command, ignorant of falling off or falling on, of doubling ranks or files, but let fly their winged shaftments without either fear or wit. Their artillery being spent, he that hath no arms to fight, finds legs to run away." Wood and other English chroniclers even compared the Indians' methods of waging war — which sometimes involved sneaking up on their enemies on all fours — to the behavior of wild animals, and pronounced them unworthy of civilized men. Like their modes of hunting, the Indians' ways of making war were seen as evidence of their backwardness and savagery.[55]

The intense ethnocentrism of most seventeenth-century English appraisals of Amerindian peoples and cultures is in itself unremarkable. Like virtually all participants in the early stages of cross-cultural contact, the English assumed that their own ways of thinking, whether about the natural or the supernatural world, and of organiz-

ing political systems or planting crops were the norm against which
those of the exotic societies they encountered *ought* to be judged. It
is not surprising that they found so much about the indigenous societies to criticize, if not condemn, and so little to recommend to their
countrymen across the Atlantic. But as sixteenth- and seventeenth-
century English responses (and those of Western Europeans more
generally) to Asian cultures in South Asia, China, and Japan suggest,
exoticism alone is not sufficient to explain the generally harsh settler
estimates of Amerindian societies.[56] Neither the great differences between European and Asian cultures nor the Christian-pagan divide
that so fundamentally separated them prevented European travelers,
traders, and missionaries from expressing wonderment at the sophistication of Asian civilizations. On the contrary, many of the Europeans engaged in cross-cultural exchanges in Asia concluded that the
civilizations they encountered there, such as those dominated by the
Mughal empire in India and the Ming dynasty in China, had much
to teach Europe about political reform or social organization, tolerance or monumental architecture.

In sharp contrast, cultures—such as those of Amerindians or sub-
Saharan Africans—that were judged to be savage were presumed by
all but a handful of English (and European) observers to have little
worth emulating.[57] And very early on, European explorers and settlers decided that the Amerindians of coastal North America had
not, despite some signs of civility, advanced beyond the savage state.
Religious differences often played a major role in this determination. But the vast gap that the English perceived between what they
deemed a civilized level of material culture and the backward state
of the Amerindian peoples was decisive. It provided key rationales
for the English appropriation of the lands that the Indians, with their
primitive technologies and backward social systems, had failed to exploit effectively. It justified the invaders' subjugation and dispposses-
sion of the indigenous peoples, and it facilitated the rather rapid

takeover of areas where the English settled. By contrast, European invasions, much less conquests, of India or China in this period were unimaginable. The North American colonizers' confidence in their technological superiority also decisively shaped their schemes for civilizing the indigenous peoples who managed to survive the epidemics and armed conflicts that had done so much to define the Amerindians' doleful history in the early years of English settlement.

❖   ❖   ❖

Decades before they established the first successful plantations in Virginia and New England, English explorers and promoters of colonization justified their ventures in terms of the civilizing influences that migration and settlement would inevitably diffuse to the savage peoples of coastal North America. Opportunities to spread the "gladsome tidings of the most glorious Gospel of our Savior Jesus Christ" were, of course, central to their sense of how the indigenous peoples would be uplifted. But as early as 1583, George Knight moved beyond obligatory references to the colonizers' responsibility to spread Christianity to an account of the material improvements that Indian societies would enjoy as a result of contacts with English settlers. Countering objections that colonization might prove harmful to and hence be resisted by the indigenous peoples, Knight argued that they would soon realize that the English had brought them means by which they might rise up from "brutish ignorance to civilitie and knowledge," by teaching them "how the tenth part of their Land may be so manured and employed, as it may yeeld more commodities to the necessary use of mans life, then the whole now doeth." Though Knight viewed proper farming techniques as the most fundamental material improvement "Christian" settlers could engender in Indian societies, he saw them as just one facet of a much more comprehensive transformation that would utterly remake the indigenous peoples' cultures and lives. His vision anticipated to a remarkable degree the objectives of the missionaries, educators, and government agents who would seek to draw the Native Americans into English and later American culture in the following centuries: "Over and beside the knowledge [of] how to till and dresse their grounds, they shal be reduced from unseemely customes to honest manners, from disordered riotous routs and companyes to a well governed common wealth, and withall, shalbe taught mechanicall occupations, arts, and liberall sciences."[58]

Knight's ambitious agenda for acculturating the Indians rested on a number of assumptions held by both the early settlers and those who later sought through educational and resettlement schemes to transform indigenous societies into something akin to the English ideal of civility. Virtually all facets of English culture were seen as superior to those of the Amerindians, and thus worthy of emulation by these backward peoples. Religious instruction and conversion were given the greatest emphasis in promotional tracts and chronicles of colonization. But the transfer of technology, the introduction of English-style agriculture, and even the fundamental alteration of the Native Americans' ways of interacting with the natural world were also central to seventeenth-century civilizing schemes.

The settlers' confidence that the Indians were eager to acquire their iron-age technology and would eventually adopt their farming techniques was premised on the conviction that advances in science and invention had elevated Europeans to a level of mastery over nature that was unimaginable for the inhabitants of the New World. William Strachey, in a 1612 account of a visit to Virginia, stressed this mastery and the potential it offered for material improvement if extended to Amerindian peoples. Both reflecting ideas then current among England's scientifically minded intellectuals and anticipating Francis Bacon's thinking in the New Atlantis (published in 1624), Strachey assumed the superior capacity of Englishmen to probe the secrets of nature and put its powers to productive and profitable use. Employing the heavily gendered language of domination favored by natural philosophers of his age, Strachey argued that the greater understanding the English possessed of the natural world would surely elevate the Indians' condition because it promised to "bring them from bodily wants, confusion, misery, and these outward anguishes, to a knowledge of a better practize, and ymproving of those benefitts (to a more and ever during [enduring] advantage and to a civeler vse) which god hath given vnto them, but envoued and hid in the bowells and womb of their Land (to them barren and vnprofitable, because vnknowne)."[59]

In the decades of settlement that followed Strachey's account, the commitment to transforming the Indians from hunters and gatherers into English-style farmers became a defining component of the Anglo-American colonizers' sense of their civilizing mission. Most English writers on the topic expressed confidence that the Amerindian peoples had the intelligence and dexterity to adopt European tools and modes of agricultural production. They had after all shown considerable ingenuity in devising their own weapons, canoes, and tools, even though these stone-age artifacts were crude by European standards. The English were also impressed by the alacrity with which Indian peoples had learned to use, repair, and even manufacture English weapons and other implements. In fact, the Indians had proved so adept at imitation that influential settlers, including John Smith and John Winthrop, warned of the dangers of teaching them English handicraft or agricultural skills, and especially of allowing them to acquire firearms. Few English observers would have disagreed with Robert Gray's assertion that "it is not the nature of men, but the education of men, which make them barbarous and uncivill, and therefore chaunge the education of men, and you shall see that their nature will be greatly rectified and corrected."[60]

Here cultural not racial factors were seen to be responsible for the

Indians' savage state. Gray and others presumed not only that Indian culture was malleable but that it would inevitably come to resemble that of the English as continuing interaction demonstrated to the Amerindian peoples the superiority of European material culture, modes of organization, and epistemologies. Gray also underscored the centrality of education in the process of acculturation, though he had little to say about how schooling for the Indians would be funded or the content of the instruction that would be offered.

The handful of missionaries and educators who some decades later actually devised schemes to assimilate Amerindian peoples to European culture viewed religious conversion and related measures to elevate moral standards as the central goals of their endeavors. But even in the "praying towns" established by John Eliot and his co-workers in mid-seventeenth-century New England, settled agriculture, the teaching of craft skills, and the transfer of technology were major objectives. The Indian settlements founded or inspired by Eliot were by far the most ambitious attempts to promote the acculturation of the indigenous peoples in the first century of contact. Underfunded, poorly staffed earlier efforts to establish schools for local Indians in Virginia had floundered, particularly after an Indian uprising in 1622. These endeavors were not seriously resumed until the early 1700s. In New England, as in Virginia, the high-minded rhetoric of conversion and uplift in promotional pamphlets usually bore little relation to the meager resources actually available for sustained missionary or educational projects. But over several decades Eliot managed to win enough financial support from a variety of sources in England and to draw sufficient resources from Indians who gathered in the praying towns to mount a sustained program aimed at turning the indigenous peoples into civilized Christians of the Protestant persuasion.[61]

Between 1651, when the first village for Indian converts was established at Nonantum, and 1675, when the outbreak of King Philip's

War began a sequence of assault and reprisal that all but destroyed John Eliot's aspiring utopias, fourteen settlements were established in Massachusetts. Central to Eliot's civilizing objectives was the propagation of agricultural skills and a sedentary way of life. Eliot and his missionary co-workers devised instructional regimens designed to put an end to the Indians' seasonal migrations and to convert them from subsistence hunter-gatherers, eking out an existence on lands that remained wild, to prosperous market farmers, working fixed fields that were legally registered as their property. What Eliot and other settlers dismissed as marginal and inefficient Indian cultivation patterns were to be replaced by English-style hoe, mattock and, somewhat later, plow agriculture. Like the Puritan communities that helped to oversee the project, and the English villages upon which they were modeled, the praying towns were increasingly defined by orchards, domesticated animals, and fenced fields. Over time the Indian residents adopted English farm implements, household tools, and methods of working the land. In addition to religious instruction, education was oriented to the inculcation of craft skills, from carpentry and masonry for men to spinning and weaving for women. Emphasis was also placed on instilling the self-discipline and commitment to manual labor that the English celebrated in their own culture and found scandalously lacking in Amerindian societies, at least among the men.[62]

If successful, the process of acculturation that the praying towns were established to promote would have brought about the effective extinction of the epistemologies, customs, modes of social and political organization, and patterns of behavior of the Amerindian peoples.[63] And the Puritans were not alone in demanding the cultural capitulation of the peoples they encountered in the New World. Assumptions of superiority were pervasive in colonizers' responses from the early days of contact. But they could not be fully acted upon as long as what Richard White has called the "middle ground" between settlers and Indians persisted.[64] In this spatial and psycho-logical zone, where neither side had yet gained dominance over the other, a rough but uneven balance was maintained in military, eco-nomic, and cultural exchanges. As the middle ground that had buf-fered the impact of the settlers' advance eroded in the late 1620s and the 1630s, and then utterly collapsed in many areas after midcentury, European arrogance and assertiveness intensified. Surviving Indian groups in coastal New England and much of the tidewater country in Virginia were left with the choice of becoming dependent clients of settler communities or of fleeing farther inland. Resort to the lat-ter option often brought them into conflict with other indigenous peoples.

Particularly in New England, the middle ground was from the be-ginning much weaker along the coast than farther west in the Great Lakes region. It was imperiled by the higher concentration of Euro-pean settlers in the coastal areas and by the devastating effects of epi-demic disease, unwittingly carried to North America by English mi-grants. It was also undermined by key technological advantages, particularly those relating to iron tools and firearms, enjoyed by the invaders. The importance of this combination in settlers' encounters with coastal peoples was explicitly recognized as early as 1606. The official instructions from the directors of the London Virginia Com-pany urged the leaders of the expedition that established the first permanent English settlement at Jamestown to employ only their best marksmen so that the Indians would be so impressed with Eng-lish weaponry that they would not dare assault the intruders. The settlers were also admonished "not to Let them [the Indians] See or know of Your Sick men" or the casualties suffered in clashes with the locals, who might conclude from this evidence that the English were little more than "Common men" and be encouraged to resist their efforts to establish an enduring plantation.[65]

Decimated and demoralized by diseases that were far less lethal for the colonizers, Indian peoples in areas where settlement was concentrated, such as Virginia and New England, were steadily drawn into webs of dependency anchored in part by the superior, iron-age technologies wielded by the colonizers. Early on, the settlers' psychologically daunting but often ineffective firearms and their more lethal swords were counterbalanced by their dependence on Amerindian peoples for native foods cultivated with indigenous tools. But as clashes between settlers and Indians grew in scale and intensity, and as the settlers' expansion exacerbated competition and conflict between different tribes and clans, increasing numbers of the Indian groups that survived at all were reduced to the status of subordinate allies of English colonies. At the same time, the Indians' demand for weapons of European manufacture, as well as a diverse array of goods ranging from tools and iron pots to blankets, mirrors, and hats, locked many of them into market exchanges in which English traders and farmers usually enjoyed the upper hand. Despite the ingenious ways Indian peoples integrated European tools and utensils into their existing material cultures, their reliance on imported goods resulted in the decline and eventually the complete loss of many traditional handicraft skills. The disappearance of such skills in turn fed an ever greater need for imported products.[66]

Christian settlements, such as those pioneered by Eliot and his coworkers, did little to preserve the middle ground that might have buffered the impact of the ever-expansive English invaders. Attempts to transmit European craft skills to Indian converts through apprenticeships floundered because of the ambivalence and outright suspicion of settlers and tradesmen. Educated Indians might well become competitors of English carpenters, masons, and ironworkers, as well as alternative sources of goods that were producing sizable profits for English merchants. Those Indian converts who took up English-style farming often found that fences and what they thought to be legal claims to the fields they cultivated offered little protection from land-hungry settlers.[67] Their displacement and the subsequent dissolution of their societies foreshadowed the policy of removal that over the next century and a half would be inflicted on the coastal peoples who had managed to survive incessant warfare and successive epidemics and had adopted the farming patterns and a good deal of the way of life of the colonizers.

Only a handful of settlers participated directly in the establishment of the praying towns, and most of the funds to support them came from Puritan donors in England rather than communities in North America.[68] But virtually all of those involved in the colonial enterprise would have affirmed the underlying assumptions that informed these projects, and few would have questioned the outcomes they were intended to achieve. Although the colonists were often deeply divided by social and sectarian differences, there was a virtual consensus across social strata, religious denominations, and communities scattered along the Atlantic littoral about the need to dominate and either acculturate or dispossess the indigenous peoples. The discourse on the necessity of converting and civilizing the Indians gradually coalesced over the seventeenth century into a hegemonic ideology that was foundational for all the English settler enclaves along the coast. Puritan elders and Virginia aristocrats consciously elaborated its key tenets in their journals and promotional tracts, and these were accepted at all levels of settler society and by those in England who wrote about or sought to shape policy for the colonies.

Even those who attempted to defend the Indians against the injustices inflicted by the settlers accepted the logic that sanctified their dispossession. The Indians had not advanced beyond savagery, and they had not devised the tools, developed the work habits, or gained the understandings of the workings of nature that would allow them to make productive use of the lands they inhabited. As a result, according to what became the consensus history of the English occupation of North America, until the arrival of the Europeans the continent had been a howling wilderness, sparsely occupied by nomadic peoples whose material culture was primitive and whose lives were occupied by precarious struggles to maintain themselves at subsistence levels. The English, having attained a higher stage of civilization, had created the scientific knowledge and devised the technology required to harness the land's abundant resources, which had been squandered by the indigenous inhabitants. The English predilection for hard work, conception of private property, and market orientation would enable them to achieve levels of material development and affluence vastly superior to the debased existence tolerated by the Indians. By example and through education, the colonizers could justify their dominance over and dispossession of these benighted peoples by conferring benefits on them. From the English invaders, the Indians would not only learn the good news of the Christian dispensation, they would also acquire the skills, tools, and knowledge to build prosperous, civilized communities.

The pivotal roles played by the railroad in the advance of the settler frontier and the displacement of the Plains Indians are captured in this 1860s lithograph. From Samuel Bowles, *Our New West* (1869).

Steam is a great civilizer . . . it will tend to the maintenance of peace, security of travel, and prosperity and strength in every department of government.

—The Reverend Samuel Wells Williams
to William Henry Seward, 1868

2

MACHINES AND MANIFEST DESTINY

Until the advent of photography there had never been a visual record of a historical process to equal the hundreds of landscape paintings, lithographs, and sketches that traced the settlement of America's western frontiers. From the early nineteenth century until the first decades of the twentieth, foreign-born and home-bred artists chronicled the nation's continental transformations. Though seldom expressive of avant-garde trends in art, frontier landscapes often captured the majestic beauty and immense scale of the American West. Almost always realistic, in styles ranging from quasi-primitivism to Romantic naturalism, the paintings provide first-hand impressions of the diverse cultures of the Amerindian peoples, the stages of Euroamerican expansionism, and the upheavals generated by the settlers' advance and the displacement of the Indians. They celebrate the self-reliance of trappers, farmers, and cowboys; the endurance of lonely pioneer women; and the quasi-mythic association of freedom with the frontier. Western landscapes also document the pivotal roles played by industrial technologies—from farm machinery and canal locks to the telegraph and the railway—in the spread of settlement across plains and mountains to the Pacific coast. And they provide the setting for innumerable violent scenes in which advancing settlers deploy lethal weaponry—from repeating rifles and six-shooters to Gatling guns—against indigenous peoples.

Many frontier landscapes recapitulate themes that emerged in the first decades of English coastal settlement in the early seventeenth century and persisted through the following century and a half of Anglo-American expansion into the interior. But in paintings of the plains and mountains of the West, there is a dramatic increase in the scale of the natural settings depicted and (often implicitly) an inflated estimate of the historic significance of the events taking place against these monumental backdrops. In Alfred Miller's *Indians Chasing a Deer* and George Catlin's paintings of Indians hunting bison in the 1830s, men and beasts are dwarfed by the hills and prairie that stretch to the even vaster expanse of the sky. The land in the scenes of the summer hunt is verdant, but uncultivated and devoid of signs of human habitation, suggesting untapped potential for agriculture and ample space for settlement. Both beckoned on the western plains, which had been egregiously misrepresented at the turn of the century in schoolbooks and on maps as "The Great American Desert," but which were widely hailed by midcentury as

George Catlin, *Buffalo Chase, Mouth of the Yellowstone*, 1832. Oil on canvas.
National Collection of Fine Arts, Smithsonian Institution.

the "Land of Promise, and the Canaan of our time . . . With a soil more fertile than human agriculture has yet tilled; with a climate balmy and healthful."[1]

In the landscapes of the more mountainous terrain farther west that were in vogue in the 1860s and 1870s—partly because of the Colorado gold rush that began in 1859[2]—the juxtaposition of raw, massive natural settings and minuscule human figures is even more dramatic. In most of Albert Bierstadt's outsized canvases of the Rockies and Sierras, for example, there is little or no sign of Indian habitation or material presence. And even though Indians, their ponies, and their encampment sprawl across the foreground of his monumental painting *The Rocky Mountains, Landers Peak* (1863), they are all but overwhelmed by the canyon walls and mountain peaks that loom into the far distance above them and cover three-fourths of the canvas. The focal point of the painting is the sunlit waterfall and its reflection in the lake, which draw the viewer's attention away from the Indian encampment that is almost wholly enveloped in shade. The community that peoples the camp is vibrant and at peace, but its bison-hide dwellings are crude, and a sense of transience and dis-organization is conveyed by the clutter of travoises, dogs, horses, and recently killed game in the lower right corner of the painting.[3] In *The Grand Canyon of the Yellowstone,* painted by Thomas Moran a decade later, what appear to be an Indian and a frontiersman sharing the view are rendered as tiny specks in the lower left center of the panoramic spectacle, while the Indian presence in Moran's *Chasm of the Colorado* (1873–1874) is reduced to bits and pieces of abandoned implements or weapons amid the rocky outcroppings that frame the lower foreground.

The exponential increase in the scale of the continental expanse captured by artists of the nineteenth century over the confined plan-tation enclaves depicted in colonial sketches and writings greatly heightened the intruders' fixation on the abundance of resources left untouched by the Indians. Western landscapes represented— with varying degrees of accuracy—the poverty of the Indians' stone-age material culture, which left them at the mercy of the forces of nature. The vagabond, improvident nature of their lives was under-scored by artistic impressions of the vast spaces they roamed. Paint-ings of Indian life made it clear that peoples of the plains and the American Southwest had domesticated horses, which European in-vaders had introduced to the land from the fifteenth century. But art also provided graphic evidence of indigenes' failure to advance be-yond the hunting-and-gathering stage of human existence to the sed-entary, farming lifestyle that Euroamericans regarded as essential for civilized life. Even more than the early colonizers, frontier settlers

Albert Bierstadt, *The Rocky Mountains, Lander's Peak*, 1863. James Smillie's steel engraving of the painting, 1866. Print Collection, Miriam and Ira D. Wallach Division of Art, The New York Public Library, Astor, Lennox and Tilden Foundations.

were dismissive of the often considerable levels of agricultural pro-duction achieved by the plains Indians and peoples farther west.

Until the 1860s, when photography began to supplant paintings and sketches, landscape exhibitions were the chief source of visual information about the western frontier. As the soon-to-be secretary of state, William Seward, proclaimed to a gathering of Minnesotans during a campaign swing in September 1860, he had been prepared for the natural grandeur of the watershed regions of the northern Mississippi River because he had read geographers' accounts and "had studied these scenes in the finest productions of art." Political notables like Seward had ample opportunities to admire vistas of the West in the outsized oil paintings that adorned the walls of the

Capitol and other government buildings, and in works that were regularly exhibited in art galleries in New York, Boston, and Philadelphia. Showings of Alfred Bierstadt's "Great Pictures," for example, were often staged theatrical events. After purchasing tickets, viewers were ushered into large rooms, in each of which the focal point was one wall-sized painting. Bierstadt's exhibitions, sometimes accompanied by dramatic performances, were enhanced by special lighting designed to accentuate the spectacle of his panoramas of the West.[4] Landscapes by Bierstadt and other frontier artists were popularized by lithographs and engravings churned out by firms like Currier and Ives that catered to popular demand. And penny dreadfuls and illustrated newspapers featured sensationalistic lithographs or ink sketches of cowboy shootouts and alleged Indian outrages. Engravings on banknotes and government bonds showing farmers felling trees and Indians slumped in despair at the spread of mills and factory towns reminded investors of the higher national purposes to which their capital was being applied.[5]

The heroic dimensions of frontier expansion were extolled in hundreds of overwrought orations punctuating Fourth of July celebrations or the opening of the canals, railways, and new territories that marked successive advances of Anglo-American domination into the western wilderness. And the spread of agrarian and then industrial civilization across the continent was lauded by poets and essayists from Walt Whitman to Ralph Waldo Emerson. In "The Young American," a lecture read to the Mercantile Library Association of Boston in 1844, Emerson emphasized the centrality of technology in the nation's rapid expansion. He explored the essential roles of engineers, and of machines such as steamboats and railroads, in opening the western frontiers and "bind[ing] them fast in one web." These ingenious contrivances, he proclaimed, made it possible for energetic Americans to extract the "boundless resources" of the continent to fuel the nation's drive for industrializa-

tion, promote the spread of its commerce across the globe, and secure for its citizenry an "organic simplicity and liberty . . . which
offers opportunity to the human mind not known in any other region."[6]

The foundational status of these representations of frontier expansion was confirmed by the master narratives of the American experience written by historians of successive generations, including some
of the most distinguished practitioners of their craft. Throughout the
nineteenth and well into the twentieth century, American school-
children imbibed the saga of the rise of their nation from textbooks
that described that process as an epic struggle by sturdy yeoman
farmers to civilize a continent thinly peopled by savage Indians.
From the 1890s, Frederick Jackson Turner's essays on the enduring
influence of the moving frontier on United States history provided
what was then believed to be an empirical, objective validation by a
professional historian for this settler perspective.[7]

These narratives also lent legitimacy to the ever more widely held
conviction that the society fashioned from the progressive mastery of
the North American wilderness ought to serve as a model of modernity for all humankind. When the Indians were included at all
in this hegemonic narrative of pioneers taming the "Wild West,"
they were almost invariably treated as obstacles to the fulfillment of
America's national destiny, and as untamed creatures of a natural
realm who were being brought into history by a uniquely inventive
culture. Missionaries and government agents sought to revive the
goal of the praying towns of colonial New England: teaching those
Indians who managed to survive the settlers' accelerating dominance how to assimilate to Anglo-American culture. But the gap that
separated settlers from Indians, widened by technology-driven transformations in "white" material culture over the course of the nineteenth century, led to pessimism among Anglo-American educators
and bureaucrats regarding the ability of the indigenous peoples to

adopt the "civilized" ways of life proffered by the continent's new masters.

Many nineteenth-century advocates of expansion envisioned the drive across the continent as part of a larger global progression that would soon encompass societies more worthy than those of the rapidly disappearing Indians who had proved incapable of fulfilling the promise offered by the emissaries from the city upon a hill. A surprising number of frontier artists drew on Bishop Berkeley's early eighteenth-century vision "Westward the course of empire takes its way" for titles of paintings and lithographs that portrayed the settlement of the West (some of which are discussed later in this chapter). But in contrast to the British philosopher, who viewed Greece as the original home of civilization, American expansionists favored China and Japan. They envisioned the ancient lands of East Asia as the ultimate destination of the moving frontier that would both join the Atlantic and Pacific coasts and provide the resources and impetus for America's mission to revitalize the decadent civilizations of the Old World. Poets, philosophers, and statesmen as diverse as Whitman, Henry David Thoreau, Thomas Hart Benton, and William Gilpin shared Emerson's certitude that America was the "country of the future," steadily progressing toward a "new and more excellent social state than history has recorded."[8]

The underlying rationale for many expansionists had shifted by the end of the nineteenth century from divine mandate to natural law. But the ideological imperatives that nature was to be mastered, resources exploited to the fullest, and technologies invented as these enterprises required had become defining sources of Americans' sense of themselves as an exceptional people, and of the ways they conceived their relationship to the rest of the world. The rapid spread of industrialization intensified the nation's commitment to science- and technology-driven growth in productivity and increasingly consumption. It fixed, for many, the conviction that American-style free-market capitalism and representative democracy provided an institutional framework for societal improvement that transcended differences among even the most disparate human cultures. Sustained material increase engendered a commitment to progressive improvement, which came to be regarded as an essential attribute of civilized society in the modern age. Americans represented themselves as energetic, hard working, inventive, adaptable, and in control of their fate, and believed that these characteristics set them apart from both the passive, backward, indolent Indians and the stagnant, fatalistic, tradition-bound societies of East Asia. The stress placed on these essentialized differences proved critical to a broad acceptance by Americans of justifications for dispossessing the Indians and for extending the nation's civilizing mission to even the most ancient and sophisticated societies on the far side of the Pacific.

It had taken nearly two centuries for European invaders—some Dutch and French, but most from the British Isles—to claim and settle the coastal seaboard and hinterlands of the area that would eventually comprise the original thirteen States of the American republic. By the decades of revolutionary turmoil in the late eighteenth century, European settlers had also begun to move into the Ohio valley, but their numbers on the frontier remained small and the pace at which they felled the forests and brought the land under cultivation was comparable to that of the colonial era. Beginning in the early nineteenth century, however, there was a steady increase in Euroamerican migration to and exploitation of the frontier lands that would ultimately comprise roughly four fifths of the continental United States. Within decades, the vast plains and mountains and the Pacific coast had been wrested from the Native Americans and the Mexicans and absorbed into the Union. The contrast between the two phases of expansion is all the more remarkable because the climate, topography, and varied environments of the West were far more formidable than those of the Atlantic coast. Prairie soils were fertile but heavy and thatched with vegetation, thus more difficult to plow for planting than those of the Atlantic seaboard. Over much of the continental interior, seasonal temperatures, winds, and other natural disturbances tended to extremes rarely experienced in coastal areas. The Rocky and Sierra mountains posed greater obstacles to travel and trade than the Alleghenies or the Appalachians.

Revolution and the founding of the first independent nation in the Americas, which prompted more political support for exploration and settlement, account for part of the acceleration of frontier expansion from the end of the eighteenth century. But an unprecedented surge of technological innovation was largely responsible for the changes in pace and magnitude. Within the lifetime of Americans born in the first decades of the nineteenth century, the United States industrialized, became a transcontinental power, and began to expand overseas on a sustained basis. And these nation-defining processes were inextricably linked in myriad ways.

The westward expansion involved not one but many frontiers. And the resource endowment, the topography, and the timing of the arrival of settlers in different areas were major determinants of which technologies would be deployed to facilitate which patterns of settlement and production. Much of the West was initially probed by hunters and trappers, whose iron-age but largely preindustrial material culture, means of sustenance, and nomadic way of life resembled those of the Indian peoples with whom they traded, socialized, and intermarried. Even the firearms, animal traps, and well-tempered knives of these first European arrivals in frontier regions were soon assimilated into Amerindian cultures. And as the paintings and sketches of Alfred Jacob Miller, Charles Deas, William Ramney, and other midcentury artists amply attest, hunters and trappers often adopted Indian techniques of pursuing game, types of clothing, and modes of constructing shelters. Such cross-cultural assimilation was also apparent on the ranching frontiers, where Anglo-American and Mexican cattle herders adapted each others' techniques of breeding, grazing, and driving cattle, and developed hybrid equipment that included such essentials as lassos, saddles, and spurs.[9]

The success of these early forays into the diverse ecosystems of the frontiers often depended on handcrafted implements and pre-industrial machinery. Watermills, for example, remained the preeminent source of power for processing agricultural produce, turning timber into lumber, and manufacturing textiles until the mid-nineteenth century. The superb American axe, which colonists along the Atlantic had wielded to clear land and transform forests into exports, remained a critical tool both for settlers in the upper Midwest and on the great plains and for those engaged in the lumber industry that flourished first in the Great Lakes region and later in the Pacific Northwest. Single-bitted axes were not replaced by long-handled, double-bitted ones and steel crosscut saws until the 1870s. The "common" plows used to cultivate the prairies in the early nineteenth century were slightly improved versions of the wooden implements of the colonial era. To turn and break up the heavy, vegetation-thatched sod found on the great plains, iron and later steel strips were added to the moldboard. But neither innovation prevented the soil from caking on the moldboard or significantly reduced the time and labor required to prepare the soil for cultivation.[10]

As steam power and industrial production spread across the young republic in the middle of the century, a revolution in transportation and communications combined with major improvements in machines for processing raw materials and cultivating soil made possible ever higher levels of resource extraction. Advances in iron-working for locks facilitated the construction of canals that linked newly settled areas with steamboat carriers on the Mississippi and centers of commerce and industry on the Great Lakes and on the Atlantic coast. Steam engines drove mechanized "donkeys" that skidded cut timber to rivers and railway lines that carried it to steam-powered sawmills where it was sawed into lumber. New technologies enabled logging companies to deplete within decades forests in the upper Midwest and the Pacific Northwest more extensive than those in the Atlantic colonies that had taken centuries to clear. By the 1870s lumbering was the nation's largest industry in terms of gross sales. Steam-powered machines drained western mines and carted the ores to railway junctions where they were transferred to trains that carried them to the smelters and rolling mills of the Midwest and the East. Increasingly specialized workshops turned out farm implements and firearms in quantities that made them affordable for farmers struggling to wrest a livelihood from oversized, drought-prone, and contested parcels of land across the plains. In the 1830s and 1840s John Deere's "singing" plow, which combined a steel share and a polished, wrought-iron (later steel) moldboard, proved highly effective for turning and cutting the humus-rich soils of the eastern prairies.[11]

Machines, such as those devised by the McCormick and Deere corporations, were essential to transforming the western prairies that Francis Parkman, the preeminent chronicler of frontier expansion, had dismissed as a "sublime waste" into one of the most productive agricultural regions of the world.[12] Through most of the nineteenth century, farm implements—from plows and planting drills to harvesters and threshers—were usually pulled by horses rather than by human farmers. This change meant that more land could be cultivated with less labor, but it did not result in significant increases in yields per acre. Particularly after the Civil War, machines that substituted—to an extent formerly unimaginable—animal traction for human sweat freed up substantial portions of the agricultural labor force for employment in industry. By 1880, thanks to these revolutionary transformations, the United States had become the first country in history in which less than half the population was able to feed the rest. Two decades later the share of the population directly engaged in agricultural production had fallen to 38 percent. The introduction of machines powered by fossil fuels, in the form of steam engines, was far more limited and problematic because of their great bulk, the amounts of water and fuel they required, and the frequency with which their boilers set fire to dry fields at harvest time. The use of farm implements driven by internal combustion engines had only begun as the frontier was officially declared closed by the compilers of the 1890 census.[13]

Widespread public awareness of the contributions of the new farm implements to the settlement of the frontier was suggested by a popular lithograph, sold in the late 1860s and the 1870s, that parodied both the expansionists' mantra "Westward the course of empire takes its way" and Emanuel Leutze's 1862 mural of that title that embellished the national Capitol. Like Leutze's painting, the anonymous parody is centered on a party of weary pioneers struggling to the top of a mountain from which they can see an expanse of western plains beckoning them onward. But rather than gazing out into the barren wilderness of the original, in the imitation the exuberant pioneers look down with admiration upon a horse-drawn McCormick reaper harvesting a verdant wheat field. Lest the viewer miss his message, the didactic lithographer appended the phrase "with McCormick's Reaper in the Van" to Leutze's hackneyed title.[14]

Of all the innovative technologies that contributed to the development of the West, by midcentury none equaled the railroad in its impact on migration, settlement, and resource extraction. And as numerous nineteenth-century depictions of trains making their way through frontier landscapes illustrate, no other material artifact could rival the railroad as a symbol for the expansive, progressive, inventive young American republic in the early stages of its ascent to global power. Although some romantics and naturalists, such as Hawthorne and more ambivalently, Thoreau, expressed misgivings about the intrusion of these massive machines into the countryside, most Americans welcomed the steam locomotive and other technological innovations with enthusiasm. They took great pride in the fact that a railroad was operating in South Carolina only a year and a half after George Stephenson's "Rocket" locomotive made the first steam railway journey from Manchester to Liverpool in England in October 1829.[15] By the late 1830s workshops in Lowell, Massachusetts, and Philadelphia were turning out American-made locomotives to rival those from England. And even though much of America's rail network had been financed by British investors, citizens and visiting Europeans alike were impressed by its extent, which by the early 1840s exceeded in mileage those in all of Europe. The Americans' zeal for new tools and machines was understandable given the urgent need for labor- and time-saving technologies on the part of what remained a small settler population relative to the vast lands they aspired to claim as their national home. And Samuel Bowles, who traveled across much of the West in 1869 just after the first transcontinental line opened, perceptively observed that isolated pioneers cherished the railway because of the connection it provided with "the completer social, moral and intellectual life of the East." With the great expansion of frontier settlement in the early nineteenth century, these imperatives grew ever more compelling.[16]

As contemporary lithographs of trains racing across the prairie or over viaducts past slow-moving covered wagons were intended to show, there had never been a technology that could match the railway locomotive for the conquest of time and space. In conjunction with surveying and mapping operations, which had begun in the colonial period but expanded significantly after the passage of the Homestead Act in 1862, the construction of railroad networks served to standardize the measurement of distances and determine the siting of frontier towns. The 30,000 miles of railway lines constructed between the early 1830s and 1860 also proved critical in opening up wilderness regions to settler migration and demarcating areas of white settlement, farming, and resource exploitation. And the golden spike that in 1869 joined railway lines from the Missouri River in the East and California in the West at Promontory Point in Utah bound the continent-spanning nation together physically for the first time. Some decades later Teddy Roosevelt, an enthusiastic chronicler of frontier expansion, proclaimed the railroad one of the "pioneers of civilization" in North America.[17]

Rarely has any machine captivated as many artists as the railroad did. Within years after the first U.S. railway lines began operations, Thomas Cole and Thomas Doughty included distant and minute, but smoke-billowing, locomotives in their scenes of the Catskill River valley and rural Massachusetts. These vistas and those of the Delaware Water Gap and eastern Pennsylvania painted in the next decades by George Inness and Jasper Cropsey suggest that the sheer scale of the natural settings into which railroads were introduced ameliorated their disruptive effects for most Americans. The viewer's eye is drawn to the trains in motion in these paintings, and yet the machines are seemingly integrated into the landscapes. Rather than dominating or disrupting the natural environment through which they pass, locomotives, viaducts, and rail lines become part of "landscapes of reconciliation."[18]

As a number of art critics have observed, however, considerable ambiguity regarding the railway can be read into some midcentury landscapes, most notably Thomas Rossiter's *Opening of the Wilderness* (1858) and George Inness's *The Lackawanna Valley* (ca. 1857). In both, roundhouses and locomotives are framed by tree stumps

and tattered forest remnants. But Inness sets the locomotive chug-
ging purposefully toward the foreground in a bright pastoral scene,
complete with languid onlooker. Rossiter's landscape is a good deal
more somber—it is a sunset scene—but the five trains and railway
trestle can be seen as the dynamic vanguard of a national crusade to
open the wilderness to civilized pursuits. Rossiter's title for the painting appears to confirm this interpretation, as did his support for the
ambitious promotion schemes of the B&O Railroad Company.[19]

No such ambivalence is evident in the two artistic representations
of the railway on the frontier that were most familiar to mid-nineteenth-century Americans—Fanny Palmer's *Across the Continent:
Westward the Course of Empire Takes Its Way* (1868) and Thomas
Gast's *American Progress* (1872). In both these paintings the celebration of technology is unmistakable. Both quickly achieved iconic
status, and their staying power as expressions of national aspirations
has been amply attested by successive editions of lithographic repro-
ductions, innumerable newspaper engravings, and their frequent
inclusion in school texts and histories of the United States, which
continues to the present day. Although they are allegorical illustra-
tions rather than landscapes in the strictest sense, each is placed in
an expansive natural setting that is clearly intended to suggest the
western frontier. Both celebrate the many roles of technology in the
conquest of the West, but Palmer gives star billing to the railway in
the form of a speeding train, which cuts the painting in half diago-
nally. The trains that make their way westward in Gast's painting are
dwarfed by the natural setting, and by an outsized, floating figure
representing the spirit of progress, who centers and dominates the
work. In a recapitulation of the stages of settlement of the frontier,
the trains are preceded by a hunter, a soldier, and a miner; a covered
wagon and a stagecoach; and farmers plowing the prairie. In this depiction the railways are upstaged by the telegraph wires that the
spirit of progress is stringing across the plains. Where Gast chroni-

John Gast, *American Progress, or Manifest Destiny*, 1872. Oil on canvas. Museum of the American West Collection, Autry National Center, Los Angeles.

cles the diversity of participants in the "winning" of the West, and isolates them from one another in his composition, Palmer focuses on a frontier community, centered on a public schoolhouse and portrayed at a specific moment. Villagers wave at the passing train, while sturdy yeomen clear trees at the edge of town and several covered wagons set off westward on a dirt path that runs parallel to the railway line.

Neither Palmer nor Gast ranked among the most talented frontier artists. But their didactic landscapes expressed the integral importance of technological prowess to American national identity. These and other artistic representations of westward expansion also embodied the civilizing ideology that provided lofty rationales for both the building of a continental empire and the beginnings of American in-

ACROSS THE CONTINENT.
"WESTWARD THE COURSE OF EMPIRE TAKES ITS WAY."

Fanny Frances Palmer, *Across the Continent "Westward the Course of Empire Takes Its Way,"* 1868. Colored lithograph, Currier & Ives. Museum of the City of New York, Harry T. Peters Collection.

terventionism overseas. Both *Across the Continent* and *American Progress* emphasized the progressive nature of railroads and other technologies: deployed by stalwart pioneers, machines propel the march of modern civilization across the frontier, spawning vigorous towns and productive farmlands. Each panorama graphically confirms technology's vital contributions to the fulfillment of America's divinely appointed civilizing mission, which would ultimately benefit all of humanity. Both artists strive to capture what Samuel Bowles called "the most indomitable energy to be found among the American people," the people whose success in spanning the continent by railway inspired Bowles to declare: "No other people than ours–daring in conception, rapid in acquirement, bold in execution, beyond any other nation—could have both educated the men for such a work and done it, too, all within five years of time."[20]

The pervasiveness in America of this faith in technology as a force for improvement was noted by mid-nineteenth-century European visitors as diverse as Guillaume Poussin, Charles Dickens, and Frederika Bremer. Foreign travelers often conceded that the Americans had some justification for believing that they had developed a unique approach to manufacturing and that frontier expansion had shaped this defining feature of their nation's exceptional historical trajectory in pivotal ways. The notion of a distinctively "American system" came to be widely accepted by both European visitors and domestic innovators, entrepreneurs, and skilled workers. Often in sharp contrast to European approaches to industrial production, the American system emphasized practical application rather than abstract theorizing; larger-scale, labor-saving production techniques rather than highly skilled, finely honed craftsmanship; and the manufacture of uniform, interchangeable parts by semi-skilled laborers reliant on standardized machine tools. And at least in the early decades of the century, many observers remained confident that the space available for expansion, the abundance of natural resources, and superior management techniques would enable the United States to minimize the urban blight, oppressive working conditions, and consequent social unrest that had evoked profound misgivings about industrialization in Great Britain.[21]

By midcentury many of the nascent industries of the United States were already competitive with those of Europe, and some had achieved significant advantages in efficiency and quality. Perhaps no American machines better exemplified the transformation of the United States into a major industrial power than the hundreds of formidable steam engines produced by George Corliss's workshop in Providence, Rhode Island, beginning in the late 1840s. At the Paris

Exposition in 1867, one of Corliss's mechanical goliaths bested a hundred rivals for highest honors. And a decade later a massive Corliss engine, weighing over 600 tons with a driveshaft more than 300 feet long, was chosen to power all the machinery in the industrial exhibition hall at the Philadelphia Centennial Exposition.[22]

With the exception of the muted unease conveyed by the ravaged forests that painters such as Rossiter and Inness placed on the margins of their railway-dominated landscapes, virtually none of the frontier art dissented from Gast and Palmer's triumphalist view of the conquest of the West. Painters and lithographers were as impressed as their fellow Americans by the power of technology to transform the wilderness into productive farmlands and vibrant towns. Like the rest of the citizenry, they assumed that the resources of the continent were inexhaustible and that the lands to be settled would never run out. The more effusive advocates of agrarian expansion went so far as to predict that clearing the forests and cultivating frontier lands would bring cooler weather to the West and increase rainfall on the prairies.[23]

By the late nineteenth century, however, early American environmentalists, such as George Perkins Marsh and John Wesley Powell, had begun to challenge this complacent optimism and assess the considerable toll that farming, mining, ranching, and other settler enterprises had already taken across North America.[24] Forests had been clear cut with little thought to replanting or the erosion that would strip soil from the scarred land. Wildlife had been slaughtered with abandon, pushing some species, such as the beavers and bison, to the edge of extinction and utterly eliminating others, including the elegant passenger pigeons that had once blanketed the skies of the upper Midwest. Steel plows uprooted vegetation and broke apart the sod that had protected the rich prairie soils for millennia, exposing them to desiccation by the sun and erosion by wind and rain. Careless farming practices, which included cutting furrows that ran parallel to the prevailing winds and neglecting to plant trees for windbreaks, compounded the loss of humus-rich soil. And late in the century settlers pushed past the 100th meridian to farm the marginal, semi-arid lands east of the Rockies. When the soil in their fields was exhausted, small landholders and especially tenant-cultivators moved on, clearing lands that had not yet been settled, using the same flawed techniques that had rendered their original farms unsustainable.

From 1889 to the mid-1890s, drought and destitution spread over much of the agricultural frontier. Expansionists blamed the weather and the lingering global economic depression, and opined that the good times would soon return with the rains. To ensure that they did, newly funded government agencies launched programs to develop better seeds, encourage the use of fertilizers to replenish overworked soils, and improve farming techniques. Combined with the widespread introduction of internal combustion engines to drive tractors and other labor-saving farm machines—and, of course, the return of the rains—these innovations carried American agriculture to unprecedented, and even more precarious, levels of specialization and productivity in the first decades of the twentieth century.[25]

Although, like many western artists, Gast and Palmer center their representations of the frontier on settler enterprises and technological innovations, both include on the margins of their panoramas the major victims of what they otherwise depict as a beneficent process. In Gast's *American Progress*, Indians flee the advancing pioneers, as do other creatures belonging to the wilderness: a herd of bison, a bear, and a deer. Palmer, in *Across the Continent*, draws a more direct connection between the technologies that facilitate expansion and the displacement of the indigenous peoples. Two Indian horsemen are engulfed in the smoke of the passing train, and the empty plains stretching into the distance suggest that the Indians have made their own defeat inevitable by doing so little to nurture the continent's abundance. Even more than their colonial predecessors, nineteenth-century promoters of expansion fixed on the Indians' failure to "improve" the lands they occupied to justify the settlers' appropriation of those lands. And no one put this proposition more bluntly than Theodore Roosevelt, whose ponderous, four-volume history *The Winning of the West* (1900) was widely regarded at the time as the definitive account of America's emergence as a transcontinental power. Writing shortly after bureaucrats in Washington decided that the frontier saga had come to an end, Roosevelt mocked the unmanly, indolent "stay-at-homes" and "do-gooders" who lamented the disappearance of the Indians or the taming of the wilderness by the "hard, energetic, practical" Anglo-Saxons. He scorned the "warped, perverse, and silly morality that would forbid a course of conquest that has turned whole continents into the seats of mighty and flourishing civilized nations." As long as the Indians were in control, Roosevelt averred, North America would remain "nothing but a game preserve for squalid savages." And because he judged that their negligence meant they "never had any real title to the soil," Roosevelt pronounced the wars waged against the Indians "ultimately righteous," even though, like all conflicts between civilized and savage races, they often turned "terrible and inhuman."[26]

Roosevelt did not categorize the Mexicans, much less the Spanish and French, with the Indians as savages. But he defended U.S. annexations of their territories or military repulses of their imperial initiatives on the grounds that America's superior technical acumen, entrepreneurial skills, and energetic laborers would develop the western regions for the benefit of all humankind. He denigrated French settlers because their miscegenation with Indian peoples had produced mixed-bloods, who were "not very industrious or thrifty husbandmen," and whose "farming implements were rude [and] . . . methods of cultivation were simple and primitive, and they themselves were often lazy and improvident."[27] Americans routinely conflated the Spanish with the Mexican oligarchy, as despotic rulers who were uninterested in improving the territories they had wrested from quasi-civilized peoples such as the Aztecs and Incas. And Roosevelt and other expansionists regarded both as degraded, indolent, and as inept as the Indians they had displaced. In this view, only the tutelage of their American conquerors would enable the Mexicans to tap the natural resources in areas such as New Mexico and California, which remained "barren wastes" after centuries of Spanish and Mexican control.[28] For the apologists for U.S. expansionism, Manifest Destiny meant more than the inevitable Anglo-American domination of two continents: it meant the spread of progress, economic growth, and civilized ways of life throughout the western hemisphere and ultimately beyond the Pacific Ocean.

Frederick Jackson Turner followed earlier chroniclers in emphasizing the frontier's inculcation of a craving for independence, individualism, and unfettered mobility. But he and other nineteenth-century observers also linked the pioneers' adaptability and mechanical ingenuity to virtues exhibited by the "composite nationality [of] the American people."[29] As they clear the forests, build towns, and till the land, the archetypical pioneers in Gast and Palmer's paintings exhibit a purposefulness, an adaptability, and a penchant for hard physical labor that are seen to set them apart from the previous inhabitants of the prairie. Expansionists disparaged the plains Indians, like the indigenous coastal peoples before them, as fatalistic, improvident, lazy, and in thrall to the forces of nature rather than striving to master them. The frontier artists depict the "hardy" pioneer folk as the Indians' direct opposites—progressive improvers, "carrying with them the indomitable Anglo-Saxon energy and the stern virtues of their fathers," and determined to render the wilderness bountiful. The settlers act decisively in the present and confidently build the basis of a better future. The Indians are immobilized, even in flight. Unable to comprehend, much less control, the vast socioeconomic transformations—physically manifested in the machines—that were driving them to the periphery of their own homelands, they were, as many white Americans feared or anticipated, destined to become extinct.[30]

The industrial processes that were contemporaneous with the high tide of frontier expansion accentuated the long-imagined dichotomy between settlers and Indians. When settlers began to migrate in ever larger numbers across the prairies, the livelihood of the more powerful Indian peoples depended primarily on hunting and gathering, although many also farmed. For most of their subsistence needs from food and housing to clothing and tools, plains tribes such as the Sioux and the Cheyenne relied on the bison. The plains Indians' domestication of horses had greatly enhanced their capacity to hunt bison. But the more plentiful supply of bison meat and hides encouraged the hyper-specialization of many plains Indian cultures, rendering them highly vulnerable to the settlers' onslaught.[31]

Through the first half of the nineteenth century, the failure or refusal of Amerindian peoples to shift from subsistence hunting to market agriculture became an ever more prominent marker of their difference from the citizenry of the American republic. Stark disparities in material culture, which intensified as the pace of industrialization accelerated, underscored fundamental differences in gender roles and relationships. Americans increasingly interpreted these discrepancies as evidence that the Indians' level of evolutionary development lagged behind that of peoples of European descent. In newly settled areas the gap in material culture was far from evident if one compared the early pioneers' sod dwellings or even their

Andrew Melrose, *Westward the Star of Empire Takes Its Way—Near Council Bluffs, Iowa*, ca. 1865. Oil on canvas. E. W. Judson Collection, New York.

hardscrabble farmhouses, such as that depicted in Sallie Ball's *Prairie Homestead*, with the lodges or tepees of neighboring Indians. But, as Andrew Melrose's imaginative variation on a familiar trope, *Westward the Star of Empire Takes Its Way—Near Council Bluffs, Iowa,* made apparent, by midcentury the contrast between established farms and Indian settlements, which settlers routinely denigrated as filthy, ramshackle, and transient, was unmistakable. With a striking painterly flourish, Melrose links a prosperous prairie farmhouse with the continent-spanning technological networks that, in Emerson's words, "shoot every day across the thousand various threads of national descent and employment, and bind them fast in one web." A locomotive, its headlamp ablaze, races straight at the viewer from the darkening sky and woodlands on the right of the canvas, scattering a small herd of deer, creatures of the wild, which—like the Indians—are unable to impede its purposeful advance.[32]

As time passed, farmers and ranchers added cavernous barns, silos, and bunkhouses for hired hands to their frontier homesteads. Wood-frame houses replaced those of sod or rough-hewn logs. And gradually, prairie homes were filled with mass-produced furnishings, rugs, curtains, wood- or coal-powered stoves and ovens, and a wide range of kitchen utensils. New England textile manufacturers found a lucrative market on the frontier for everything from work trousers to suits and dresses fit for Sunday services. And neighbors vied to "surround themselves with the comforts and amenities of metropolitan life." The material comforts enjoyed by respectable settler households contrasted sharply with the deteriorating living standards of their Indian neighbors as worn hides could no longer be replaced, woodlands were fenced off by settlers, and fine pelts were traded for cheap clothing, blankets, and tin tubs. Pioneers sought to observe an idealized, gendered division of farm labor, although the hardships of frontier life meant even more blurring of the boundaries between women's and men's work than in colonial times. Contemporary memoirs have nothing but praise for pioneer wives who joined in clearing fields, plowing and planting, and tending livestock. When illness, drought, or the death of a spouse threatened the homesteader family's livelihood, women—and children including girls—took over these tasks, or men assumed a greater role in childrearing and domestic routines. These images of collaboration and purposeful activity were the antithesis of the long-prevailing stereotypes of the lazy Indian male and his female counterpart, the squaw drudge. The widening disparities in material culture, coupled with the Indians' presumed inability to negotiate the transition from hunting and gathering to farming, were seen by many Americans as further evidence that their extinction was inevitable, and perhaps imminent.[33]

Ironically, the same industrial technologies that Gast and Palmer depicted as aiding the westward movement were undermining the landownership of yeoman farmers, whom both artists valorized and whom Thomas Jefferson had considered essential for the success of an agrarian republic. On the one hand, inexpensive farm implements, and canal and railway networks, did make it possible for small landholders to cultivate relatively large plots of land and to produce marketable surpluses.[34] But on the other hand, railway companies, banks, and speculators came to control a substantial portion of the land that the U.S. government confiscated from Amerindian peoples and then surveyed and parceled out to settlers and railways. Large estates—many of them devoted to raising cattle or sheep—came to dominate farming in many areas. And large and small landholders alike were dependent on capital-intensive transport, processing, and marketing facilities controlled by often distant companies and entrepreneurs. Over the next century these forces would prove pivotal in the transformation of an idealized agrarian republic, peopled by independent yeomen farmers, into a highly centralized capitalist society in which agricultural production was dominated by agribusinesses.[35]

When farm families forfeited property rights, they lost not only their livelihood but also an important marker that distinguished them from the Indians. By the mid-nineteenth century the Indians' resistance to adopting what the invaders deemed a higher stage of human development had become a major arguing point for those who contested the Jeffersonian view that the Amerindian peoples were capable of full assimilation to Euroamerican culture.[36] That view came under increasing assault as settlers in the southern Appalachians and the eastern Mississippi valley pressed state and federal officials for the right to claim Indian-occupied lands. In the early 1830s advocates of "removal," including President Andrew Jackson, argued that the failure of southeastern tribes to develop the territory they occupied justified relocating them across the Mississippi River. Contrasting the "country covered with forests and ranged by a few thousand savages" with "our extensive Republic studded with cities, towns, and prosperous farms, embellished with all of the improvements which art [technology] can devise or industry execute," Jackson intimated that Providence had decreed that the Indians should yield their "unused" lands. A supportive congressman later declared that if the U.S. government did not allow enterprising settlers to supplant the Indians, it would "obstruct the march of science—stay the works of art, and stop the arm of industry."[37]

The fact that the Cherokees and the other tribes of the Five Nations, whom many Americans regarded as the most civilized of the Indian peoples, were the victims of the most infamous removal operation boded ill for all indigenous peoples. The Cherokees and to varying degrees neighboring tribes, such as the Choctaws and Chickasaws, had assimilated to Anglo-American culture as thoroughly as any Indians. They enjoyed considerable success as commercial farmers, intermarried with European settlers, resided in European-style houses, and dressed in the Anglo-American fashion. Yet, to justify the seizure of their lands, their extensive acculturation was denigrated or denied. Refused redress by the Supreme Court and the federal government, they were dispossessed, humiliated, and in the winter of 1838–39 forced to march to "Indian Territory" in what would later become the state of Oklahoma, under appalling conditions that resulted in the deaths of thousands, many of them women and children.[38]

The view that Indian peoples lacked the capacity to be fully assim-

ilated into American society appeared to be given scientific validation by the research of two highly respected early American ethnologists, Samuel Morton and W. H. Morgan. Morton, a physician
educated in the United States and Europe, was by the early 1830s
an avid collector of skulls and a somewhat ambivalent advocate of
phrenology. His comparative measurements of human cranial capacity were published in a lavishly illustrated, highly influential
tome entitled *Crania Americana* in 1839. On the basis of disproportionate samples of skulls of the five races that he claimed made up
the human species and dubious mathematical calculations,[39] Morton found that the mean skull size of "barbarous [North American
Indian] tribes" was just slightly larger than that of "negroes," and significantly smaller than those of Caucasians, Mongolians, and Malays. Since Morton's purpose was to establish correlations between
skull size and innate intelligence, he concluded that the Indians'
"wild ways" and low state of societal development were biologically
rather than culturally based. In accord with other racial determinists
of his time, he averred that phrenological deficiencies explained
the Indians' lack of contributions to invention or scientific investigation, their stunted imitative capacity, and their failure to progress to
higher levels of human attainment. All of which suggested that efforts to promote their assimilation into settler society, a branch of the
most highly civilized Caucasian race, would prove futile.[40]

W. H. Morgan's extensive anthropological investigations of Amerindian societies yielded less overtly racialized findings than Morton's
cranial indices. Nonetheless, Morgan's theories regarding the dynamics of evolutionary progress in human societies, set forth in his
1877 book *Ancient Society*, appeared to buttress the arguments of
those who doubted the Indians' potential for assimilation to "civilization." Unlike Morton, Morgan believed that cultural constructions
—especially political and familial organization—were the major factors impeding or facilitating human social evolution. But ascending

the evolutionary ladder from savagery to civilization, Morgan argued, depended on transformative technological innovations, "such as the domestication of animals or the smelting of iron ore, [which] gave a new and powerful impulse forward." And he gave a physiological twist to this progression, arguing for a causal relationship between increases in the mental capacity of human groups and those groups' capacity to fabricate innovative technologies and build more sophisticated institutions. Unfortunately, the lengthy time span he deemed necessary for increases in intelligence sufficient to turn barbarians into civilized peoples rendered this evolutionary possibility all but irrelevant to the Amerindians.[41]

In their representations of the frontier, neither Gast nor Palmer includes even a suggestion of the violence that was pervasive in final stages of the dispossession of the Indians. This omission may help to account for the acclaim which contemporaries and succeeding generations accorded both landscapes as mythic allegories of the processes of national expansion. But a countervailing trend was evident at a more popular level: graphic depictions of aggression were staple fare in the illustrated magazines and penny adventure stories of the Wild West published in the postbellum decades. Sketches of cowboy shootouts were interspersed with lithographs of Indian attacks on stagecoaches and engravings of cavalry skirmishes with bands of agile Indian riders.[42] And artistic impressions of American expansion, particularly into the Great Plains, chronicled the devastating effects of even nonmilitary industrial technologies on both the indigenous peoples and the bison herds upon which their survival depended.

By the last half of the nineteenth century, the plains Indians were acutely aware of the importance of technology to the settlers' en-

Theodore Kaufmann, *Railway Train Attacked by Indians*, 1867. Oil on canvas. John F. Eulich Collection, Dallas, Texas.

croachment on their lands and threat to their cultures. Assaults by tribal war parties on railroads, stagecoaches, and telegraph lines were reported in travel accounts and narratives of the settler-Indian wars published from the late 1860s to the 1880s. And paintings of Indians uprooting railroad tracks, such as Theodore Kaufman's nightmarish *Railway Train Attacked by Indians* (1867), reflected the near obsession of pioneers on the frontier and publicists in the eastern cities with the dangers posed for these networks of expansion by tribes that had not yet been confined to reservations. On the other side of the cultural divide, Indian leaders lamented the proximity of their peoples' hunting grounds to the routes opened by wagons and railways because settlers in transit slaughtered game animals for food and hides, cut down or burned scarce stands of trees, and allowed their livestock to overgraze the prairie.[43] Newbold Trotter's 1897 painting *Held Up*, a panorama of a passenger train waiting for large herd of bison (by then rare) to get out of its way, captures metaphorically the myriad collisions of cultures and biosystems that were occurring across the frontier. The painting also expresses the American expansionists' long-held conviction that the bison herds, and implicitly the Indians who depended on them, were little more than obstacles impeding laudable efforts to colonize and develop regions the Indians had done nothing to improve.

The direct way in which the railroads contributed to the destruction of the bison herds—and thus of plains Indian culture—was amply illustrated by wood engravings published in widely read periodicals and books such as *Frank Leslie's Illustrated Newspaper* and Richard Dodge's *The Plains of the Great West* (1877). *A Lively Scene on the American Prairie*, printed in the former in 1868, depicts rifle-toting hunters (and perhaps ordinary passengers) who have climbed on top of railway cars halted in the middle of the plains to slaughter hapless bison as the herd scatters in all directions. In a comparable scene in Dodge's travel narrative, the imminent killing is merely suggested by the descent of passengers sporting top hats and rifles from a train as terrified bison stumble over one another and the railway tracks. However reprehensible, the killing of bison for the amusement of travelers probably had far less impact than the excessive hunting of some of the plains tribes. Eager to meet the strong demand for buffalo hides in eastern markets—due in part to their suitability for leather belts used in industrial machinery—increasing numbers of Indians were drawn into the unrestrained hunting of the bison that contributed to the precipitous fall in their numbers, from 20–30 million before the Civil War to a single herd of approximately 800 sheltered in Yellowstone Park in 1895.

Excessive hunting by Indians magnified the deleterious effects on the bison of drought, disease, shifts in prairie vegetation, and competition with domesticated livestock for pastureland. But more than any other factor, the relentless slaughter by tens of thousands of Euroamerican hunters, which accelerated from the early 1870s, drove the bison to the edge of extinction by the mid-1880s. For the Indians this systematic—often quite deliberate—attempt at biocide was all the more appalling because once the tongues (a culinary delicacy for many Euroamericans) had been severed and the hides stripped from the carcasses, the rest was left to rot. The Crow medicine woman Pretty Shield could not forget the stench of the putrid flesh of skinned bison in the Judith River Basin, nor could she understand why the "white man" would kill *all* of the great beasts even though he had no use for their meat. Perversely, Albert Bierstadt's melodramatic attempt to capture the anticipated end of the species on canvas, *The Last of the Buffalo* (1889), indicts the Indians—who had everything to lose from the bisons' demise—as those responsible. Surrounded by skulls, carcasses, and dying bison, a warrior on a rearing white horse slays a charging bull with a spear.[44]

Between the late 1860s and the 1880s the destruction of the bison was paralleled by sporadic but brutal warfare that completed the dispossession and displacement of the Amerindians. The end of the Civil War in 1865 did not result in an appreciable increase in the number of soldiers assigned to frontier posts. But it meant that the poorly trained volunteers who had been recruited locally during the war could be replaced by veteran fighters or newly enlisted personnel led by seasoned commanders. And these forces were armed with weapons that had been improved or introduced during the long conflict between North and South: breech-loading artillery, Gatling guns, repeating rifles, and Colt revolvers. Yankee heroes, including most famously George Armstrong Custer and William Tecumseh Sherman, made good use of these in the wars on the western frontiers. (In June 1876 Custer made the fatal mistake of leaving the Gatling guns behind in his haste to engage an Indian war party congregating on the banks of the Greasy Grass or Little Bighorn River.)[45]

The arsenals where many of these weapons had been invented or refined, and where the "American system" of manufacture had originated, came to excel in their mass production by specialized workmen using machine-tooled, interchangeable parts. The Gatling machine gun, named after the Philadelphia physician who invented it in the early 1860s, was a special source of pride because in conception and construction it was "entirely American."[46] And though Indian tribes could obtain rifles and other small arms from gun runners or capture them in battle, most of the weapons they could trade for were obsolescent, if not defective, and they had difficulty replenishing ammunition. The American government's capacity to supply its widely dispersed cavalry, in contrast, was steadily enhanced as railways were extended westward across the plains to join with those snaking through the mountains from the Pacific coast.

Initially the military's main assignment was to patrol routes to the West and protect migrants from Indian bands fighting to preserve tribal hunting grounds that the government had recently promised would be theirs in perpetuity.[47] And despite the persistence of the American commanders' and soldiers' centuries-old disdain for the Indians' reliance on ambushes, hit-and-run attacks, and other ruses in waging war, they could not deny the plains warriors' manly courage, determination, and fighting skills. Even Teddy Roosevelt was forced to admit that the Indians were "the most formidable savage foes" the Europeans had to contend with on any frontier. American cavalrymen were disconcerted by the rapidity with which Indian fighters adapted their tactics to offset their enemies' technological advantages. Indians on racing ponies charging in irregular order, for example, made poor targets for field artillery, a point driven home by the reactions of prominent Indian chiefs who visited Washington in the 1860s. After witnessing a demonstration of massive coastal guns that was intended to intimidate them, the chiefs scoffed, since none could imagine that experienced warriors would be so foolish as to stand still while the stationary guns were firing.[48]

American disparagement of the Indians' ways of waging war as treacherous and cruel was in any case hypocritical. Government promotions of the cavalrymen as righteous crusaders and defenders of innocent civilians were blatantly at odds with their infamous winter campaigns against the plains peoples in the mid-1870s. In surprise assaults on Indian villages, the soldiers often massacred women, children, and men too old to fight or flee. They also destroyed villagers' shelters and food stores in the midst of the harsh winter season when the scarcity of game and edible vegetation made their staples all but impossible to replenish.[49] The legendary bravery and tactical acumen of Indian bands slowed the settlers' encroachments and gained a number of stunning victories, such as the "massacre" of Captain William Fetterman and 79 of his men near what is now Sheridan, Wyoming, and the elimination of Custer's units on the Greasy Grass River. But in addition to the superiority of their enemies' weapons, discipline and coordination, and communications technologies, rivalries within tribes and between different Indian peoples, logistical limitations that made extended campaigns unsustainable, and the depletion of their numbers meant that the Indians could not halt for long, much less reverse, the loss of their lands and autonomy.

By the late 1870s or the 1880s, depending on which frontier struggles were at issue, the combination of the destruction of the bison herds and relentless military aggression forced Indian peoples across the West to choose between internment on government-run reservations and annihilation. Some remnants of tribes sought refuge in Canada or northern Mexico, but most resigned themselves to confinement on the reservations. In the years just after the Civil War, American officials offered manufactured goods ranging from steel

William Fuller, *Crow Creek Agency, D.T.*, 1884. Oil on canvas. Amon Carter Museum, Fort Worth, Texas, Acquisition in Memory of Rene d'Harnoncourt, Trustee, Amon Carter Museum, 1961–1968. 1969.34.

knives, iron pots, and textiles to firearms to entice Indian leaders to sign treaties that committed them to settle their people in specified areas, thereby renouncing claims to far larger ancestral hunting grounds. By the 1870s, once proud peoples were forcibly reduced to dependence on government-supplied food, clothing, shelter, and periodic subsidies. The distribution of these necessities was controlled by federal Indian agents, who were often corrupt or incompetent (or both), and Christian missionaries, who were not above withholding delivery if the Indians in their charge resisted their civilizing initiatives.[50] The broken Indians who stare out somberly from William Fuller's painting in the primitive style *Crow Creek Agency, D.T.* [Dakota Territory] (1884) are a world apart from the exuberant hunters in Catlin's landscapes of the 1820s and 1830s, or even the animated villagers in Bierstadt's *Rocky Mountains*, painted just two decades earlier.

Government policy regarding Indian peoples confined to the reservations was premised on the assumption that they would either adopt the sedentary, agrarian lifestyles of Euroamericans or perish, a fate which those who refused to be "civilized" were widely thought to deserve. By the late 1850s, Indian agents from disparate locales reported that peoples living on the reservations had begun—at times quite successfully—to adopt the settlers' farming practices and tools. Plains Indian women, who like their counterparts on the Atlantic seaboard had traditionally been in charge of agricultural production, often defied the wishes of missionaries and Indian agents by becoming active participants in the transition to market agriculture. But owing in large part to official determination to turn Indian warriors into yeomen farmers, technology transfers and commercial exchanges were usually channeled to Indian men, who usually worked with their wives in plowing and planting. In many instances, however, the "destitute" state of confined tribes meant that they could not make productive use of the lands they had been allotted. Some peoples lent these out to settler farmers, others left them uncultivated and sought to revive their former nomadic ways. Blaming the persistence of Indian indolence and savage customs, such as communal landholding, for these failures to improve their lands, government agents supported legislators who pressed for the privatization of reservation allotments. The enactment of that policy in the 1887 Dawes Act opened the last areas left to the indigenous peoples of the West to speculators and bank foreclosures, which contributed mightily to the impoverishment and marginalization of the remaining Amerindian population in the century that followed.[51]

As in the praying communities of the seventeenth century, in both the reservation schools and special institutes such as the one founded by Richard Henry Pratt at Carlisle, Pennsylvania, education and work were quite consciously gendered. In the best-run schools, boys received instruction in Christian teachings, English (speaking and reading), and elementary science and mathematics, and were put to work on model farms or in blacksmith shops where they could learn vocational skills. Usually separated from the boys in classrooms or schools exclusively for girls, young Indian women were also taught to read English and count, but the focus of their vocational "training" was on inculcating "proper"—that is, American—approaches to domestic tasks and hygiene. Students of both sexes were urged to renounce "backward" tribal practices, from hunting and sweat cures to modes of rearing children and preserving food (drying, for example, rather than canning, which the rapid spread of canneries from the 1870s made the preferred method of preservation). These educational programs marked a resurgence of the once-embattled faith that the Amerindian peoples could (eventually) be fully assimilated into American society.

In all its variants, American-supervised education centered on pedagogical maxims and disciplinary regimens that were deliberately designed to erase tribal cultural identity. Students were forbidden to speak Indian languages, which their teachers rarely bothered to learn. The children's long hair was roughly cropped; and they were made to dress in American style. Whenever sufficient funding was available, they were taken from their families and boarded at school in order to insulate them from the degrading practices and superstitious beliefs that the authorities deemed pervasive sources of corruption in tribal life. Though missionary education made it possible for exceptional individuals, such as Ely Parker, a full-blooded Seneca and Ulysses S. Grant's private secretary and confidant, to aspire to professional careers in mainstream America, they remained a tiny minority of the Indian population well into the twentieth century. Most graduates of the Indian schools either returned to the reservations and struggled to readjust to tribal ways they had been taught to despise, or became misfits in mainstream society, where most of the citizenry considered them racially inferior and incapable of sustaining the civilized ways that had been grafted onto their savage natures.[52]

Long before the West was won, American expansionists had set their sights on a much wider global mission. The drive across the continent was paralleled by ever more ambitious overseas ventures that were in many ways an extension of the moving frontier. As early as the 1780s, U.S. merchant ships—freed from the mercantilist restrictions that had contributed to the outbreak of the American Revolution—had crossed the Pacific and established a direct trading link with China. The profits garnered by those who invested in the 1784 voyage of the *Empress of China,* the first U.S. ship to trade at Canton, fed expectations of great fortunes to be made in the "China market." Until the early 1840s, however, the bureaucrats of the ruling Qing dynasty, which was almost as reclusive as the shogunate in Japan, frustrated these ambitions by restricting commerce with Westerners to two ports, Canton and Macau on the south coast. And in the decades before and after Great Britain routed the Chinese in the Opium War of 1839–1842, British traders and diplomats dominated Western dealings with the vast empire. Furthermore, fifteen thousand miles or more separated Boston, New York, and other U.S. ports on the Atlantic coast from the "Far East." On a good run, the round-trip voyage across the Atlantic, past the Cape of Hope, and through the Indian Ocean took sailing ships from eight months to a year. These formidable obstacles notwithstanding, numerous American mercantile firms and families, including the Browns of Providence and the Astors of New York, amassed or added to great fortunes with profits from the China trade.[53] But even though commerce overwhelmingly dominated the first half-century of American intercourse with China, and strongly motivated Perry's expedition to "open" Japan, it was from the outset linked to higher purposes and an intensifying sense of national mission.

Until well into the nineteenth century, cultural influences—like mercantile goods—flowed mainly from China to the United States. But the late eighteenth-century infatuation of Americans like Thomas Jefferson and Ben Franklin with China's achievements in agriculture and technology was increasingly overshadowed by exposés of the corruption, decadence, and intense conservatism of the scholar-gentry officials who controlled Chinese contacts with foreigners.[54] By the early 1800s these often scathing critiques bolstered a resolve by the Western powers to compel—forcibly if necessary—political and social changes in China, as well as a commitment by American expansionists to a civilizing mission that for the first time explicitly targeted societies beyond North America. As the "star of empire" moved inexorably westward, the conviction that the United States had become the highest repository of civilization took hold among politicians, social commentators, and missionary organizations. And frequently these disparate groups linked the nation's original mission to tame the North American wilderness with a global vision of a progressive, energetic young republic reaching across the Pacific to revitalize the ancient but petrified societies of the Far East.

Most advocates of this grander civilizing mission viewed America's exceptional technological and material achievements as proof that the United States was destined to take the lead in the quest of "civilized" Westerners to uplift the peoples of Asia. In the 1780s the poet and essayist Philip Freneau predicted that the agricultural abundance generated by the settlement of the frontier would prove a catalyst for peace among the nations of the world. More than a century later, the same promise of American beneficence lent a scant hint of hope to the conclusion of *The Octopus,* Frank Norris's muck-raking assessment of the railways' impact on the development of the American West. After surveying the human casualties of the railroad magnates' machinations, the narrator of the novel declares that the fields of ripened wheat remain "untouched, unassailable, undefiled" and will yield a great harvest that will "roll like a flood from the Sierras to the Himalayas to feed thousands of starving scarecrows of the barren plains of India."[55]

In the century between Freneau and Norris, advocates of expansionism from Thomas Hart Benton and William Gilpin to William H. Seward and Teddy Roosevelt reveled in the prospect of American teachers and entrepreneurs marshaling their superior scientific knowledge, technical acumen, and industrial technologies to revitalize backward Asian societies. These visionaries viewed the railroads spanning North America as thoroughfares to the entrepôts on the Pacific coast, which were rapidly becoming rivals to the centers of trade, shipping, and military power in the eastern United States. The nation's unofficial poet laureate, Walt Whitman, proclaimed the railway and the steamship keys to a "Passage to India" that would transform America into a major force in global commerce and render it the mediator between the continents of Europe and Asia. In this vision of global rejuvenation, even the pursuit of economic profit was elevated to a force for improvement. Expansionists confidently predicted that the spread of American-style capitalism into global market networks would inevitably undermine benighted regimes, such as those in China and Japan, that sought to preserve stultified societies by closing them off from the rest of the world.[56]

Proponents of this global vision considered the conquest and consolidation of the "magnificent empire"[57] on the western frontiers to be essential for the projection of American influence overseas. For much of the nineteenth century the West served as an internal colony, providing staple foods, minerals and other raw materials, investment outlets and markets, and lands for settlement for the increasingly urbanized, industrializing East—and by the last half of the century the upper Midwest. Ironically, the region's rapidly growing surplus in agricultural production gave rise to pro-imperialist lobbies among the grain merchants, farmers, ranchers, and miners of the quasi-colonized West.[58] But to some extent these impulses in support of overseas interventionism were offset by the need to devote resources to the development of the frontier regions themselves. In an era when American entrepreneurs often relied on British investment capital and technological innovations for critical infrastructural projects such as canals and railways, funding for overseas commercial enterprises was in short supply. Except during the Civil War, America's army was minuscule compared to those of the great powers of Europe, and its soldiers' training and generals' skills were generally (and not always fairly) disparaged by their European counterparts. And though the U.S. naval fleet was a force to reckon with by the end of the century, it remained smaller than those of most of America's industrial rivals, especially Great Britain, and it was spread over a daunting expanse of coastlines and sea lanes.[59]

America's actual influence overseas until the end of the nineteenth century was usually far more limited than the inflated expansionist rhetoric and self-congratulatory memoirs of the era might suggest. The colonization of South and much of Southeast Asia by the European powers left only China, Japan, and Korea open to significant U.S. political overtures, and even in East Asia the Americans usually had to cede the leading role to the British and vie with the French, Germans, Russians, and eventually Japan for secondary parts. The Perry expeditions to Japan in 1853–1854 provided an exceptional, and brief, moment in the international limelight for the United States. And aside from Commodore Perry, Townsend Harris, and a few others, most U.S. diplomats serving in East Asia, at least until the 1860s, lacked the language proficiency, professional training, and social skills to distinguish themselves.[60]

These shortcomings provided opportunities for missionaries to enhance their status among Americans residing in foreign countries and their influence on American policy. From the time they began to arrive in the early 1830s, Protestant missionaries assigned to China sought to acquire at least a modicum of familiarity with written Chinese and the local dialect of the spoken language. Some who ministered to the Chinese for long periods, such as Peter Parker and Samuel Wells Williams, became fluent in the language. Able to communicate without interpreters, and far more familiar with Chinese society and customs than virtually any American diplomats or merchants, they emerged as cultural brokers in the middle decades of the century. Williams joined the Perry embassy as an interpreter (even though he was proficient in Chinese rather than Japanese). At roughly the same time, Peter Parker served for over two years as the official American envoy in Beijing.[61]

Given the merchants' almost exclusive concern with turning a profit and the rather confined lives led by most diplomats, it is not surprising that by the 1840s the missionaries were widely regarded as the best-informed Americans in China. For decades they would be the main conduit of information about things Chinese for the citizenry back home. Because the Japanese remained more firmly in control of foreign access and their own internal affairs, the missionaries never attained the influence in the island kingdom that they enjoyed in China. Also, Japan's industrial growth and emergence as a military power sparked intense scrutiny by Western diplomats, engineers, and military personnel, whose views tended to carry more weight than those of the missionaries.

By the late nineteenth century Americans' assessments of their nation's civilizing mission, and of the capacity of the Chinese and Japanese to revitalize their societies by mastering the ways of the West, were informed by a new surge of technological innovation, material increase, and social change that cumulatively made up a second industrial revolution. Occurring in all the industrial nations, but most pronounced in Germany and the United States, this new stage of mechanization was driven by a growing substitution of steel for iron and petroleum for coal. It was also marked by a dramatic expansion in the manufacture of chemical products, and by the emergence of a panoply of industries connected to electrification. The market-driven international corporation was yet another component of this new phase of industrialization, as was the shift from an orientation toward work and community toward an emphasis upon mass consumption and self-fulfillment in the United States and, to a lesser extent, the other Western industrial societies.

From agriculture to manufacturing, the succession of inventions and scientific breakthroughs that contributed to America's industrial transition were worked into increasingly complex, interconnected, and interactive technological systems.[62] These innovations were systematically integrated by the corporate enterprises that had come to dominate the U.S. economy—and in many respects society and polity. As industrial production became reliant on applied scientific research, laboratories and experimental facilities became essential components of American manufacturing complexes. Electrification and innovative manufacturing techniques required ever greater quantities of raw materials. Their delivery depended on expanding communication and transport networks. Manufacturers were also dependent on transportation networks for distribution of their products, advertising agencies and retail distributors for domestic sales, and shipping lines for export. The increasingly mechanized agricultural sector also relied on communications and transport networks as well as on farm implements supplied by the industrial sector and synthetic fertilizers, insecticides, and livestock feed concocted by corporate and university scientists and tested on auxiliary experimental farms.

These transformations driven by complex and intertwined scientific-technological systems played major roles in a succession of demographic shifts that fundamentally refashioned American society. New machines, weapons, and transportation systems had initially facilitated the dispersion of settlers, making it possible to settle and farm areas at ever greater distances from the eastern seaboard. By the last half of the nineteenth century, however, industrial technologies were contributing to countervailing demographic trends that made for the concentration rather than the dispersal of the American populace. Though railway networks and steamship lines had initially enhanced geographic mobility, the increases in the scale of industrial production, and the complex technological systems on which they depended, required the concentration in urban areas of both working and managerial classes, whose growing numbers ensured the multiplication of clerical, retail, and service workers to support them. The development of mass urban transportation, sewage disposal, affordable housing, and electrification made possible unprecedented metropolitan growth. By 1910 nearly 38 percent of Americans lived either in city centers or on their outer rings.[63]

Diverse and interconnected technological systems provided the impetus for dramatic—and often unsettling—changes in the physical and social contours of America's towns and cities. From the 1880s onward, paving, electrification, telephones, trolley lines, and later subways facilitated substantial increases in the size of urban centers and gave rise to middle-class suburbs as key locales for the concentration of population and wealth. Elevators and steel-frame and concrete construction allowed urban areas to grow up as well as out, as monumental skylines became the defining architectural feature of American cities. Parks and sports stadiums catered to the craze for physical culture and competitive games that swept American (and

European) society. Department stores ostentatiously displayed an abundance of consumer goods that was perhaps the most obvious manifestation of America's advances in mass manufacturing. From flush toilets to cast-iron stoves, most new consumer products, including a bewildering array of what were advertised as labor-saving gadgets, were intended for use in the home. These much-touted inventions often added to rather than reduced the burden of women's household work. But they literally brought home to the middle- and upper-class Americans who could afford them a sense of the nation's technological proficiency and productive capacity.[64]

Between 1876, when the Centennial Exhibition at Philadelphia displayed the wonders of American science and technology, and the turn of the century, the nation's manufacturing capacity more than quadrupled. By 1900 the urban-industrial sectors had decisively eclipsed the rural-agrarian, despite the fact that mechanization had helped make American agriculture the most productive in the world. Between 1880 and 1900 railway companies added 100,000 miles of track to the 163,000 miles already constructed. But in these same decades, expanding urban markets replaced railroads as the chief component of American economic growth. Production of iron and steel for bridges, trolley lines, and metal-framed buildings, in addition to that for railroad expansion, had outstripped America's industrial rivals by century's end. In 1880 the United States produced less than 4 million tons of pig iron, ahead of Germany's 2.5 but only half of Great Britain's output. By 1900 America's iron production had soared to 14 million tons, nearly equaling the combined 16.5 million tons of Germany and Great Britain. The rate of increase in American steel production in these decades from 1.3 to 10 million tons was more than matched by Germany's jump from 0.7 to 6.7 million tons. But American steel manufacture, roughly equal to that of Great Britain in 1880, more than doubled that of its rival by 1900.[65]

Overall, at the turn of the century the value of American manufactures was equal to that of the combined manufactures of its three main competitors, Britain, Germany, and France. This great leap in industrial prowess gave credence to the widely held conviction that the United States had become the most progressive and prosperous society in all human history. Appropriately employing a railroad metaphor in his paean to America's *Triumphant Democracy*, Andrew Carnegie compared the republic's unprecedented economic growth to the "rush of an express" past European nations that seemed to progress "at a snail's pace." Mixing machine and social evolutionist imagery, Carnegie insisted that only sustained growth and technological innovation would permit the United States to maintain its hard-won dominance in an ever more competitive and dangerous international arena.[66]

Carnegie's conviction that America had surpassed its industrial rivals implicitly affirmed material advance as a measure of a society's level of civilized development. And this assumption was widely shared by diplomats, merchants, and missionaries engaged in America's efforts to revitalize the ancient societies of China and Japan. Knowledgeable observers conceded that both these cultures had historically displayed considerable aptitude for technological innovation. But in the Chinese case in particular, Americans and Europeans stressed that these breakthroughs had been achieved by a more vibrant society of the distant past. Missionaries' and travelers' reports on the degraded state of Chinese society appeared to confirm American stereotypes that were shaped by encounters with Chinese migrants— overwhelmingly impoverished laborers—who had begun to arrive in California by midcentury. Composite representations of Chinese public spaces as filthy, of Chinese homes and public works as dilapidated, and of the Chinese themselves as dishonest, deceitful, and enervated by addiction to drugs and unimaginable immoral indulgences were used to justify racial discrimination, communal violence, and exclusionary policies in the United States and disdain for and segregation from the "teeming masses" in China itself. Contradictory censures of the Chinese as either indolent or so hardworking that they threatened to displace American workers also appeared in reports from overseas and in the domestic press.[67]

The negative impressions of a minority of Americans living in China were at cross purposes with the much more favorable assessments of most Americans who resided or traveled extensively in the Qing empire. These visitors who ventured beyond the foreign enclaves were impressed by "the magnificent Chinese achievement" of forging a civilization that had not only endured, but excelled in most human endeavors, for millennia. Whether missionaries, engineers, or retired army officers, those who wrote informed accounts of China in the late 1800s countered racist characterizations of the Chinese by emphasizing China's long history of pioneering inventions as fundamental to human progress as printing and gunpowder. The missionaries usually attributed China's stagnation and deplorable contemporary condition to the close-mindedness, isolation, and despotic control of the Qing rulers and their Confucian functionaries. Breaking from the precedents set by past dynasties and even the early Qing rulers, the dowager empress Cixi (Tz'u-hsi) and her allies had for decades repressed innovative solutions to the empire's mounting problems, blocked the overhaul of its obsolescent infrastructure, and all but excluded progressive influences from abroad. With their long gowns and carefully manicured fingernails, the "mandarins" were frequently characterized as effete — but highly obstructive — relics of a bygone age.[68]

In contrast to the mostly critical appraisals of the current state of China, Japan continued to receive the high praise from American visitors that had been bestowed by Perry and most of his entourage.

The samurai elite's admiration for the guns, machines, and disciplined organization of Westerners impressed even those travelers or military attachés who considered the architecture of the Japanese quaint at best, their traditional implements and weaponry crude, their scientific understandings fantastical, and their statistical compilations careless or indecipherable. Like most Asiatics, the Japanese were also deemed by some Americans to have little regard for truth, accuracy, or plain dealing and an irritating obliviousness to clock time.[69] But these criticisms were almost invariably overridden in accounts of Japan by admiration for the cleanliness, frugality, industriousness, and orderliness of the Japanese people.

Of special interest to more seasoned Japan watchers, such as Townsend Harris and William Griffis, was the contrast between the Japanese openness to foreign ideas, inventions, and modes of production and the Chinese reluctance—and often outright refusal—to borrow from outsiders. Those who had traveled in both countries, for example, found the railways in Japan far more efficient and comfortable than those in China, and in some cases the equal of any in the United States. And Japan's resounding victory in the 1894–1895 war against its far larger Chinese neighbor reinforced the opinion of Americans (and Europeans) that the manly, vigorous, receptive Japanese were far ahead of the Chinese in their quest to attain levels of political cohesion and military power comparable to those of Western nations.

The remarkable ability of Japanese craftsmen to adopt and soon produce imported innovative technologies, from surveying equipment to steamships, was repeatedly singled out by American observers as evidence of the Japanese capacity to turn their backward, feudal society into a progressive, industrialized one. By the 1880s visitors marveled at the rapid pace of this transformation as attested by the Western-style hotels, universities, and office buildings springing up in coastal cities such as Tokyo (formerly Edo) and Yokohama. Japan's leap into modernity was brought home to millions of spectators at the 1904 St. Louis World's Fair, who were dazzled by the combination of the refinement of Japan's traditional handicrafts and its industrially produced machines, transport vehicles, and consumer goods. Surviving commentaries on the Asian pavilions at the fair were far less kind to what were viewed as the overwrought productions of traditional Chinese artisans. If manufactured goods from China were exhibited, they were apparently not deemed worthy of appraisal.[70]

Although American representations of the Chinese were in general far more unfavorable than those of the Japanese, both peoples were widely considered capable of rejuvenating their societies through extensive borrowing from the industrialized nations of the West. The Chinese and the Japanese were routinely lumped together by Western racial theorists as "Mongolians" or "Mongoloids," and usually ranked just beneath the various branches of the "Caucasian" race. In the United States, immigrants from both societies were subjected to discrimination on the basis of their presumed racial differences from the "white" majority, which at various times provided the rationale for humiliating legislative measures aimed at constricting or blocking immigration from East Asia. But almost all reasonably informed American observers were confident that both the Chinese and the Japanese could successfully adopt Western ways of thinking, organizing, and relating to the natural world.

The level of development the Japanese had attained within a matter of decades after the intrusion of the Perry embassy provided ample proof of their ability to master Western science and technology and to employ these to build a modern society. The success of the Japanese in these endeavors prompted a number of specialists on East Asia to judge them the equal of Americans or Europeans in intelligence, thereby further complicating the daunting task of those who sought to establish precise racial classifications for the human species. Racist categories were also muddled by the efforts of some "theorists" to claim the Japanese as "white Asians" or to rank them at the same level as Caucasians. Even Teddy Roosevelt was forced to admit that, though they were not Anglo-Saxons or even Caucasians and their immigration to the United States should be restricted, the Japanese were "as remarkable industrially as in warfare." And he averred that by combining what was "strongest in [their] ancient character and traditions," they had "assimilated with curious completeness most of the characteristics that have given power and leadership to the West," and thereby positioned Japan to "take its place as a great civilized power of a formidable type."[71]

Like many of his contemporaries, Roosevelt held a far lower opinion of the Chinese. He found them weak, disorganized, and unmanly in their lack of self-control, discipline, and martial prowess. It is not clear whether he believed these qualities were racially inscribed or products of a moribund social system. But the collective judgment of the most distinguished late nineteenth-century American writers on China clearly favored a cultural rather than a biological explanation. Missionaries who had lived in China for decades, such as W. A. P. Martin and John Nevius, dismissed the notion that the Chinese were mentally inferior to Americans or any other "race" as too ignorant to merit refutation. They saw the Chinese as uninterested in punctuality or accuracy, and as unable to think abstractly, but they blamed these shortcomings on the primitive, pictographic nature of the Chinese language and on the decrepit educational system. Martin argued that the Chinese were nevertheless "inventive, civilized, and changeable," as demonstrated by the "long list of the most useful discoveries" they had made during the long history of their civilization. Nevius pointed out that the comparison between Western and Chinese technological development that critics used to demonstrate China's backwardness would prove precisely the opposite if one compared the two civilizations just a few centuries earlier.

And despite the social turmoil that had thrown the Qing empire into anarchy, most well-informed China watchers would have concurred with Major-General James Wilson's estimate that the "natural intelligence" of the Chinese people "was quite as great as those of other races." Chester Holcombe, secretary of the American legation at Beijing in the mid-1890s, echoed this view when he predicted that the "high average intelligence of the Chinese" and their renowned capacity for learning would enable them to compete successfully with Western students in acquiring the technical proficiency that was critical to their country's rebirth.[72]

From the outset, those who sought to extend America's civilizing mission across the Pacific realized that reviving Chinese civilization involved very different challenges and would require rather different approaches from those worked out for Amerindian peoples. As had been the case on the great plains earlier in the century, Protestant missionaries became major agents for the transfer of the American version of modern civilization to China, and in roughly the same time span, to Hawaii. In the decades before the Civil War, more than 40 percent of American Protestant missionaries were serving on Indian reservations (10 percent), in Hawaii (20 percent), or in China (nearly 13 percent). By the end of the century, more than 80 percent of the Protestant missionaries assigned overseas were proselytizing in China.[73] But the missionaries who began to arrive in China in the early 1830s soon recognized that the higher level of civilization they encountered there, the far greater numbers of the indigenous population, and the open hostility of the Qing bureaucrats would compel them to accommodate to local conditions far more than they had on the American frontiers. In contrast to the authority they assumed on Indian reservations and the considerable influence they exercised over many Hawaiian notables,[74] missionaries in China found that beyond tiny bands of converts their influence over the Chinese was nil, and that even their nonproselytizing activities were restricted to very limited areas. Notwithstanding Western military victories against overmatched Chinese forces that enabled the missions to expand their operations to port cities beyond Canton and later into the interior, Christian preachers remained reviled outsiders. The number of converts was still minuscule at the end of the century—100,000 in a population of approximately 500 million in 1900[75]—and the missionaries soon realized that the cultural erasure that they and others had pursued (albeit with limited success) in efforts to assimilate the Indians and Hawaiians was clearly impossible with the Chinese.[76]

Although the missionaries came to rely heavily on the presumed superiority of Western science and technology to gain access to and converts among the Chinese, they concluded that the level of both had to be more sophisticated than what had impressed the Indians or the Hawaiians. Chinese handicraft skills were a match for those of any Western nation, and missionaries in the field acknowledged that they had little to teach the Chinese about hard work, tilling the soil, or marketing the products of their labors. The missionaries were certain, however, that Chinese men could learn a great deal from the example of the relationships between Christian spouses about how to treat their wives and raise their daughters, whose lowly status and condition were deemed a measure of how much Chinese civilization had regressed. Missionaries were also sure that Chinese women would benefit from the lessons in proper approaches to homemaking, including hygienic practices, that dominated the curriculum for girls. But excepting small numbers of converts, most from the lowest social levels, missionaries—men or women—ordinarily had little access to either Chinese women or their households. American missionaries soon discovered that social work, particularly the provision of medical treatment, was the most effective way to establish personal interactions with the Chinese. Until the communists shut down the Christian missions over a century later, medical clinics and hospitals as well as schools that offered medical training and instruction in Western sciences were the major focus of missionary efforts in China.[77]

Despite some misgivings that a commitment to social work might detract from the winning of converts, there was strong support in most denominations for medical care and science education. Like other Americans, Protestant ministers and their congregations viewed the material improvements that were transforming U.S. society as signs that the young nation was indeed exceptional and its citizenry truly God's chosen people. And for those determined to share the gospel with "heathen" peoples, the telegraph and the railway were "noble inventions" by which "civilization, republicanism and Christianity" could be disseminated. These technological wonders and the scientific discoveries associated with them were further validated for the faithful by the largely uncontested assumptions that they had been devised exclusively by Christian peoples and that they demonstrated the unique receptivity of Christian civilization to creativity, innovation, and critical thinking.[78]

Advocates of the medical missions as "auxiliary" activities were confident that the immediate benefits this form of social work would offer the Chinese people would prepare the way for widespread conversion. Missionaries with the broader American civilizing mission in mind were also convinced that the superiority of Western medicine and science would impress upon reform-minded Chinese the advantages of American approaches to social improvement. In this view the prescriptions, instruments, and healing techniques deployed by missionary physicians would lend them the aura of "miracle workers" in the eyes of the Chinese, enhancing their authority in situations where so many forces conspired to undermine it. Successful healing was also seen as an antidote to Chinese superstition and credulity, which were judged to be pervasive and debilitating, and as a means of weakening reliance on traditional herbal medicines and acupuncture, which most missionaries thought ineffectual if not harmful.[79] The medical missionaries viewed their clinics and hospitals as models of orderliness, cleanliness, and proper sanitary practice. The late nineteenth-century determination of city planners and health boards to provide at least the middle-class residents of America's towns and cities—from whose ranks most of the missionaries were recruited—with filtered water, sewage systems, and regular garbage collection served to fix these amenities as defining features of modern, civilized societies.[80]

Although the schools for girls emphasized "domestic science" and childrearing, American female physicians were deliberately assigned to Chinese mission stations to underscore the contrasts between the possibilities for educated women in the United States and the degraded condition of women in China. The training of Chinese medical assistants, and by the 1880s of both male and female doctors, enhanced the social status of Christian converts and also demonstrated that it was possible for the Chinese themselves to harness Western science and technology to build a prosperous, secure nation. Medical and scientific training were also seen as effective ways of inculcating a penchant for empiricism, self-discipline, and systematic investigation. As official restrictions on (though not always popular hostility to) missionary activities in China weakened in the last decades of the nineteenth century, medical care was increasingly matched by instruction in Western sciences in missionary schools for boys—and in some cases for girls—from the elementary to the college level. And throughout the decades of civil strife and nationalist agitation in the following century, these activities, along with related projects for the translation of key Western and Chinese works, particularly in the sciences, were central not only to the proselytization of Christianity in China but to the wider American ambition of establishing the United States as the protector of and tutor for that beleaguered country.[81]

Commodore Perry's expectation that America would become Japan's main tutor as the island society sought a place among the modern nations was repeatedly frustrated in the last decades of the nineteenth century. American teachers, technicians, and entrepreneurs were engaged—at times in significant ways—in the transfer of the Western technologies, science, and institutional models that made possible Japan's rapid transition from an isolated feudal kingdom to an industrializing, expansion-minded regional power. But the relatively bloodless dismantling of the obsolescent components of the Tokugawa political and social order, and the emergence of highly skilled, diplomatically astute leadership dominated by the lower samurai, meant that the development process in Japan was controlled by the Japanese themselves far more than was possible in China, which was racked by internal strife and threatened with dismemberment by predatory imperialist powers.[82]

In its early stages the Japanese drive to modernity was highly dependent on *yatoi*, Westerners who were paid to reside in Japan and share their scientific, technical, and organizational expertise with students, bureaucrats, and factory managers. In contrast to China, however, which also grew reliant on Western assistance, Japanese government officials systematically recruited outside advisors, paid them handsomely, and politely sent them home when their assigned tasks had been performed. Missionaries and other foreigners whose activities in Japan were not officially sanctioned found that their access to the people and ability to undertake even modest projects for "improvement" were rigorously constricted by direct government

regulation. Government controls were reinforced by intensifying nationalist sentiments among the Japanese people. This led Japanese
Christians, whom the "unequal treaties" imposed by the Western
powers after Perry's visits permitted to practice their faith openly
again, to assert control over Christian churches in the islands, further marginalizing American missionaries.[83]

The introduction of the American sport of baseball provided a
striking example of the contrasts between Japanese and Chinese responses to cultural imports. The Japanese enthusiastically embraced
the game, incorporating its stress on virtues gendered manly and
prized by the samurai, such as teamwork, physical prowess, and
group competition, and soon developed their own teams. By contrast, baseball was more or less forced on both boys and girls attend-
ing Christian missionary schools in China. Although some Chinese
nationalist thinkers stressed the importance of physical fitness to
China's "self-strengthening" efforts, schoolgirls found calisthenics
"very strange . . . almost ludicrous," and contrary to feminine mod-
esty. Chinese boys found the physicality and aggressiveness of American sports antithetical to Confucian ideals of male demeanor, and
thought basketball more work than play, which did little to recommend it to the scholar-gentry and professional classes who attended
Christian schools to learn English and familiarize themselves with
Western scientific thinking.[84]

Americans involved in what they viewed as the "Westernization"
of Japan were increasingly cognizant of the fact that, like the Chinese, the Japanese were determined to retain as much of their pre-
contact culture as possible. As Sidney Gulick, a missionary serving
in Japan at the turn of the century, observed, the Japanese conceded
that science and technology were essential "tools" for attaining what
Americans and Europeans deemed modern civilization. But they
were also clear that civilization, even in its modern incarnation, was
much more than material things and the ability to unlock the secrets

of nature. Even though Japan's leaders concluded that they had little choice but to master, and improve upon whenever possible, the levers of power fashioned by the West, they refused to sacrifice their vaunted aesthetic sensibilities, time-tested patterns of social organization, and reverence for the natural world.[85]

Japan's policymakers found the American path to national prosperity and stability less and less appealing as the pace of change in the islands accelerated. Not only did influential Japanese intellectual and political leaders—many of whom had either studied in or traveled extensively in the United States—become increasingly critical of American culture, they chose European models for "improving" their own society. The Japanese political system was modeled on the parliamentary systems of Europe, particularly that of Wilhelmine Germany. Japan's new army was patterned on that of Prussia; its navy on Great Britain's. And even its system of compulsory mass education—though adhering to American precedents and drawing on the experience of American academics for a time—was reorganized along French lines, and was later to receive an infusion of German influence. These critical choices with regard to the kind of capitalist democracy Japan aspired to be were doubly unsettling to American expansionists because American-Japanese relations were amicable through most of the last half of the nineteenth century.[86]

By the mid-1890s, however, Japanese industrial exports were becoming competitive with those of the United States and other producers, and Japan's imperial designs on China and Korea were dramatized by its stunning victories over China in the mid-1890s and Russia in 1904–1905. Those triumphs bolstered Japan's primacy as a model of successful Asian self-improvement in response to challenges from the Western powers, and elevated the status of its universities and technical schools, which became major training centers for the Chinese, Vietnamese, and other colonized peoples from South and East Asia. Resentful of what they perceived as America's collusion with the European powers to deny Japan the spoils that were its just deserts from the two wars, Japanese leaders more openly pursued policies of national self-interest. Increasing numbers of American Japan (and in some cases, China) watchers concurred with their European counterparts that industrialization and nationalism were feeding the "Yellow Peril" that had begun to coalesce in East Asia. Fantastical projections of mechanized Asian hordes marauding throughout Eurasia, and of cheap manufactured goods flooding the markets of Europe and America convinced some prognosticators that the American mission to rejuvenate the ancient societies of East Asia had gone awry.[87] These alarms also implicitly called into question the assumption that the American model was as appropriate for the Japanese and Chinese as for the Indian peoples of the great plains. But lessons that might have been instructive in shaping U.S. policy were scarcely noted by contemporary observers and largely lost on succeeding generations.

There were no opinion polls to verify popular sentiment, but it is likely that the great majority of nineteenth-century Americans adhered to the paradoxical assumptions that, on the one hand, their exceptional technological aptitudes and modes of socioeconomic organization had been critical to the forging of a prosperous and powerful nation, and, on the other, that it was their destiny to guide the rest of humanity down the same path to societal development. The consensus on these contradictory propositions was so strong that few thought to point out that a sparse population, vast frontiers, and an abundance of natural resources gave the United States advantages in its quest to become a highly developed industrial nation that few other societies possessed. Almost all of those engaged in the

Japanese woodblock print of the Chinese surrender in the Sino-Japanese War, 1895. Library of Congress.

expansionist enterprise refused to explore the broader implications of the disappointing results of the civilizing offensives extended or launched over the course of the nineteenth century. Even the threat of extinction had not been sufficient to compel most Indians to abandon their resistance to assimilation into American culture. Some individuals did acculturate with varying degrees of success, but most Indian peoples struggled to retain as much as possible of their customary ways, even when doing so meant living isolated and often impoverished lives on the margins of American society.

The Japanese and the Chinese exhibited a similar reluctance to adopt American ways, though they saw the necessity of selective borrowing from industrialized Western societies. What advocates of America's civilizing mission in East Asia ought to have realized by the turn of the century, however, was that those who offered American technology transfers, educational initiatives, and diplomatic support could not control the ways in which the Chinese or Japanese would make use of them or the impact they would have on the transformations occurring throughout the "Far East." A Japanese woodblock print, produced to commemorate Japan's victory over China in the 1894–1895 war, might have served as a powerful caution regarding outcomes. Akin to the lithographs and line drawings that sought to convey the saga of westward expansion to a mass audience in the United States, the print depicts the surrender of the Chinese aboard a Japanese warship. The Japanese military commanders are the paragons of manliness that so impressed Theodore Roosevelt as he surveyed the virtues and flaws of the different "races of man." Clad in tight-fitting, Western-style uniforms, they stand ramrod straight, disciplined, and proud as the Chinese emissaries move toward them across the clean, uncluttered deck. Dressed in flowing Confucian gowns, with the long "pigtails" they were required to wear as a sign of submission to the Manchus, the Chinese are portrayed as effeminate, submissive relics of a dying civilization. Behind them their British advisors witness the surrender that seals a defeat that is profoundly reconfiguring the world order. Like the Americans, the British were unable to avert China's humiliation, and it could well be occurring to both that the coercive effort of Western powers to "open" Japan to international intercourse was yielding outcomes that were neither intended nor conducive to the persistence of Euro-American global hegemony.

This 1915 photograph of Emilio Aguinaldo, former leader of the Filipino insurrection, standing with Frank L. Crone, the American director of education, in front of a prizewinning field of corn, was touted as emblematic of the "peace and prosperity" that American colonization had brought to the islands. From Dean Worcester, "Making Over the Philippines," *World Outlook*, April 1915.

Our political sway has not been imposed upon the [Filipino] people to any greater extent than was necessary; and by the very fact of our superiority of civilization and our greater capacity for industrial activity, we are bound to exercise over them a profound social influence.

—FRED W. ATKINSON,
General Superintendent of Education in the Philippines, 1905

3

ENGINEERS' IMPERIALISM

William Howard Taft was nonplused when President McKinley offered him a position on the five-man commission that was to oversee the establishment of a civilian government in the Philippine islands. Although the telegram summoning him to a meeting with McKinley in Washington had given no indication of what the president wished to discuss, Taft might well have guessed that it had something to do with judicial appointments, which were the focus of his own ambitions. Taft had served with distinction as the nation's solicitor general and later as a circuit judge. For a decade support had been building in the Republican party for his elevation to the Supreme Court. But he had been passed over when McKinley filled a vacancy on the Court in 1898, and there were no openings in late January 1900, when he met with the president and the secretary of war, Elihu Root, at the White House. McKinley's offer took Taft by surprise. "He might as well have told me that he wanted me to take a flying machine," he was later to remark. Taft had shown little interest in foreign affairs to that point in his career. And though he had opposed the war with Spain and the annexation of the Philippines, he had paid little attention to either and knew virtually nothing about the islands that had recently become America's largest colonial possession.[1]

Taft's lack of interest and ignorance had evidently been matched by those of President McKinley himself in the spring of 1898, when Admiral Dewey's rout of what passed for Spain's Asian fleet in the battle of Manila Bay belatedly thrust the United States into the scramble among industrialized powers for colonies in the Pacific. As the ultimate arbiter of the congressional debates over whether the United States should colonize the Philippines, barter them off to another imperial power, or allow the leaders of the Filipino insurrection against Spain to establish an independent republic, he had been obliged to study a globe to determine the location of the "darned islands." Jacob Schurman, the academic who had headed the First Philippine Commission in 1899, was no better informed than the president. Schurman admitted that he was "very ignorant regarding the Philippines," and he believed such ignorance was "the normal condition of the great majority of the people of the United States" at the time.[2]

By the last decades of the nineteenth century the Philippines were being factored into the calculations of military strategists, advocates of overseas expansion, and businessmen who were interested in the markets of the Far East. But it is fair to say that most Americans knew very little about the islands when Congress ratified their annexation in February 1899. By one estimate there were fewer than ten Americans in the Philippines when the Spanish-American war broke out.[3] Most of these were merchants who rarely ventured beyond the capital and main port at Manila. Americans with substantial knowledge of Philippine history or Filipino society were even more scarce, as Dean Worcester's abrupt emergence as an "expert" on the islands amply attests. Worcester, an academic from the University of Michigan, had spent considerable time in the Philippines, but he was a zoologist, who had devoted his fieldwork there to non-human fauna. He had spent little time in Luzon, where resistance to Spanish rule and the Filipino nationalist movement were concentrated. Consequently, Worcester had no first-hand knowledge of the Filipino revolutionaries or the social conditions that had given rise to their struggle for independence.[4]

Given an absence of alternatives, Worcester was recruited as one of McKinley's chief advisors on issues relating to the proposed annexation. Soon after joining the inner circle of executive policy-making, Worcester was asked for his views on McKinley's "benevolent assimilation" proclamation, which set forth the general principles of colonial administration that were to be followed for nearly half a century of American rule in the Philippines. Compared with more established advisors in McKinley's administration and most of the congressmen who became embroiled in the controversies over whether or not to annex the islands, Worcester seemed a very knowledgeable fellow indeed. As McKinley confessed to General Elwell Otis, who badly mismanaged both negotiations with the Filipino leadership and the subsequent war against the Filipino people, politicians in Washington were "ignorant of . . . conditions in the islands incident to our occupation." This ignorance was reflected in the Americans' early reliance on foreign "experts," particularly the British journalist John Forman, who was touted as a "prime authority," and the acting British consul in Manila, Frederic Sawyer. It led also to an often uncritical acceptance of the views of compliant representatives of the Filipino elite in the Manila area. The shortcomings of these sources of information were illustrated by the often vague and contradictory impressions that passed for facts in the accounts of early American observers of the islands, including estimates of the Filipino population that varied between six and twelve million.[5]

Despite active lobbying by Filipino leaders in Manila and Washington, congressional debates over annexation were usually only marginally concerned with the situation in the Philippines, about which most representatives were either unabashedly indifferent or appallingly misinformed.[6] With rare exceptions, the key players on both sides of legislative clashes over the fate of the islands addressed that issue from the standpoint of whether imperialist annexations would strengthen the United States in international affairs or undermine its democratic and republican traditions. For most participants, including the congressmen themselves and the journalists and lobbyists, domestic issues were paramount. These ranged from celebrations of colonial expansion as a prerequisite for America's rise to great-power status and continuing prosperity to warnings about competition for U.S. produce by Philippine agricultural imports and the perils of admitting additional millions of dark-skinned peoples to the Union.[7] If the Filipinos were mentioned at all, it was only in terms of platitudes stressing the need to rescue them from their savage or barbaric state and to the duty of America's manhood to take up its share of the civilizing mission.

If Taft had been aware that his ignorance of the Philippines was shared by most of America's political elite, McKinley's request might have come as less of a surprise. McKinley confessed that he had originally opposed annexation, but the Philippines had been colonized and now it was America's obligation to rule them effectively. Without mentioning the paucity of qualified competitors, McKinley assured Taft that his qualifications had been carefully considered, and that everyone consulted agreed that he was the ideal person for the post. Root appealed to Taft's patriotism and sense of duty, pointing out that one who had enjoyed such a "fortunate career" ought to rise to the challenge of the "risk and sacrifice" of service to his country in its time of need. Once McKinley had assured Taft that such service would only enhance his prospects for nomination to the Supreme Court, the appointment was all but secured. Taft's subsequent request that he be made head of the Commission government was readily granted. And although his knowledge of the Philippines improved little until after his arrival in the islands in June 1900, he plunged with enthusiasm into the role of imperial proconsul that had been so abruptly and unexpectedly thrust upon him.

Judging from Taft's correspondence and the reactions of those working with him to establish a civilian colonial administration in the islands, he seldom questioned his ability to succeed at tasks for which he had little preparation. Equally noteworthy was his confidence that U.S. colonial rule would bring peace and prosperity to a society about which he and most other Americans knew next to nothing. In public pronouncements and personal letters, Taft exuded an ebullient optimism about everything from his own health and that of the American soldiers stationed in the islands to the prospects for the civilizing mission that he was eager to launch there. But, as his correspondence also attests, the hearings that Taft and his fellow commissioners began soon after their arrival made evident both the complexities of Filipino politics and society and the devastating impact American colonization efforts had thus far had on the islands.

Widespread hostility to the American invaders had been the inevitable consequence of the patronizing attitudes and often duplicitous negotiating tactics adopted by U.S. diplomats and military leaders in their dealings with the *ilustrado* leaders who had directed the struggle for independence from Spain. American miscalculations regarding the motives and depth of support for these revolutionary leaders arose in large part from a general indifference to political and social conditions in the islands and the aspirations of their diverse peoples. American ignorance also contributed to the bungled diplomacy and arrogant military maneuvers that led to the outbreak of warfare between the formerly nominally allied Filipino and American forces in February 1899. The four-year "insurrection" that followed resulted in the deaths of tens of thousands of Filipino soldiers and more than 700,000 civilians. The latter perished mainly as a result of epidemic diseases spread by the warring parties; the destruction of crops, farm implements, and housing; and direct assaults by American units on villagers, especially during the later guerrilla phase of the conflict. Over 4,000 United States soldiers were killed in military campaigns to crush Filipino resistance, which drained hundreds of millions of dollars from the American treasury. Particularly on Luzon and several of the Visayan islands, where the fighting was concentrated, the war also took a heavy toll on the communications infrastructure, which had already been heavily damaged in the struggle to oust the Spanish, the islands' original colonial overlords.[8]

The shambles that America's venture into colonization had made of the Philippines must have at times led Taft to regret the decision to venture from his comfortable life as a circuit judge into the perilous realm of colonial administration. And the Second Commission's hearings soon made it clear that costly errors arising from ignorance were compounded by the Americans' lack of experience as colonial rulers in areas beyond the continental United States. Although merchants, planters, and missionaries had established an increasingly dominant American presence in Hawaii in the 1870s and 1880s, those islands were not formally annexed and administered until after

1898. Rapid victory in the Spanish-American war meant that the United States rather abruptly assumed responsibility for the governance of a diverse and widely dispersed island empire, consisting mainly of Hawaii, Puerto Rico, the Solomons, and the Philippines. President McKinley admitted that when he decided to annex the Philippines he had no idea what to do with them.

Elihu Root, a corporate lawyer from upstate New York who in 1899 was appointed head of the war department, which at that point administered the Philippines, sought to compensate for his ignorance of colonial affairs by undertaking a personal crash course in British imperial history. But like numerous other American officials then and later, Root soon concluded that the paternalistic British approach was inappropriate for the United States because it stressed the maintenance of law and order rather than the progressive uplift of the colonized. Luke Wright, who succeeded Taft as governor of the Philippines in 1903, shared Root's view. He declared that the Americans' aim of preparing subject peoples for self-government rendered British precedents irrelevant. Because the British were convinced that colonized peoples were incapable of self-rule, Wright maintained, the "natives [were] employed only in the most subordinate positions" and "no attempt [was] made to develop and build up in the native, by practical experience, a capacity to govern." In contrast to later officials who proudly proclaimed the superiority of U.S. policy, however, Wright intimated that the inapplicability of British precedents was problematic for American administrators. It left them with "no experience in the government of Oriental peoples" in a situation in which they were confronted by "the hostility of a large part of the inhabitants of the [Philippine] islands."[9]

Inexperience and the paucity of reliable information about local conditions did not dampen the American colonizers' enthusiasm for sweeping reforms and ambitious public works projects. Within weeks of their arrival, the commissioners began to propose bold solutions to the prodigious problems they attributed mainly to centuries of benighted Spanish rule. The confidence with which they set about remaking the colony in spite of the calamities that U.S. intervention had already visited upon the islands can in part be explained by their conviction that American society provided an ideal model for civilizing the rest of the world. Because they assumed the superiority of the material culture, ways of thinking, and modes of social organization of their progressive and democratic republic, the colonizers rarely questioned their ambitious, and highly ethnocentric, goal of remaking Filipino society in accord with precedents drawn from the American experience. Determined to approach their task in a systematic, empirical fashion, the members of the Second Commission renewed and extended the inquiries initiated by their predecessors. From ethnologists and newspaper reporters to military officers and nonresisting members of the Filipino *ilustrado* elite, they compiled hefty volumes of information regarding the peoples and cultures whose future they were committed to shaping. The Americans' sense of themselves as the most scientific of modern colonizers complemented their certitude that they could overcome any obstacles by drawing upon American engineering skills and industrial technologies. These more than any other presumed assets bolstered their assurance that, despite the unpromising beginnings of their civilizing mission to the islands, they were destined to transform the Philippines from the backward colony the United States had wrested from the Spaniards into one of the most progressive and prosperous nations in Asia.

The aspiring colonizers assumed that their superior technology and training would allow them to make short work of Filipino military resistance. In fact, some thought that the demonstration of the killing power of America's weapons and the technical proficiency of its engineers was sure to convince the recalcitrant Filipinos of the advantages of acceding to U.S. tutelage. Jacob Schurman wrote that General Antonio Luna, one of the more audacious Filipino commanders, was "filled . . . with amazement and dismay" when the Americans in pursuit of his forces bridged a river "he had supposed impassable." Schurman reasoned that because "nothing [was] so impressive to the Oriental mind as power," victory could be won only by the utter "annihilation of [rebel leader Emilio] Aguinaldo's army as a fighting machine" by the superior U.S. forces.[10] But American casualties rose disturbingly in the following months as Filipino forces shifted to guerrilla warfare to counter the invaders' superior firepower. The colonizers sought to explain away their military reverses by characterizing the conquest as a fiercely fought contest between the forces of civilization and those of barbarism. As Theodore Roosevelt had frequently asserted, the most brutal (and by implication the dirtiest) wars were those on the fringes of expanding civilized societies. However cruel, these conflicts with "barbaric" or "savage" peoples determined who would control valuable lands and resources and whether great civilizations would continue to grow and hence flourish. Roosevelt insisted that "native" resistance left the civilized Western powers with no choice but to wage war until the "natives" acquiesced to colonial rule.[11]

In the case of the Philippines this characterization of the adversaries as barbaric or savage was bolstered by the "rebels'" resort to guerrilla tactics. American reporters and military men routinely referred to Filipino soldiers as "landrones" (bandits), and Taft dismissed their resistance as cruel, treacherous, and little more than the "terrorism of marauding bands." For many Americans this underhanded mode of warfare justified the same sort of harsh "reprisals" that for centuries had been inflicted on Amerindian adversaries. But Roosevelt argued that it also made it impossible for Americans to withdraw from the islands. If they retreated, he predicted, not only would the Philippines revert to a state of feudal backwardness and "savage anarchy," but American honor and manhood would be called into question. It is likely that Roosevelt was sensitive to the charge, made by a number of observers, that the failure to respond forcibly to alleged affronts to Americans perpetrated by Filipino soldiers and civilians before the outbreak of open warfare had been taken by the "rebels" as evidence of the occupiers' cowardice. Clearly viewing the situation as a test of the manly virtues of the American "race," he chastised those who opposed annexation for "their unwillingness to play the part of men," and warned that if the Americans did not fulfill their duty as a civilized people to colonize the Philippines, a more manly Western nation would take up the challenge.[12]

The destruction caused by the forcible occupation of the Philippines and reports of atrocities committed by American forces sustained anti-annexationist sentiment in the United States for years after Congress voted to colonize the islands, and even raised doubts among some expansionists about the viability of America's imperial experiment. But the badly mismanaged conquest appears to have had little effect on the military leaders and soldiers engaged in the fighting or the civilian officials (and missionaries) charged with administering captured provinces.[13] As the latter debated long-term policies to be pursued in "pacified" areas and strove to establish enduring institutions even as the war continued, they evinced scant concern that they and their fellow Americans had, as Wright warned, "no experience in the government of Oriental peoples." In part, American administrators plunged ahead because in their view they had no other choice. The damage done by the insurrection and its repression had to be repaired, and the authority of the colonial regime appeared to be the only antidote for the chaos engendered by the continuing struggle. But the ebullient optimism of official reports and personal memoirs in the years during which military campaigns laid waste some of the richest and most heavily populated regions of the islands suggests that the unacknowledged contradictions of America's approach to the Philippines had much more to do with developments within the United States than with the conditions encountered in the "unknown islands in far-off seas."[14]

The Spanish-American War was in many ways the culmination of the decades of economic growth and westward expansion that had propelled American merchants, missionaries, and sailors across the Pacific to Hawaii, Japan, and China. The officers who commanded the armies that conquered the Philippines, the administrators who formulated colonial policy and directed the resulting projects, and the missionaries and teachers who carried America's civilizing vision to the villages were drawn overwhelmingly from the middle classes that had been the main agents and beneficiaries of America's late-century surge to power. They were predominantly male, though women went to the Philippines and other colonial territories in significant numbers as schoolteachers, missionaries, and wives of colonial or corporate employees.[15] They grew up in households in which new technologies had altered work patterns and gender roles and had contributed to more diversified and nutritious diets, better lighting, improved hygiene, and greater opportunities to amass consumer goods. They came of age in towns transformed or brought into being by new systems of transport, communication, sanitation, and energy. In school as well as in professional careers, they were continually reminded that they were privileged to live in an age of such wondrous improvements, and to be citizens of the world's most advanced and democratic society. And they were told by popular authors, such as Edward Byrn, George Morison, and Josiah Strong, that America's achievements would inevitably be disseminated to all humankind.[16] They journeyed to the Philippines with the self-assurance befitting representatives of a higher civilization. They set forth to uplift the savages and to vanquish barbarism, material backwardness, and superstition by establishing American institutions, exporting American machines, and inculcating American ideas and attitudes among less fortunate peoples. America's civilizers still descended from the city upon a hill, but now the church steeples were dwarfed by skyscrapers and smokestacks, and they were more likely to carry slide rules and T-squares than the King James Bible.

A downside to the accelerated technological innovations and social changes of the late nineteenth century also influenced the Americans' attitudes and approaches toward the Philippines. The maldistribution of the material rewards of the new systems of production intensified longstanding social divisions in the United States and created new ones. It led to periodic conflict between social groups that gradually hardened into endemic hostilities along class lines. Class divisions were compounded by a heightened awareness of ethnic and racial identities fed both by the flood of immigrants from Eastern and Southern Europe and by the growth of the African-American population in urban areas. Beginning in the 1870s these converging tensions were exacerbated by a series of severe depressions in the agrarian and manufacturing sectors that generated radical challenges from agricultural and industrial workers. These upheavals alarmed not only the entrepreneurial and managerial classes, who were the major beneficiaries of the second industrial revolution, but middle-class professionals and petty bourgeois clerks and shopkeepers, who were determined to distance themselves economically and culturally from the laboring classes and in many instances from ethnic and racial minority groups.[17]

Many Americans viewed the decades around the turn of the century as marked by a pronounced decline in community identity and social commitment. This "crisis in the communities,"[18] which was often linked to a heavier emphasis on competition, consumerism, and the movement of population from small towns and rural areas to urban centers, provided the context for challenges to patriarchal authority within the family and to paternalistic managerial styles in the workplace. The suffragette movement and other feminist initiatives profoundly threatened prevalent assumptions regarding masculine and feminine attributes and social roles. These challenges, along with shifts associated with corporate and government bureaucratization, led to widespread fears of social decay and physical degeneration. In this view, sedentary living and the loss of the financial independence traditionally associated with property ownership and self-employment were eroding the manly qualities that did much to account for the West's domination of the rest of the globe. At the same time, feminist transgressions appeared to deny the maternal instincts and spousal sensibilities that were deemed essential to the healthy reproduction of the "civilized races."[19]

Fin-de-siècle prophets of decline cited diverse trends as signs of degeneration. These included the widespread incidence of neurasthenia and other nervous disorders among the middle and upper classes, the effete lifestyles of sexually ambiguous personages such as Oscar Wilde, and rising numbers of the physically and mentally unfit in Western societies. Remedial responses ranged from eugenics campaigns and the promotion of sports and other forms of physical culture to efforts to revive traditional handicrafts and expeditions to the "exotic East" in search of spiritual and aesthetic fulfillment. Boisterous guardians of American manliness advocated both military action and roughing it in frontier areas like the "Wild West." In such locales a man could restore his independence and virility by taking up Teddy Roosevelt's "strenuous life," which was touted as the key to national salvation and to the regeneration of Western civilization.[20]

But far greater numbers of aspiring youths at all social levels, particularly young middle-class men, preferred to emulate the inventors and engineers who had contributed so much to America's rise to national prosperity and global power. Many modeled their lives on the careers of inventors, especially Thomas Edison and Alexander Graham Bell. Inventors and engineers were celebrated not only as exemplars of American ingenuity but also as innovators, managers, and planners who could find solutions to the manifold social ills that beset the republic as it entered its second century.[21]

The centrality of engineers' contributions to America's transformations and the professionalization of their work had raised their status and their financial remuneration to levels unimaginable in earlier decades. Whether engaged in constructing canals and railways, designing machines and production processes, or building bridges and roads, engineers had made critical contributions to the nation's early industrialization. As the chemical and electrical sectors of the economy expanded with the second wave of industrialization, new engineering specialties developed to take up the research and development, and in many instances the managerial tasks, that were central to the new systems of innovation, production, marketing, and distribution. Engineers were also essential to the emergence of the twentieth-century American city. They were indispensable in virtually all facets of urban planning, from sewer and electrical lines, transport systems, and new industrial plants to sports stadiums and amusement parks. In the military sphere, they contributed to the innovations in weapons design and production, ship construction, and logistical systems that revolutionized warfare in the industrial West. It is little wonder that they were widely regarded as role models for citizens of modern democracies, and that their skills and vision were deemed vital to America's continuing preeminence in a tense and highly competitive international arena.[22]

The process of professionalization established engineers as experts in command of esoteric knowledge, a specialized jargon by which to communicate that knowledge, and the technical skills to turn that knowledge to practical advantage. Professional organizations such as the American Society of Civil Engineers and the American Institute of Electrical Engineers, and technical schools and colleges specializing in engineering education, devised formal standards and procedures by which mastery of engineering knowledge could be certified, access to engineering positions limited, and the profession as a whole defined.[23] These organizations also articulated the achievements, ethics, professional objectives, and social responsibilities of the profession for a national audience. This articulation was vital to the construction of the engineer as the cultural hero of the age.

The engineers' stress on the scientific nature of their work was central to their appeal as men to be emulated for both youths at home and colonial bureaucrats overseas. In the American context, the fact that the science in question was applied rather than abstract or theoretical was of great importance. Engineers could translate their esoteric knowledge into technological innovations that enhanced humankind's material well-being and mastery of nature, and thus they were of direct benefit to society. As George Morison proclaimed in his presidential address to the American Society of Civil Engineers in 1895: "We are the priests of material development, of the work which enables other men to enjoy the fruits of the great sources of power in Nature, and of the power of mind over matter. We are priests of the new epoch, without superstitions." Engineers had been instrumental in establishing material improvement and increase as defining attributes of modern civilization. In so doing, they had earned popular recognition as critical agents of progress and social advancement, and as arbiters of the future of humanity.[24]

Membership in the engineering profession was overwhelmingly middle-class and male, and the scientific and technological pursuits of engineers were also largely white male preserves. Participation in both these subcultures valorized engineers as (masculine) masters of nature, which continued to be gendered feminine. Andrew Carnegie, for example, mused in the mid-1880s: "Man is ever getting Nature to work more and more for him. A hundred years ago she did little but grow his corn, meat, and wool. Now she cuts the corn, gathers, binds, threshes, grinds, bakes it into bread, and carries it to his door . . . she will carry him whithersoever his lordly desire may lead him. Across continents and under seas she flies with his messages. Ever obedient, ever untiring, ever ready, she grows more responsive and willing in proportion as her lord makes more demands upon her."[25]

Literature and popular imagery depicted engineers as physically active, authoritative—if not domineering—individuals. Their manly demeanor and self-assurance made them more than a match for disgruntled and ill-disciplined laborers, whether newly arrived immigrants, African Americans, or the semi-savage peoples of distant lands.[26] Fredrick Taylor, a mechanical engineer by profession, strove to elevate the disciplining and supervision of workers to a science. His elaborate schemes for "scientific management," which presupposed that white, middle-class men—preferably engineers—would be in charge, served to reinforce status distinctions and spatial distance between solidly middle-class, overwhelmingly white engineer-managers and the working-class employees, both men and women, they supervised.[27]

The American interlude in the Philippines can best be understood as a vast engineering project. The colonial era in the islands was inaugurated by the chief engineer (and cartographer) of the war department, who, acting upon McKinley's instructions, added them to the territories of the United States on a large map in the president's office. From the outset, it was assumed that engineers would take a leading role in the colonization of this distant and "exotic" land. Their mission was set forth in some detail in *Duty of the American People to the Philippines* (1898) by an anonymous writer using the

pseudonym "Publicola": "Our railroads and great iron establish-
ments will be found not bad schools from which to graduate administrators for such provinces and districts. There are hundreds of
young engineers with some practice in the field who would go out
for minor positions with the same sense of duty and something
very nearly the same education as that taken to India by Lawrence,
Edwardes and Roberts."[28]

Although none of the early governors was an engineer, members
of the profession pervaded all other levels of colonial administration.
From the time of the shift to civilian government in 1901, it was stip-
ulated that the supervisors of the colony's twenty-seven provinces
must be civil engineers. The engineer-supervisors were thus pivotal
members of the three-person boards in charge of the day-to-day administration of the provinces. Engineers were also in charge of the
land revenue and geological surveys, the mapping expeditions, and
the public works projects aimed at rationalizing space across the islands, sorting out property claims, and assessing the colony's resources. In addition, they served on the provincial boards of health.
In that capacity, engineers emerged as key strategists in the medical
campaigns against the succession of epidemics that struck the Philippines in the first years of American rule. An engineer also headed
the committee appointed to investigate the potential of the Baguio
area as the site of a hill station where the colonists could vacation
and escape the coastal heat. There were never enough engineers
to staff all the positions assigned to them, a situation which in part
accounted for the high salaries and other special incentives offered
to recruits from the United States. The shortage of engineers is a recurring concern in the many letters written by Taft in his years as
governor urging prominent friends back home to nominate qualified
young men for engineering posts in the Philippines.[29]

Perhaps no one valued the engineers' contributions to the colonial project in the Philippines more highly than the fourth governor,

W. Cameron Forbes. Reflecting on the first decades of American rule, Forbes declared:

> In the main it may be said that of the many creditable and useful services rendered by Americans in the Philippine Islands that given by the Bureau of Public Works has been among the foremost. The engineers, from those in the highest positions down through the service, brought to their work true American enterprise and the will that succeeds . . . Too high praise cannot be given to the devotion revealed and service rendered by the band of pioneers who, as builders and planners, did their part in the fine work of civilization the United States undertook to carry out in the Philippine Islands.[30]

As the qualities Forbes lauded suggest, the influence of the engineering profession extended far beyond the large number of its members who took up administrative positions or served as advisors for U.S. corporations in the islands. Engineers, particularly civil engineers, set the standards for Americans engaged in the civilizing mission. This meant that the ethos of the engineering profession came to define that mission, to set its goals and parameters, and to determine the policies and instruments through which it would be implemented. The esprit and can-do attitude that predominated in the American engineering profession at the time do much to resolve the paradox of the colonization of the Philippines. When subject peoples were viewed in terms of a set of technical problems that could be addressed by institutional adjustments and technological inputs geared to material increase, American ignorance of the Filipinos' customs, beliefs, and modes of social interaction appeared unimportant. The colonizers' indefatigable faith in their ability to rule the Philippines successfully and to establish prosperity and representative government there had little to do with the nature of the societies they proposed to transform. Rather it was inspired by their witness to and participation in the dissemination of technologies, modes of organization, and technical procedures that they were convinced had made the United States the most productive, technically advanced, and materially endowed nation in human history.

The capacities to stimulate material advance and to put unexploited resources to productive use were definitive for the commissioners who formulated colonial policies and the officials who enacted them. Employing some of the terminology that would later be favored by modernization theorists, they explicitly identified their purpose as the "development" of the "backward" or "undeveloped" islands. Even before annexation, the lexicon of economic development had been invoked by congressmen and journalists seeking to justify colonization.[31] Some Americans viewed development initiatives as the only way to compensate for the death and destruction caused by the suppression of the Filipino insurrection. Carl Crow, a reporter who visited the islands a decade into the colonial era, went so far as to boast that the colonial regime's improvements in sanitation alone had "saved ten times the lives that were lost through the bullets of American soldiers."[32]

During and after the conquest, the promise of material improvement was invoked to justify the perverse logic of forcibly subduing the Filipinos in order to bring them peace and prosperity. The pseudonymous Publicola, a fervent advocate of annexation, argued in 1898 that "without the expectation of marked industrial improvement," the United States would not be justified in retaining the islands.[33] As governor, Taft viewed material improvement as the key to the success of pacification efforts. He urged the secretary of war to hasten the transition from military to civilian government so the public works projects that were essential to reconciling the Filipinos to American rule could be undertaken. Theodore Roosevelt, who as president provided strong support for Taft's initiatives in the Philippines, declared: "Nothing better can be done for the islands than to introduce industrial enterprises. Nothing would benefit them so much as throwing them open to industrial development. The connection between idleness and mischief is proverbial, and the opportunity to do remunerative work is one of the surest preventatives of war . . . Congress should pass laws by which the resources of the islands can be developed . . . the vast natural resources of the islands must be developed."[34]

Although roads were built and sewage systems installed in the early years of the commission governments, public works were elevated to the highest priority under the governorships of James Smith (1906–1909) and W. Cameron Forbes (1909–1913). In his 1908 message to the commission and the Philippine assembly, Smith prefaced his inventory of improvements to the islands' transport and communication infrastructure, sanitation systems, and harbors with a rather overwrought paean to America's "disinterested benevolence": "Then for the first time since the world began did a nation, flushed with victory and mistress of the fate of conquered millions, turn her face from earth to heaven, and, catching some of that divine charity which inspired the Good Samaritan, set herself to lift a subject people to a higher plain of progress." Forbes, an even more enthusiastic advocate of material development, promoted public works projects designed to attract corporate investment and spur industrialization. At a banquet in his honor near the end of his term as governor, Forbes defended his emphasis on public works and economic development in the strongest terms:

> I have made material prosperity my slogan while governing
> these islands. I have worked for material development. I
> have done it on the principle that a chain is no stronger
> than its weakest link, and our weakest point was the back-
> wardness of our community from a material point of view,

due to uneconomical and unscientific methods, lack of adequate means of transportation, lack of capital, lack of education on the part of the people, lack of incentive for labor, lack of physical strength on the part of the laboring and directing classes the result of poor nutrition and hygiene. I have set myself to seek out the fundamental reasons for this backwardness and remedy. I have not done this because I believed material development was the whole thing, but because it was the thing most needed here.

As Forbes made quite explicit elsewhere, in his version of the civilizing mission, public works, from roads to schools, needed to be fully developed before responsible colonial officials could even begin to think seriously about turning governance over to the subject peoples.[35]

The American colonizers' sense of themselves as agents of a higher civilization whose task was to render the undeveloped Philippines productive and prosperous, was, of course, grounded in tropes that had informed earlier encounters between settlers and the indigenous peoples of North America. But these familiar themes took on somewhat different meanings in the Philippines. In part the differences can be attributed to the influence of broader fin-de-siècle intellectual currents. They also owed much to significant differences between American interactions with the colonized Filipinos, particularly the *ilustrado* elite, and the earlier exchanges with Amerindians. Furthermore, the perceptions and policies of Americans in the islands were tempered by encounters with other Asian cultures, particularly those of China and Japan. As in other colonial societies in this period, the very meaning of civilization was much more reductively equated with superior science and technology than it had been in the pre-industrial era.[36]

Americans were even more prone than the Europeans or the Japanese to see engineering feats as both definitive gauges of the level of development achieved by other societies and essential components of their own mission to civilize. Charles Elliott, for example, who served under Governor Forbes as commerce and police secretary, believed that the quality of the roads in a society was a "fair index" of its peoples' "intelligence and enterprise." Forbes himself considered roads the best indicator of the quality of government and the caliber of the men in charge. Members of the 1901 Philippine Commission declared that peoples without roads were necessarily savages. American bureaucrats graded the diverse peoples of the Philippines on a scale that ascended from "savage" Negrito hunters and gatherers to the semicivilized, highly Westernized *ilustrados*. The ranks allotted to particular peoples depended largely on estimates of the level of their material culture—most prominently their state of dress or undress, their tools and weapons, their dwellings, and the degree to which they had transformed their natural environment. The social evolutionist thrust of these gradations was also reflected in the Americans' consensus that a key part of their mission was to develop Filipino society to the point where its peoples could survive in a predatory international arena without the protective intervention of the United States.[37]

The correspondence between levels of material culture and stages of evolutionary development was popularized for Americans back home by the ways in which nonwestern societies were portrayed at the expositions and world fairs that were in vogue around the turn of the century. The design of lavish spectacles, such as the World's Columbian Exposition of 1893, held in Chicago, and the Louisiana Purchase Exposition of 1904, in St. Louis, were intended to impress visitors with the technological and scientific wonders that had car-

ried "modern" Western societies to the pinnacle of human development. The focal points of the expositions were clusters of neoclassi-
cal buildings, such as those in the Court of Honor in the White City
at the 1893 fair, or monumental structures, such as the Column of
Progress in St. Louis in 1904. The exhibits in these orienting spaces
not only featured technologies that exemplified scientific and technical advances but quite explicitly celebrated them as triumphs of
white, male, middle-class ingenuity. These marvels were deliberately juxtaposed to exhibits in the "midways" or "reservations" that
radiated away from the Western industrial clusters. The outlying
exhibits rather dramatically displayed the more primitive material
cultures of "racial" groups identified as less advanced, if not savage
or barbarous. The *tableaux vivants* devoted to nonwestern cultures
were arranged hierarchically from Japanese shrines and Egyptian
street scenes to grass-hut villages meant to depict "native" lifestyles
in such "exotic" locales as West Africa and the Pacific islands.[38]

The Filipino peoples chosen to be displayed in these glorified
sideshows, and in the natural history museums that were also in
vogue at the time, were those, such as the Igorots and Negritos, who
were designated as "tribals." Often displayed in seminakedness, representative "specimens" of these ethnic groups both satisfied the voy-
euristic impulses of fairgoers and vividly promoted the contrasting
images of the backwardness of colonized peoples and the progressivism of the American citizenry that the organizers of such exhibitions
intended to project. But the material culture and social organization
of these minority groups were neither typical of the Philippines as a
whole nor even remotely representative of the urban *ilustrados* or
the peasant villagers who made up the great majority of the population. *Ilustrados* and peasants, not the highly publicized tribals, were
the main groups from which those actually administering the islands
constructed representative images of the Filipinos. The Western-
educated elite and the agrarian classes were also the principal targets

of American projects for social reform and economic development. This focus was illustrated by the fact that the colonial organizers of the massive Philippine Exposition at the 1904 St. Louis fair, which was spread over forty-seven acres, took care to contrast scantily clothed primitives, who as yet had not been redeemed, with Western-clad Filipino children in American-style schoolhouses and several well-drilled, smartly uniformed Filipino "Scout" units and marching bands.[39]

The contrast between the colonizers' representations of themselves as civilizing engineers diffusing advanced technology and techniques and their representations of the Filipinos as ignorant, backward, and materially impoverished informed virtually all aspects of American policymaking and projects in the islands. Americans often made exceptions for the rather refined lifestyle they discovered among wealthy *ilustrado* families. Consequently, in their assessments of Filipino attainments—or lack of same—and capacity for improvement, many commentators distinguished between the small Western-educated elite and the peasant majority. Most conceded, for example, that the *ilustrados* were quite well educated, adept at professional pursuits (especially law), well mannered, hospitable, and fond of music and dance. The fact that many of the *ilustrados* (literally, the enlightened ones) shared with progressive American administrators and educators a reverence for the Enlightenment ideals of rationalism, empiricism, and scientism did much to enhance mutual respect as well as the colonizers' estimates of Filipino potential for advancement. But Americans also disparaged the *ilustrados* as self-serving, deceitful, vain, and prone to long-winded speeches. And though the Filipino peasants were typically seen as ignorant, credulous, and childlike, numerous Americans remarked on their comely physiques, their polite and law-abiding natures, and their aptitude for military service, if commanded by competent leaders.[40]

In addition to distinctions between the elite and the peasantry, the colonizers codified ethnolinguistic differences in the Filipino population, which were usually expressed through racial categorizations (Malays versus Negritos, for example) or designations suggesting levels of evolutionary attainment (semicivilized, barbarian, savage). The diverse array of "tribal" minorities and the heavily Islamicized Moros of Mindanao posed significant classificatory challenges, but Americans constructed an essentialized Filipino with reference to the Tagalogs and related ethnolinguistic groups who occupied the plains of south-central Luzon and the Visayan islands. Some made distinctions between these quite varied, yet predominately agricultural, lowland-dwelling peoples, but most lumped them together as an advanced or intermediate type of the Malay race. In general estimates of the Filipinos' level of and capacity for development, the colonizers tended to gloss over differences between elite and peasantry in their attempt to distill essential attributes consistent with their policy recommendations and their assessments of the impact of colonization on the islands.

Like their representations of the Amerindian peoples, the colonizers' essentializations of the Filipinos were inextricably connected to their impressions of the lands the latter inhabited. The islands were seen as fertile and rich in natural resources but woefully underdeveloped, with vast tracts of rain forest wild and uncharted and inhabited only by "savage" peoples whose slash-and-burn agricultural patterns were deemed primitive and destructive.[41] Many commentators gave the colonizers' fixation with developing the islands' resources a familiar racial twist. Theodore Roosevelt shared the view of the Asian "expert" William Griffis that the Americans, as the most advanced strain of the Anglo-Saxon race, had become key agents of the expansion of civilization from the highly developed temperate to the undeveloped tropical regions of the globe. Countering the warnings of Benjamin Kidd, who feared that the "white" race would be corrupted by residence in the tropics, Griffis praised the vigorous and technically proficient Anglo-Saxons for finding ways of making use of the resources that less advanced tropical peoples had failed to exploit. And Roosevelt agreed with Griffis that the Anglo-Saxons' tropical projects had contributed significantly to the material advancement of Asian and African societies. Some colonial officials explicitly contrasted the energetic and dominating Anglo-Saxons with passive "Orientals" mired in fatalism. Like the Amerindians before them, the Filipinos were seen as exerting little control over their environment because they accepted that "nature should be propitiated, not controlled." They allowed their societies' potential for growth and development to "be cramped by [their] physical surroundings" and never dared "to adapt [their] surroundings to their desires."[42]

Many Americans attributed the underdeveloped state of the Philippines to the Spanish, who had controlled the islands for centuries but had failed to build the communications and transportation infrastructure that was essential to the extraction of minerals and timber and the creation of market outlets for local products. Even the meager improvements that had been introduced had been allowed to deteriorate by the lethargic and decadent Spanish overlords. American policymakers pointedly contrasted their own scientific procedures and sophisticated technologies with the "medieval" attitudes and crude fabrications of the Spanish. Boasting of the level of development that would be or had been achieved under U.S. rule, Americans chastised both the Spanish and the Filipinos for allowing "one of the richest portions of the earth [to lie] dormant since the earth took its form" because neither had the tools or the knowledge to harness the forces of nature.[43]

In the American view, the repressive Spanish colonial regime had impeded the development of the islands through its stunting and corrupting effects on the character of the Filipino population. The essential Filipino who emerges from official reports and correspondence and published accounts of the early years of U.S. rule is in many ways the polar opposite of the equally essentialized American engineer. As a product of scientific training, the engineer exemplifies empiricism, critical testing, and technical expertise. By contrast, the Filipinos are depicted as superstitious, credulous, impractical, and lacking a "sense of reality." They are said to pay little attention to empirical proofs or the relationship between cause and effect. Instead they stubbornly adhere to fantastical folk beliefs and explanations grounded in supernatural interventions.[44] These irrational predispositions were seen as most distressingly manifested in Filipino resistance to American campaigns to eradicate epidemic diseases such as cholera, as well as to projects aimed at providing sewage systems or unpolluted water. American officials reported their dismay that Filipinos held evil spirits responsible for the spread of disease or that those afflicted with cholera believed they could be cured by drinking the water of a "sacred" spring near Manila, which turned out to be fed by a broken sewer pipe.[45]

Routinely deploying tropes that had long pervaded European imperialist discourse, American colonizers characterized the Filipinos as lacking initiative and needing steady supervision—in sharp contrast to the prototypical American engineer, who took charge and excelled in carrying out a vision or design. The image of the active, progressive, self-made engineer was pitted against the image of fatalism, passivity, and lack of drive that Americans associated with the Filipino peasantry, though not always with the *ilustrado* elite. Americans judged the Filipinos as imitative rather than innovative, deeply committed to their customary ways, incapable of solving their own problems, and utterly dependent on American administrators and technicians to lead them into the modern age.[46]

Americans also commented on the Filipinos' proclivity for tall

tales or outright dishonesty, and viewed this failing as a manifestation of their childlike nature. This trait, too, contrasted with the image of engineers, revered for their honesty and strict adherence to
the truths revealed by science. Engineers were sticklers for precision
and accuracy; the stereotypic Filipino was vague, careless, and indifferent—if not hostile—to statistics, whether used to count people,
measure property, or design bridges. Beyond suspicions that Filipinos habitually cheated on their taxes and were incapable of reliable
measurements, American colonialists feared that the combination of
childlike dependence and congenital dishonesty would make it difficult for the Filipinos to develop the political ethics and equitable,
self-governing institutions that, the Americans repeatedly asserted,
were the ultimate goals of their experiment in colonization.[47]

If some Americans blamed Spanish oppression for the Filipinos'
supposed propensity to cheat and lie, others argued that they had an
aversion to work, particularly manual labor, and that this was another legacy of Spanish rule. Like their counterparts in all tropical
colonies, American colonial officials generally agreed that the enervating climate and the ease of obtaining food and other necessities
in the islands' lush environment were the most basic sources of the
Filipinos' indolence.[48] But a number of observers traced the Filipinos' dislike of hard work to the excessive exactions of government
and landlords as well as the paltry wages they believed to have
been standard under the Spanish regime. In this view, these abuses
had left the peasantry malnourished, weakened by disease and parasites, and with little incentive for serious or sustained exertion. The
ilustrados had imbibed a disdain for manual labor that Americans
associated with the culture of the Spanish nobility. Whatever the
explanation, the predominant view of the staunchly middle-class
American colonizers was that Filipinos were lazy and careless workers who were oblivious to clock time and evinced high rates of ab-

senteeism. As improvident "citizens of the land of mañana," they gave little thought to their future advancement.[49]

Although many characterizations of the Filipinos were unquestionably racist in both the language employed and the categorical sweep of the traits assigned, American attitudes toward the islands' population were a good deal more ambiguous than these pronouncements might lead one to believe. This ambiguity was exemplified by the striking disjuncture between the colonizers' social interactions with the *ilustrados*, which both sides acknowledged were often racially charged, and the estimates of Filipino aptitudes held by Taft and other American administrators as well as travelers, teachers, and social commentators. The latter were highly essentialist, but they often (at least implicitly) contradicted the pervasive racial rhetoric of the age. It is not surprising that many American soldiers, coming as they did from a society where strict social separation of the races was ascendant and racist epithets were quite respectable in most circles, viewed their efforts to suppress Filipino resistance in racial terms.[50] For most, the conquest of the islands was an extension of the wars against the plains Indians or vigilante assaults on assertive African Americans. Like their British, French, and Dutch counterparts, American colonizers created rather closed communities in areas where they were concentrated, such as Manila. Though they mingled with prominent *ilustrados* at official gatherings, they preferred to socialize at virtually "whites only" clubs and in their homes, where, with notable exceptions, the only Filipinos likely to be present were servants. Whether at work in Manila, traveling in the countryside, or vacationing in the hill station at Baguio, Americans sought to demarcate clearly—through dress, manner of speaking, and behavior—the social and physical space that separated them from even the most distinguished of their colonial subjects.[51]

Despite their preference for social segregation, many Americans were quite explicitly open to the possibility of significant, if not unlimited, Filipino self-improvement and advance as a people. They almost invariably assumed that progressive uplift would be fostered mainly through the islanders' adoption of American ways. And though this presupposition was clearly ethnocentric, those who espoused it at least implicitly called into question the notion that racial limitations would prevent the Filipinos from becoming thoroughly civilized in the imperial, evaluative sense of the term. But even among those who were sanguine about the Filipinos' capacity for improvement, there was considerable disagreement as to how long the process would take. Some doubted that "Orientals" could advance as far as the Anglo-Saxons, and many considered the "natives'" prospects in terms of the racial categories and lexicon in use at the time. Nonetheless, prominent Americans frequently cited the Filipinos' highly developed intelligence as evidence of their potential for education and advancement. Working on these assumptions, colonial policymakers in the islands strove to establish the physical infrastructure, the institutions, and a public educational system that would transform what they viewed as a backward, feudal society into an economically developed, and ultimately independent, democracy capable of full integration into the international capitalist system. During and just after the First World War, retrospectives on the first decades of the colony proclaimed that the Filipinos had realized these expectations in ways that were unprecedented for a colonized people.[52]

These assessments were accepted far beyond the inner circles of colonial policymaking, as is confirmed by the angry rebuttal of Mrs. Campbell Dauncy, a British resident of the Philippines, against the arguments in an anonymous article criticizing "Arbitrary Race Distinctions," which had appeared in the *Manila Times* in June 1905. Incensed to read what she called the "hundredth article" in the American press in which the intelligence of Filipino students was compared favorably with that of American children, Mrs. Dauncy countered with a rather truncated version of the "arrested development" thesis. This theory, elaborated by numerous racial thinkers in the nineteenth century, claimed that whereas African or Asian children might be as clever as Europeans, when the former reached adolescence their intelligence was stunted—for reasons that varied from author to author. As Mrs. Dauncy put it, "It is as though a veil were drawn over the brightness of their minds, and they not only progress no further, but even go *backwards!*"[53]

It is likely that Mrs. Dauncy's view was widely shared by her contemporaries, including most Americans both in the Philippines and at home. And yet the civilizing rhetoric and practice of American policymakers in the Philippines in the early 1900s implicitly problematized, and often explicitly contravened, the racist beliefs of the age. Some officials and commentators revealed deep ambivalence when estimates of Filipino capacities for civilized development were pitted against notions of biologically based racial difference. Notable in this regard was Francis Burton Harrison, the fifth American governor of the islands. Harrison condemned race prejudice as "one of the most poisonous growths of modern times," and was convinced that the Filipinos were eminently improvable. He also averred, however, that they "are in no single respect, as far as I can observe, like the negro race." He acknowledged that Americans living among "large Oriental populations" displayed instinctive "prejudice against color," which he feared might prove an insurmountable obstacle to their ability to govern "vast colored populations." He praised the Filipinos for everything from their refined manners to their intelligence, and promoted them in the colonial service and legislature far more energetically than any of his predecessors. But Harrison also viewed the low incidence of intermarriage between Filipinos and Americans as a "fortunate" trend, particularly since he was certain that both sides opposed these unions.[54]

If most American officials shared the racial prejudices of the overwhelming majority of their compatriots, their development projects and promotion of able Filipinos to ever higher positions provide ample evidence of a generalized belief in the potential of the colonized for advancement. U.S. administrators were committed to establishing a system of mass education, and also to nurturing universities and medical and technical schools, which they considered essential for a self-reliant Philippine nation of the future. At a time when the engineering profession was so highly regarded in American society, it was striking evidence of colonial officials' high estimate of Filipino abilities that just over a decade after annexation the number of Filipino engineers in government service was nearly double that of their American counterparts. A decade later a Filipino was appointed the director of the public works department.[55]

A comment in one of Taft's letters as governor captured the generally unexamined contradictions between American racial convictions and official confidence in the Filipinos. The letter was written to introduce two Filipino students to the captain of the ship on which they were to travel to the United States to study engineering at the University of California. Taft assured the captain that the students were "good boys, imbued with a proper spirit to learn English, to become educated Engineers and to become good Americans."[56] Though the students were in fact youths, Taft's reference to them as "boys," rather than students or young men, may indicate unreflective acceptance of one of the most pervasive tropes of racialized discourse and social interaction. But his confidence that they would succeed in the rigorous engineering program of one of America's most distinguished universities as well as "become good Americans" defied the racist strictures of the age. It also presaged the assumption of the later modernization theorists that there were no inherent, or racial, barriers that would limit the capacity of non-western peoples to follow the liberal-capitalist American path to societal development.

Although Americans' representations of themselves and of the Filipinos were often not racist in the sense of emphasizing attributes that were seen to be innate and biologically determined, they were frequently highly gendered constructs. The traits assigned to Americans epitomized manliness as it was understood in the mainstream middle class and much of elite culture in the United States at the turn of the century. The essentializations of the Filipinos were somewhat more variable and a good deal more ambivalent. In the gendered distinctions Americans made between themselves and Filipinos, engineering imagery and values were pervasive. The predominance of men in the engineering profession complemented the masculine ambiance of the patriarchal society in the Philippines. Colonial engineers were celebrated as rational *men* in control of both themselves and their less technically proficient subordinates. In direct opposition, Americans who sought to distill "Filipino characteristics" stressed an emotional and unpredictable temperament. Filipinos were seen as normally polite, friendly, and law abiding (thus amenable to colonization), but they were said to be prone to jealousy and vengefulness. When provoked, they were thought to be highly excitable and given to "alarms and intrigues."[57]

These inclinations, although not necessarily viewed as exclusively feminine, certainly clashed with the image of rationality and control that epitomized manly behavior. They were also associated with the childlike qualities that had been ascribed to the Filipinos and other colonized "races" as well as to American women. Men might display controlled anger on the playing field or when disciplining lackadaisical workers or intimidating dissenting colleagues. But it should be tempered by purposeful reason and never give way to the loss of emotional control.[58] The Filipinos' propensity for the latter was captured in accounts of, for example, a "frenzied" Malay native running "amuck through crowded streets, killing indiscriminately all he meets until slain himself."[59]

Both implicitly and explicitly, the colonizers represented themselves as self-made, active, disciplined, hardworking, and engaged in a hands-on manner in strenuous and creative pursuits that would benefit the societies they strove to uplift. Many of these qualities had heroic connotations, particularly when associated with engineers creating monumental construction projects in backward, overseas locales.[60] As these activities and the Americans' avid promotion of sports among the peoples of the islands suggest, physical fitness and activity were becoming central components of manly identity in middle-class American culture. With exceptions for members of the *ilustrado* elite, contrasts with the passive, undisciplined, indolent, and unambitious Filipinos were consciously drawn in virtually all American representations of themselves and their colonial subjects. Some emphasized qualities of the Filipinos that were unmistakably intended to suggest effeminacy: the diminutive size of Filipino men, their sentimentality and vanity, and their fondness for ostentatious clothing (the first report of the Taft Commission mocks the "natives'" preference for "patent-leather and other showy shoes"), jewelry, and flamboyant display in everything from manners to oratory. In this context, even references to the affectionate nature of Filipinos or their emotional attachment to music, dancing, and other leisure activities were likely to be ambivalent.[61] A number of American writers remarked on the strong position of women in Filipino society, both in economic roles and in behind-the-scenes influence on decisionmaking within the family.[62] This observation was intended to suggest that, at least with regard to one gauge of civilized development long employed by Westerners, some Filipinos had advanced significantly. But what appears to be praise can also be read as further evidence of the weakness of Filipino men.

It is not by chance, then, that much of what was written about the Philippines in the first decades of colonization gives a strong impression of childlike and effeminate "natives" at play in their rich but untamed environment, while manly, industrious Americans build roads and open schools that give promise of a progressive civilization in the making. It is also not coincidental that the colonizers' civilizing projects were aimed at improving Filipino men. Even though colonized (and indeed American) girls and women were included in educational, health, and athletic initiatives in larger proportions than in any other colonial society of the time, males predominated in numbers, in resources devoted to their training, and in opportunities to advance to the highest levels. In fields such as science, engineering, and law, American development schemes almost invariably and quite unambiguously buttressed the paternalistic status quo inherited from indigenous and Spanish cultures, a pattern that was also much in evidence in the United States.

The Americans' self-representations as hard-working, disciplined, and technically proficient agents of material and social development served to obfuscate the underside of their presence in the Philippines. One aspect of the latter was the virtual epidemic of sexually transmitted diseases in areas occupied by American soldiers and administrators in the early 1900s.[63] The focus on construction projects and public education also helped both the colonizers and the Filipinos move beyond the killing, rape, and destruction that had dominated the first years of their encounter. And it was critical to the recruitment of ambitious and talented American youths into the colonial administration. The appeal to stereotypic manly virtues is obvious in the sloganeering of recruiters who insisted that "only strong

Americans [ought to be] at the helm of the colonial government,"
that "brains, industry, perseverance, and patience were absolutely essential and good judgment indispensable."[64] Some candidates for
positions as district engineers may have been drawn to Theodore
Roosevelt's challenge to take up the strenuous life. As administrators
from the bureau of public works emphasized, these posts involved
arduous tasks that only the most responsible "men of strong character" were capable of performing. Like many of the soldiers who
served in the campaigns to put down Filipino resistance, some of the
teachers and administrators who went to the islands did so out of a
sense of adventure, a need to assert their independence, or a wish to
escape the tedium of a desk job back home.[65]

It is improbable, however, that service in the colonial bureaucracy
proved any less stifling of individuality or creativity than its corporate
counterpart. And dysentery, heat and humidity, and myriad insects
soon disenchanted those who had gone to the Philippines in search
of an exotic paradise. There is little direct evidence that Americans
sought to overcome nervous disorders or restore their virility through
contact with the peasants of Luzon or the tribal peoples of the highland regions. The peasants were routinely described as degraded and
impoverished by Spanish oppression. On some occasions missionaries and ethnologists characterized the most admired of the tribal
peoples, particularly the Igorots, as noble savages and compared
them favorably with urban-dwelling Filipinos. But at other times
these same observers conflated them with the rest of the highland
and forest peoples as filthy, primitive, and animalistic. The prevailing impression conveyed in fairs and natural history exhibits in the
United States was that, however exotic, Filipino tribals were backward rather than noble savages. They were partially evolved peoples
threatened with extinction and urgently in need of colonial projects
that would result in their material and moral uplift.[66]

Male American administrators, schoolteachers, and missionaries may also have hoped that service in the Philippines would provide a paternalistic refuge from the class and ethnic conflicts, status anxieties, and feminist challenges they encountered in the United States. They were certainly prone to represent the islands as a vast tabula rasa on which they could implement their grand designs for infrastructural and resource development with little resistance from the locals. The colonial administrators, mainly Republican and conservative but reform-minded and progressive, deemed the colony far more malleable and manageable than U.S. society itself. It was an arena where capitalism, self-governing institutions, mass education, and industrial technology could be put to work in the service of manly ideals that were highly contested, if not in peril, back home.

Neither the Americans' sense of their civilizing mission nor their reliance on its scientific and technological moorings was unique. British, French, Dutch, and German colonial officials and missionaries went forth to Africa and Asia with similar convictions and aspirations in the late nineteenth century.[67] But American and European approaches to formal colonialism differed in significant ways. Most of the differences were rooted in the Americans' assumption that the Filipinos could and ought to adopt the colonizers' institutions, material culture, and ways of life. Though some Europeans shared this assumption, the extent to which colonized peoples could assimilate Western culture was a controversial and racially charged issue. In fact, in the years before the First World War, there was a decided retreat in the European empires from policies and projects fostering assimilation in favor of shoring up indigenous elites who were *not* Western-educated and promoting revived or invented "traditions" as counters to nationalist challenges to European rule.[68]

In McKinley's "benevolent assimilation" proclamation and other early pronouncements of American colonial policy, considerable care was taken to assure the Filipinos that their new overlords would respect, in the words of Secretary of War Elihu Root, "their customs, their habits, and even their prejudices." But these assurances were contradicted by McKinley's central notion of assimilation—presumably to American ways—and were explicitly retracted by Root immediately after his half-sentence promising respect. Root reminded the members of the Philippine Commission of "the great principles of government which have been made the basis of our [American] governmental system," and insisted that these must be "established and maintained . . . however much they may conflict with the customs or laws of procedure with which [the Filipinos] are familiar."[69] As the remainder of the instructions to the commission made clear, the prime concern of American officials—especially in the critical years just after annexation—was enunciating these principles and formulating policies deemed consistent with them, rather than attempting to take into account the complexities of Filipino society and culture.

From the outset the colonizers aimed to remake the islands and their peoples in accord with American models and ideals. As one writer quipped in a 1920s retrospective on American rule, the colonizers strove to turn the Philippines into "a sort of glorified Iowa."[70] This meant not only that the colonizers dressed, built their houses, and dined as they had back home, but that from education and sports to democratization and highway construction, they aimed to Americanize the Filipinos as well. Because universal elementary schooling had facilitated social mobility and leveling in the United States, the colonizers reasoned that it would gradually undermine the "feudal" dominance of the *ilustrado* class. Capitalist competition and free enterprise had made America productive and prosperous; surely they would do the same for the Philippines. In American accounts from the early colonial decades, roads, buildings, sewage systems, and even sporting events are frequently compared to their counterparts in the United States. Filipino pastimes, especially the men's obsession with cockfights, and indigenous customs, including the *ilustrados'* penchant for flowery oratory and the scant clothing of the common folk, are almost invariably disparaged. Americans often champion baseball as a substitute for the bloody cockfights, and photos of *ilustrado* notables in Western suits, cravats, and starched collars are deployed (rather incongruously) in supposed before-and-after comparisons with G-string-attired "savage" peoples. These sentiments and intentions defy official rhetoric proclaiming respect for Filipino culture and early admonitions about the difficulty of transferring American ideas and institutions because of the islands' exoticism, complexity, and diversity.[71] And they help explain the Taft Commission's determination to pursue "development along the lines which American ideals require."[72]

As early colonial administrators often emphasized,[73] development along American lines not only conflicted with policies long pursued by the established colonial powers, it was likely to prove subversive to the Western imperialist enterprise as a whole. From the time of the arrival of the Taft Commission in the islands in June 1900, American officials cultivated ties to the *ilustrado* elite. Those notables who were willing to cooperate were soon granted powers and positions that contemporary British, Dutch, and French proconsuls were determined to deny to the upstart Western-educated classes of colonized peoples in their overseas possessions.[74] And the U.S. agenda in the Philippines was disruptive of Western global hegemony in other ways as well. The American rhetoric of colonialism often called into question dominant racist presuppositions in its assumption that there were no innate barriers to the Filipinos' adoption of American institutions, learning, and modes of social organization. Material development, which had long been a major concern in the European colonies, was more central to the American vision of the civilizing mission. But European engineers seeking to implement technical schemes for colonial improvement often found their plans altered or frustrated by "Orientalist" scholars, ethnographers, and bureaucrats, who prided themselves on their superior understanding of colonized societies. In the American colonies this counterpoise was either feeble or nonexistent. As a consequence, the engineers and their allies had only to contend with indigenous elites, recalcitrant peasants, and the intermittent interference of tight-fisted congressmen.

As control of the Philippine colony shifted from the military to the civilian commission government in the early 1900s, infusions of technology and the public works projects they made possible were increasingly regarded as the most critical of the forces that would bring modern civilization to the islands. As early as 1899, Jacob Schurman argued that the construction of roads was the key to ending Filipino armed resistance, adding: "Not even schools are so important." A year later Taft began to champion railroads as "the one thing more needful than anything else in these islands," a view seconded by the members of the commission governments he led. Luke Wright, who followed Taft as governor in 1903, declared railroads more important than schools in civilizing the islands, an opinion heartily endorsed by his successors James Smith and especially William Cameron Forbes, who headed the colonial administration until 1913.[75]

In the first decades after annexation, substantial portions of the colonial budget were devoted to public works projects. In 1913 Frank Strong, a mechanical engineer and owner of a machinery company in Manila, celebrated the capital's transformation into a "model American city" and a showplace of Asia, complete with "pure water from the mountains twenty-five miles away piped to every house; sewage picked up all over the city by electrically operated pumps and delivered one mile into the harbour; electric cars in the principal streets and far into the suburbs; electric lights in stores and houses; gas delivered to all parts for cooking and lighting; concrete sidewalks, paved streets, wide boulevards lined with trees, public parks, added to the single one we found, and two thousand automobiles." Strong also catalogued the improvements across the archipelago:

> Four hundred and fifty miles of railways, 1,140 miles of high grade macadamized roads, and 1,340 of lighter surfaced roads; 5,170 steel and concrete bridges and culverts, serving the needs of 2,000,000 people whose means of intercommunication formerly were by the crudest of roads or mere trails; telegraphic communication to all parts of the islands; 600 artesian wells supplying pure water to towns, thereby very greatly reducing the death rate; several thousands of miles of coast accurately surveyed and the 57 lighthouses of Spanish days increased to 144, rendering navigation safe.[76]

Before-and-after comparisons celebrating these transformations were commonplace in travel accounts and official publications. Some took the form of paired photographs, as in a 1915 article appropriately entitled "Making Over the Philippines" by Dean Worcester, who continued to be regarded as one of the foremost experts on the islands. These photographs and their captions contrast "frail" bamboo-raft ferries, primitive roads, and crude plows from before annexation with graceful concrete bridges, well-paved highways, and steam-driven tractors—all products of American ingenuity and technical prowess. American officials not only contrasted these improvements with the deplorable record of the Spanish with regard to public works, they confidently proclaimed the new bridges and roadways built in the Philippines superior to any in the territories of other colonial powers, and often as good as or better than those in the United States itself.[77]

The administrators who launched America's elaborate social engineering project in the Philippines regarded public works as both the

impetus and the infrastructural grid for their schemes of socioeconomic, political, and cultural transformation. Roads, railway lines,
and telegraphs were hailed as the best educators and the key to developing the islands' resources and market economy. And they were
touted as the instruments by which the age-old isolation and hostility
of the different ethnic groups of the islands could be overcome and a
viable nation eventually created.[78] Public works were the material
manifestation of a larger project of rationalization that was aimed at
reconfiguring spatial and temporal perceptions and social relationships through census operations, geological and land tenure surveys,
and legal reforms. They were seen as complementary to other development initiatives, including campaigns to eradicate epidemic diseases and improve health and hygiene, to "modernize" Filipino agriculture, and to establish elementary schools with American-style
sports programs throughout the islands.

Beginning in 1902 a cholera epidemic ravaged many areas in the
Philippines. Malnutrition, brought on in part by the destruction of
food supplies during the years of warfare, contributed to the rapid
spread of the disease. The epidemic ensured that measures for improved health and sanitation would be given high priority by the
colonizers. Owing to both ignorance of local conditions and a
refusal to take Filipino concerns seriously, American measures to
combat cholera provoked a great deal of resistance. Though Americans interpreted the rumors that proliferated in the wake of their efforts against the disease as examples of Filipino superstition and credulity, they had much more to do with heavy-handed, even arrogant,
official responses. The burning of the houses of cholera victims, for
example, and the isolation of the sick in makeshift hospitals from
which few returned, prompted speculation that the colonizers were
promoting the epidemic by poisoning wells or were poisoning those
who had been taken to hospitals. These measures also contributed

to widespread refusals by Filipinos to report cases of infection and attempts to conceal the bodies of those who died of the disease.[79]

Official frustration and Filipino resistance were perhaps inevitable in a situation that pitted the aggressive, allopathic American approach to disease against a populace heavily reliant on homeopathic remedies. The colonizers' warlike campaigns to eradicate pathogens were deeply unsettling to people who "recognised the importance of a patient's state of mind, [and] attributed disease to a lack of harmony between man and his environment."[80] But the American doctors who served as advisors for the colonizers' campaigns against disease were staunch believers in germ-theory etiologies, which from the 1870s had grown increasingly dominant in Western medical thinking. They had little knowledge of or regard for indigenous medicines and healing techniques. Their ethnocentrism was countered by the suspicions and hostility of the Filipinos, natural responses given that the Americans were still waging a brutal war of conquest and were also forcibly implementing anti-epidemic measures that violated the Filipinos' sense of community, their notions regarding the treatment of illness, and their very bodies.

The mishandled campaign against cholera gradually gave way to health and hygienic measures that colonial administrators deemed better administered and clearly beneficial to the subject population. In the official view, American rule facilitated the diffusion to the islands of advances in tropical medicine and sanitation from Europe and the United States. The colonizers also established laboratories and medical centers in the Philippines to pursue research on tropical diseases, and they founded a medical school, expanded existing hospitals, and built new ones. A board of health was created to oversee the assault on disease-carrying germs. Food-inspection regulations were imposed, sanitary ordinances promulgated, and local officials taught the rudiments of "modern" medical theory and disease control. American and later Filipino engineers were enlisted to design, construct, and supervise Manila's first sewage system and the reservoirs and pipelines needed to provide the city with filtered running water. Many of these measures were, of course, intended to promote healthy conditions in areas where Americans were concentrated. But colonial administrators were explicit about their commitment to provide at least urban areas with the sanitary facilities that they regarded as essential for truly modern cities.[81]

Although funding constraints often forced the cutback or cancellation of ambitious sanitary schemes, serious efforts were made to improve conditions in rural areas by digging hundreds of artesian wells and offering instruction in new techniques of waste disposal.[82] Despite continued resentment, if not open resistance, extensive inoculation programs were launched to control smallpox and other contagious diseases. Priority was given to incoming Americans, but the efforts were soon expanded to encompass all but the most remote of the Filipino population. Swamps and stagnant ponds were drained in attempts to eradicate malaria in settled areas, and government-sponsored researchers tested ways to combat the parasites that debilitated the majority of Filipinos.

Perhaps the most intrusive measures were those taken to promote cleanliness. Filipinos, especially those living in coastal or riverine areas, bathed frequently, but the colonizers usually interpreted these regular ablutions as a response to the tropical heat rather than as prompted by a concern for cleanliness. This reasoning appeared to be borne out by the facts that Filipinos often bathed in highly polluted water and that only members of the elite classes used soap. These practices gave added urgency to projects aimed at improving waste disposal and water supply. Missionaries tackled the problem of the "dirty natives" at another level by extolling the "power of soap and water." They also urged American capitalists, such as William Proctor (of Proctor and Gamble fame) to donate large quantities of soap to cleanse their "tribal" charges. Carl Crow, an American journalist who traveled widely in the islands, believed that the adoption of Western eating utensils by Filipino adults would lead younger generations to emulate this more sanitary practice. And an official delegation reporting on the impact of colonization in the 1920s considered the promotion of Western notions of hygiene by Filipino women's clubs a "most encouraging sign of the times."[83]

U.S. officials ranked health and sanitation projects among the most beneficial and enduring legacies of American colonial rule in the Philippines. Again stressing American exceptionalism, some incorrectly asserted that the United States was the only colonizing nation that had introduced health measures not just for the benefit of expatriates but for the welfare of the subject population. Victor Heiser, head of the colony's bureau of health, declared that American improvements had turned Manila into one of the world's cleanest cities. He proclaimed the Philippines a model for the rest of Asia and predicted that the American example would compel less enlightened colonial and Asian governments to introduce "modern" systems of sanitation. Taft lauded the advances in sanitation and disease control in the islands as among "the great discoveries of the world" and averred that they would be seen by future generations "to have played as large a part as any in the world's progress in the current hundred years."[84]

Health and sanitation measures served as the entering wedge for the colonial regime's efforts to expand its agenda for development into the thousands of rural communities where most Filipinos lived. Asserting military control over resistant areas and then combating the cholera epidemic preoccupied administrators during the first years of the occupation. But beginning in some areas as early as 1898, the recurrence of an epizootic of rinderpest became the focus of projects for rural improvement. Rinderpest had first ravaged the water buffalo herds, the main source of nonhuman power employed by the peasant population, in the late 1880s. Mortality rates, as high as 80 percent of the carabaos in some areas, led to sharp declines in agricultural production, food shortages, and malnutrition. The decimation of the herds also contributed to locust and malarial infestations on lands no longer cultivated or used to pasture water buffalos.[85]

Replenishing the herds and destroying locust swarms or draining the breeding grounds of anopheles mosquitoes delayed the introduction of programs designed by the newly established bureau of agriculture to modernize farming. But the bureau's initiatives soon came to center upon agricultural experiment stations, university-based research laboratories, and test plots patterned after the ones that had contributed to the transformation of agricultural production in the United States. Whether appropriate or not for the tropical soils and monsoon climate of the Philippines, mechanized plows and tractors, chemical fertilizers, and hybrid seeds were introduced into the islands. Colonial administrators and visiting consultants emphasized the scientifically grounded and conservation-minded aspects of the American approach. But official rhetoric and the nature of the programs funded leave little doubt that the colonial regime's priority was building a flourishing export economy on the basis of a revitalized and market-oriented Filipino agricultural sector.[86]

As a number of administrators stressed, the regime's intent was to create a rural population consisting largely of landowning cultivators committed to production for the market.[87] Not coincidentally, if fulfilled, this vision would transform Filipino peasants into a class that very much resembled the idealized, independent, and enterprising farmers who had turned North America into one of the granaries of the world. But colonial policymakers insisted that private investment, capitalist enterprise, and market competition, rather than extensive government intervention, were the keys to molding the Philippines into something similar to a "glorified Iowa." Thus it was acceptable for the government, drawing on precedents from agrarian development in the United States, to provide research and technical support and to build or fund transport and communication networks, as well as to launch homesteading schemes to encourage settlement of uncultivated areas.[88] But beyond setting easily circumvented restrictions on the amount of land individual families and corporations could amass, the colonial regime eschewed serious efforts to carry out land reform or regulate peasant indebtedness, tenancy arrangements, or land alienation. The negligible level of government involvement in these critical facets of rural life meant that the landholding class could steadily enhance its wealth and power at the expense of the peasantry. This allowed the persistence of an essentially feudal social structure in the rural Philippines, despite the considerable expansion of market production and capital investment under American rule.

American officials were well aware that rural society on Luzon scarcely resembled its counterpart in Iowa or anywhere else in the North American farm belt. They explicitly acknowledged the possibility that economic expansion in a free-enterprise, competitive mode could serve to entrench the *ilustrado* class, frustrating hopes of creating an independent, landholding, politically empowered peasantry.[89] Unwilling or unable to intervene directly to alter these outcomes, many of those committed to America's civilizing project assuaged their anxieties by accepting official rhetoric that fixed upon mass public education as the guarantor of gradual social reform and the enfranchisement of the peasants. But others doubted that popular education would prove a panacea for the islands' social inequities and highly skewed distribution of wealth and political power.[90] Given serious funding shortages and more critically the content and orientation of the elementary curriculum, these doubts were well founded. After several shifts in educational philosophy in the first years after a bureau of education was established in January 1901, the focus of public elementary instruction became training in the industrial arts for boys and the colonial variant of home economics for girls.

The strong influence of Booker T. Washington's vocational orientation for education, particularly on Fred Atkinson, the first director of the bureau of education, suggests that these policy choices were at least in part made with racial considerations in mind. Atkinson and later officials insisted that the underemployed and impoverished Filipinos were more in need of practical skills and steady jobs than of a grounding in the liberal arts. There was a good deal of truth to these assertions. But the emphasis on craft skills and vocational preparation left ordinary Filipinos with few options other than subordinate positions as skilled laborers serving the colonizers and the indigenous elite. At the same time, the access to private schools enjoyed by the children of the elite ensured their continued dominance of the professions and indigenous entrepreneurship. It also meant that they could monopolize the positions opening up in the colonial bureaucracy and in emerging Filipino political parties.

Ironically, the very qualities of American educational programs that enfeebled their potential for genuine social change rendered them compatible with the ambitious engineering project that was at the heart of the colonial enterprise. Public education at the elementary and secondary levels not only supplied trained mechanics and craftsmen, it aimed to inculcate values affirmed by the colonizers' self-representations and to disparage habits and attitudes associated with the "tradition"-bound Filipino culture. Such prominent policymakers as Taft, Worcester, and Elliot insisted on the need to impress Filipino youths with the "dignity of labor," thus counteracting centuries of Spanish and *ilustrado* disdain for physical work. Instruction was organized in ways that accustomed Filipino students to industrial time and the importance of punctuality and self-discipline. The curriculum was designed to build character, promoting honesty, accuracy, thrift, cleanliness, and self-control in matters sexual as well as occupational. Housekeeping and hygiene were core subjects for girls and young women. Some educators euphemistically conceived these as aspects of a broader curriculum in "domestic engineering" that emphasized the importance of women's roles as mothers and homemakers in the colonizers' efforts to civilize and modernize the Philippines.[91]

Through organized sports this process of character building was extended beyond the classroom to the students' leisure time. The same diverse virtues ascribed to athletics in late nineteenth-century American society were trumpeted by administrators, missionaries, and schoolteachers in the Philippines. Americans repeatedly identified the introduction of Western sports—especially the uniquely American game of baseball—as an effective measure to improve the physical health, self-assurance, and moral probity of Filipinos. Although both young men and young women were encouraged to participate in sports, anticipated benefits were gender specific. For example, athletic competition was expected to enhance the aggressiveness, physique, and virility of Filipino men. By contrast, sports, particularly softball and tennis, were seen to make Filipino women more active and outgoing, partly by compelling them to set aside "unsuitable" traditional clothing that had restricted their movement. The colonizers linked the improvement of the population's physical condition through athletics to more fundamental cultural transformations, including those relating to sexual mores, teamwork, and individual exertion. They believed that such changes would heighten the Filipinos' sense of civic responsibility and national identity, and thereby enhance the colony's chances of surviving, even flourishing, when it eventually became an independent state.[92]

❖　　❖　　❖

The civilizing projects launched by Taft and his successors in the Philippines would inform American policies in other nonwestern so-

cieties throughout the twentieth century. Relative to other areas occupied and administered by the United States in the early 1900s,
such as Puerto Rico and Haiti, the Americans encouraged far higher
levels of active participation by the Filipinos in all phases of development initiatives, including planning, design, and management. Both
the colonizers' assumption that the Filipinos were capable of undertaking these tasks and their conviction that the advancement of the
colonized depended on assimilating to American ways anticipated
key tenets of modernization theory, which in subsequent decades
would become the dominant discourse on U.S. development assistance for nonwestern peoples and societies.

American administrators acknowledged that the popularity of
public works projects and educational programs among both the
ilustrados and the peasants was a strong reason for pursuing them. In
apparent contradiction to official platitudes about promoting self-
governance and nation-building, some administrators were candid
about their expectation that colonial development programs would
engender long-term Filipino dependence on American technical expertise and technology.[93] The fact that engineering projects and
technical education programs were the sorts of initiatives that the
U.S. Congress was the most likely to approve (if not fund adequately) made them doubly attractive. In addition, because most of
those in the colonial administration in the early years were staunchly
middle-class and largely Republican, they viewed building highways
and public schools as more appropriate tasks for government than
land reform or other state-sponsored alterations in the islands' paternalistic patronage and social systems.[94] But colonial policymakers
such as Taft, Forbes, and Worcester were Republicans of the progressive persuasion epitomized by Theodore Roosevelt. Accordingly,
they were committed to what has been characterized as a moral variant of capitalism (rather than the cutthroat strain so evident in the
robber baron era), and to the government checks essential to main-

tain it. In this regard, it is noteworthy that the level of state involvement in development projects in the Philippines was often determined by precedents established by similar projects in the United States, particularly those associated with the opening of the frontier.

Given these constraints, serious alternatives to an approach to colonization inspired and largely directed by engineers failed to materialize. Despite protracted debate over the annexation of the islands and the conduct of the war of occupation, dissent regarding the narrowly defined developmental policies rarely addressed the ideological orientations underlying them. Rather, it focused on specific policy questions, such as the extent to which public elementary education ought to be vocational, whether railroads should be funded by the government or private investors, and the pace at which Filipinos ought to be put in charge of political affairs. The engineers' imperialism that came to dominate the colonial project in the islands sparked only limited contention and little overt resistance, particularly from American administrators, teachers, and businessmen actually stationed in the islands. Dissent was indirect, diffuse, and often bound together with lingering misgivings about the colonization of the islands in the first place.

The "modern" society that the engineers' imperialism was intended to fashion in the Philippines fell short of its goal in part because neither public funding nor private investment was remotely adequate for the ambitious projects envisioned. A suspicious, poorly informed, and tight-fisted Congress forced colonial officials to levy unpopular taxes for projects as basic as road construction and repair.[95] Private investment never reached early expectations, to some extent because of uncertainty regarding how long the Philippines would retain colonial status. But perhaps more critical were the suspicion of and restrictions applied to American investors by both colonial administrators and *ilustrado* politicians. The former were wary of a colonial reprise of the robber baron era; the latter were vehemently opposed to the exploitation of the islands' peoples by capitalists from overseas.[96]

Capital shortages notwithstanding, the emphasis on development driven by industrial technology impeded the fulfillment of American aspirations to build a modern, democratic nation in the Philippines. By focusing on public works and material increase within a market-driven, laissez-faire economic structure, the colonizers avoided addressing the underlying maldistribution of power, wealth, and access to resources in Filipino society. Even before annexation, American officials were forging alliances with prominent *ilustrados*—a process that was inevitably intensified as a result of the "insurrection" led by other *ilustrado* factions. As American dependence grew on *ilustrados* like Emilio Aguinaldo, former leader of the Philippine revolution and president of its short-lived republic, and other members of what has been aptly labeled the Filipino *compradore* classes, the possibility of significant reforms, such as extensive land redistribution or measures to curb the rampant corruption and intimidation in the patron-client systems that dominated politics in Manila and the provinces, faded inexorably. The Filipinization of the political system, which was all but complete by the end of Francis Harrison's tenure as governor-general in 1921, further entrenched obstacles to meaningful reform.[97]

Not surprisingly, members of the *compradore* elite were content to support American development initiatives that provided them with skilled laborers and transport facilities for the marketing of their crops. America's great engineering project created the infrastructure with which Filipino politicians could forge a nation from the diverse peoples of the islands. But a good deal of evidence suggests that by and large the *compradores* accepted and internalized the colonizers' hegemonic rhetoric about development. They shared with the Americans a vision of modernity fashioned in accord with Enlightenment precepts, including critical enquiry, empirical scientism, and rational approaches to material improvement. More critical perhaps was their realization that cooperation regarding public works projects need not compromise their political demands or delay independence.[98] Consequently, Filipino journalists felt free to complain about the arrogance and insensitivity of Victor Heiser and other health officials while conceding that the Americans' work constituted a "valuable factor of progress and of material development." Most members of the elite may not themselves have deigned to engage in manual labor, but some lamented that an aversion to hard work was turning the Philippines into a nation of "Señoritas." Americans commented on Filipinos' fondness for flashy American automobiles, and on the readiness of members of the elite to concede the backwardness of the masses and their need for American technology and training.[99] Filipino writers catalogued the virtues of the "Filipino women of our days" who under the "influence of Saxon education" and the example of American women had become agents of "progress," "improvement," "vigor," and "health." Some *compradore* commentators extolled the benefits of American sports to Filipino health and character in much the same terms used by colonial administrators. These writers called on the Filipino people to adopt the pioneering and enterprising spirit of their colonizers.[100]

Just how fully some *compradores* internalized the colonizers' developmental rhetoric is illustrated by a "Pageant of Culture" that was written by Eulogio Rodriguez, assistant director of the (Philippine) National Library, and performed by students of the Philippine Women's College in early 1930.[101] The purpose of the pageant was to show what "countries" from India and China to Spain and the United States had contributed to the "culture and progress" of the islands. In a sequence of vignettes, costumed students appeared bearing gifts representing each country's influence. While they paraded in, a herald announced their contributions, ranging from "mother" India's philosophies to England's Magna Charta. America, "the second Rome in grandeur and power," was lauded as having bestowed "democracy—the modern school system, sanitation, agriculture, American industries, and the like up-to-date things." Allowing for a somewhat different order of contributions, W. Cameron Forbes could not have dictated a more fitting summary of the colonizers' achievements.

Warnings that the *compradores* might monopolize political power and enhance their domination of the inequitable socioeconomic system had been voiced by American officials, such as Schurman and Taft, from the early days of commission government. Nonetheless, in their rhetoric at least, colonial policymakers remained confident that more machines, better roads, and mass education would in time avert this outcome and render the colonial experiment a resounding success. In fact, however, material increase without social reform exacerbated the divisions and tensions within Filipino society. Persisting undercurrents of peasant resistance were channeled into local uprisings, quasi-millenarian religious movements, and endemic everyday and avoidance protest. If there was any doubt that America's mission to bequeath its developmental brand of civilization to the Philippines had fallen far short of its stated goals, it would be clearly resolved by the revolutionary struggles of the decade after the Second World War and the intensifying civil strife that culminated in the overthrow of the Marcos regime in the mid-1980s.[102]

President Teddy Roosevelt enjoying a photo op at the controls of a steam shovel during his much-publicized visit to the construction site of the Panama Canal in 1906. Library of Congress.

No English interpreter is needed when a Chinese or a Peruvian
sees this series of working dams, or electricity flowing into a
single farmhouse, or acres that phosphate had brought back
to life.

—David E. Lilienthal,
Chairman, Tennessee Valley Authority, 1944

4

FOUNDATIONS OF AN AMERICAN CENTURY

During the first visit of the Perry expedition to Japan in 1853, American emissaries were pleasantly surprised by Japanese inquiries about the railway a U.S. company had been building across the Isthmus of Panama since 1850. Sixty years later a Japanese dignitary, visiting the gigantic construction site that Panama had become after nearly a decade of excavation for an inter-oceanic canal, did not mute his admiration for what American engineering ingenuity had achieved. When asked if he found the project as impressive as he had anticipated, he exclaimed: "Oh, very much greater . . . it is stupendous, magnificent, colossal! No nation but the great, rich American nation could build this canal. Japan has much to learn from you."[1] On all counts, his remarks were tailor-made to please his American hosts. His hyperbolic estimate of the scale and success of the canal project accorded perfectly with those that had begun to appear in books and newspaper articles in the United States as the waterway neared completion. For those directly involved in building the canal, his assessment amounted to a satisfying public retraction of the highly negative appraisal of a team from Japan that had inspected the site in 1905—just as work got under way—and recommended that Japanese laborers not be allowed to work there because Panama was so unhealthy and working conditions so hazardous. More broadly it acknowledged that American material triumphs had indeed made the United States the worthy tutor of Japanese people, a mission that a half century earlier Perry and his entourage had been certain their countrymen were destined to fulfill. But above all, the distinguished visitor's comments left no doubt that in his view America had arrived as a global power. And international acceptance of that reality was precisely why President Theodore Roosevelt and many others, including Alfred Thayer Mahan, had expended so much energy and political capital to win congressional approval for the construction of a canal through Panama, where the French had ignominiously failed just over a decade earlier.[2]

Excepting the inveterate traveler and America watcher Viscount Bryce, who also visited the Isthmus as the canal was nearing completion and famously called the project "the greatest liberty Man has ever taken with nature," no foreigner matched the Japanese dignitary's estimate of the magnitude of the U.S. achievement in Panama. But foreign approbation was hardly needed, and was very likely intentionally muted in reaction to the bombastic claims made by the

Americans themselves. Much of the American posturing was about sheer size and excelling the most renowned previous human creations. Many who praised the magnitude of the Panama undertaking agreed that it was a "story without parallel in the annals of man." They exulted in the project as "the greatest single achievement in human history," which had "smashed into smithereens . . . all of the world's engineering records." Visiting the canal zone in 1906, when the undertaking appeared to be floundering, President Roosevelt had set the tone for this sort of claim in a rousing speech in which he declared the project "the biggest thing of [its] kind that had ever been done." Later enthusiasts would point out that in the decade of excavation at Panama, the equivalent of all of the sand and soil cleared at Suez was removed every fifteen months, which made the digging of the latter seem "a small enterprise." By the time the "great ditch" was finished, one recordskeeper declared, enough dirt would be removed to fill the flatcars of a train 96,000 miles long. Another estimated that the "total amount of material handled . . . ranged somewhere around 260,000,000 [cubic] yards," and averred that "the building of the pyramids, for five thousand years or more the wonder of the world, was play in comparison." Estimates of the magnitude of every conceivable aspect of the project were offered, from the number of loaves of bread baked for the laborers in a year to the size of the locks in comparison to the Washington Monument, the *Titanic*, and the circumference of the earth. Even the throw weight of the outsized artillery pieces forged to protect each end of the canal was quantified ("a ton of steel propelled by a quarter-ton charge of smokeless powder") to dazzle the citizenry of the United States and interested onlookers around the world.[3]

In an age when strident nationalism and jingoism prevailed among the industrial powers, these excesses of patriotic pridefulness were not unique. But the emphasis on technical mastery in these comments suggests that the belief that the United States was "technology's nation" had become even more central to American identity by the early 1900s than it had been in the preceding century of industrialization, territorial expansion, and U.S. emergence as a major player in the global economy. The ability to overcome the environmental, financial, and above all engineering obstacles in Panama that had defeated Ferdinand de Lesseps and the French, who had earlier linked Europe and Asia with a sea passage through the desert at Suez, was seen as proof that the United States had surpassed even the established industrial powers of Europe in the aptitudes that would ensure power and prosperity in the modern age. The canal's rise "phoenix-like out of the ruins of the French enterprise," its celebrants concluded, could not fail to impress America's rivals with its "sufficiency in the face of the largest demand upon man's engineering acumen."[4]

Unbounded energy, inventiveness, adaptability, and faith in technological solutions had enabled the Americans to work through the successive crises that had ultimately reduced the French enterprise at Panama to a literal shambles. American managers were depicted as steadfast and foresightful, and their modes of deploying and disciplining a motley, mostly non-European labor force as superior to those of the lackadaisical, chronically disorganized French. The youthful American engineers, foremen, and supervisors were praised for their manliness, work ethic, and moral rectitude, which some writers contrasted with the character of their French counterparts, whom rumor mongers had stigmatized decades earlier for their dissolute ways.[5] And many saw larger purposes in the American achievement. They were confident that the canal had contributed immeasurably to bringing "the nations of the earth into closer intercourse with one another," and thereby "done a service to the cause of universal progress and civilization, the worth of which the passage of time will never dim." And all the while, the canal boosters assumed, "The world is looking on," and realizing that America was now prepared to take up the global mission that had been foreseen centuries earlier by John Winthrop.[6]

The contrast between the congratulatory indulgence of many popular writings about the canal and the accounts of those who actually designed the locks, found ways to control tropical diseases, and supervised construction is striking. William Gorgas describes his campaigns to eradicate malaria and yellow fever in measured, clinical prose. He acknowledges that his successes would not have been possible without his earlier work with Walter Reed and Carlos Finlay against yellow fever, which was ravaging Americans in Cuba, and the pioneering work of Ronald Ross, a British doctor stationed in India, who had meticulously traced the etiology of malaria. The papers assembled in a collection edited by George Goethals, the autocratic governor of the Canal Zone during the critical years of construction, are bland, nuts-and-bolts descriptions of engineering solutions to problems large and small.[7] Authors such as William Sibert, who supervised the building of the Gatun locks and dam, and John F. Stevens, the second and decisive chief engineer on the project, refrained from disparaging their French predecessors. They credited the French for the excellence of their surveys, the substantial areas they had excavated, and the channels they had dredged. Sibert and Stevens reported that "immense quantities of material and machinery was found distributed along the entire line [of construction]," much of which was of "splendid workmanship." The Americans were also able to repair and use more than fifteen hundred buildings left by the French. Cognizant of the grand tradition of French engineering, American engineers and canal officials seldom faulted French technical proficiency. In explaining the French failure they focused instead on the lack of government support, the inadequacy of the excavating equipment available in the 1880s, and the fact that the French attempt had predated the breakthroughs in tropical medicine that allowed the Americans to succeed.[8]

Hyperbolic acclaim notwithstanding, the building of the Panama Canal was a remarkable undertaking and an engineering masterpiece. Success hinged on the Americans' ability to dominate some of the most inhospitable terrain for human habitation on the face of the earth. A vast swath of dense rainforest—teeming with tarantulas, scorpions, poisonous snakes, and innumerable insect pests and wild animals—had to be cleared and constantly secured from reclamation by the fecund vegetation. The Chagres River, which became a raging torrent in the rainy season, had to be contained, and a passageway cut through the mountainous region at Culebra, where massive mud slides had frustrated every French excavating stratagem. Swamps and fetid pools contributed to an environment that could be lethal, particularly for migrants who lacked immunities to yellow fever, malaria, and other endemic diseases. Panama was in sum an extreme manifestation of the tropical environment that Europeans had long considered impossible for peoples from temperate regions to settle in and improve. At the time there was little dissent from the maxim that "whites" could not sustain physical labor in the heat and humidity of tropical locales. The barriers of climate and disease were particularly frustrating because European and American advocates of imperial expansion, most famously Benjamin Kidd, had become increasingly vociferous in their insistence that the fertile lands and abundant natural resources of tropical dependencies were essential to the continued dominance of the Western industrial powers.[9]

The success of American efforts to build a canal at Panama brought global recognition that had not been accorded to U.S. accomplishments in the Philippines. From the perspective of European colonizers, the overseas empire the United States had so hastily patched together in the late 1890s consisted mainly of tiny, not particularly desirable areas left over from the fierce intra-European competition for colonies. And Europeans had been largely dismissive of claims that the American approach to governance in the Philippines was without precedent in the extent to which mass education, science, and technology were enlisted in projects to uplift the subject peoples and to forge a modern, independent nation. American achievements in Panama, in contrast, were all but impossible to marginalize. Gorgas's campaigns against yellow fever and malaria, for example, soon put an end to centuries-old miasmic theories that blamed these diseases on exposure to poisonous tropical vapors, and overcame the obstacle that more than any other had frustrated the earlier French project.[10]

In light of turn-of-the-century notions about the impact of environmental difference on migrating "races," contemporaries judged the epidemiological breakthroughs at Panama as a watershed advance for "temperate" peoples. They were seen to open the tropics to European and American conquest, settlement, and capitalist enterprise as never before. And this influx of "Caucasian" or Anglo-Saxon energy and invention ensured that societies in tropical regions would someday be developed as fully as those in industrial temperate lands. Tropical peoples might even create civilizations of their own, though contemporary pundits, including the widely quoted geographer Ellsworth Huntington, assured their readers that civilization had flourished much more rarely in hot, enervating climates, and then at levels of accomplishment far below those in temperate regions, particularly Europe and North America.[11] These assumptions about "racial" aptitudes meant that it never occurred to the French or the Americans that any of the peoples of Latin America might themselves be capable of building an isthmian canal. That left no alternative than for more energetic peoples to do it for them.

The public works projects that were the pride of American officials in the Philippines paled in comparison with the engineering feats accomplished at Panama. After considerable debate, the architects of the Panama project were determined to build a lock canal rather than the sea-level waterway the French had attempted. A lock design would still require massive, difficult digging in the mountainous interior, where topographical and geological configurations produced innumerable mud slides. But by using multiple locks at both ends of the "great ditch," American engineers created in effect a bridge of water with steps at each end that lifted ships up and across the Isthmus and then back down to sea level. This approach considerably reduced the depth of excavation required where elevations were the most extreme. The huge locks themselves were engineering wonders, and they were flanked by extensive facilities for provisioning ships at the Atlantic entrance and repair yards on the Pacific end. An oversized railway was designed to carry away excavated rock and dirt—some of which was used to build protective barriers at the entryways to the canal on each side of the Isthmus. Joseph Bishop, a secretary of the Canal Commission, reported that visiting engineers and foreign dignitaries (including the one he later quoted from Japan) were impressed by the incomparable "efficiency, perfection of detail, precision, and smoothness of operation, unity of spirit and enthusiasm" of those engaged in the Panama project. Clearly, Bishop intimated, the manly engineers of America deserved the acclaim that had been bestowed upon them by experts "from all quarters of the globe."[12]

The generous praise of the canal by the eminent Japanese visitor was all the more remarkable because it came at a time of rising tensions between Japan and the United States. By the end of the nineteenth century, Americans had become disconcerted by the determination of the Japanese to chart their own course to industrial modernity. Japan's surprise victory in the Sino-Japanese War of 1894–1895, and its even more stunning rout of the Russians a decade later, left the

Western powers with little doubt about its expansionist designs and ambition to become the paramount power at least in East Asia. Theodore Roosevelt's insistence from the 1890s that a canal across Central America, whether at Nicaragua or Panama, was essential for the United States to coordinate its transcontinental bases of commerce and project its naval might in both the Atlantic and the Pacific had become an article of faith for American expansionists.

To the Japanese, American motivations for the construction of a fortified canal were all too obvious. The U.S. capacity to exert military force, particularly through naval actions, from Hawaii across a bridge of Pacific islands to the Philippines was already a blatant challenge to Japan's imperial ambitions. In the 1890s Japanese resentment had been directed against European rivals, particularly France, Germany, and Russia after they had blocked Japan's annexation of China's Liaotung peninsula at the end of the 1894–1895 war. In the following decade the United States became the focus of Japanese anxiety, not only because of the threat posed by the canal project but because the Japanese believed they had again been denied their just rewards after a major victory, this time over the Russians, by the Treaty of Portsmouth, which Roosevelt had negotiated to resolve Russo-Japanese conflict in 1905. In the years that followed, Japanese expansionists were exasperated by American diplomatic moves aimed at curbing their influence in Manchuria and China proper, even though they had been granted a free hand in Korea. The tensions of international competition were compounded by deep insult in 1907, when the Japanese were outraged by California legislation that mandated segregated schools throughout the state for "Orientals," including Japanese.[13]

Given the heightened aggressiveness of U.S. foreign policy at the turn of the century, there was good reason for Japanese concern. And the Japanese could not miss the hypocrisy of American measures to curb Japan's imperial designs at a time when the United

States itself was building an island empire in both the Caribbean and the Pacific. In the aftermath of the Spanish-American War, the United States had secured control over the Caribbean approaches to the proposed isthmian canal by annexing Puerto Rico and assuming control over Cuba through successive military occupations and the Platt Amendment, under which Cuban insurgent leaders were coerced into ratifying American supervision of Cuban foreign relations and internal affairs. The large U.S. military base established at Guantanamo on the eastern tip of Cuba, and — as part of the same burst of imperialist aggrandizement — the formal annexation of Hawaii with its superb naval facilities at Pearl Harbor, anchored a formidable defense perimeter around Panama that the United States systematically pieced together while the canal was under construction.

Although concern for the defense of the canal was more often a pretext than a major motive for these unilateral — at times preemptive — interventions,[14] they inaugurated a decided shift to more aggressive, expansionist American policies toward formerly colonized societies overseas, particularly those in Latin America. Moves to assert U.S. domination had been evident in Roosevelt's schemes to win control over the corridor through the middle of Panama where, after a good deal of controversy, Congress had voted to construct the canal. A motley assortment of Americans (obsessively abetted by Philippe Bunau-Varilla, who had played a major role in the French attempt to build the waterway) had fomented the Panamanian "revolution" that put an end to Colombia's sovereignty over the Isthmus. And Roosevelt did not hesitate to use gunboat diplomacy to ensure that the bloodless coup succeeded and a compliant regime was installed in Panama. In 1905 the president sought to legitimize all of these — and later — expansionist schemes by setting forth a corollary to the Monroe Doctrine, which gave the United States the right to intervene in neighboring countries where social unrest or political upheaval threatened.[15]

The thoroughly colonial regime installed in the Canal Zone marked a decisive retreat from the relatively progressive U.S. approach to civilizing uplift in the Philippines. From the workforce that built the waterway to the insular community that developed at the construction site, segregation was as pervasive as in the American South in the Jim Crow era. Administrators, engineers, supervisors, and even clerical personnel were all "white" and male, and overwhelmingly Americans. Popular accounts of the canal "saga" returned again and again to heroic depictions of the youthful "khaki heroes of Panama," and the predominantly "Anglo-Saxon" engineers, "whose genius and resourcefulness [had become] the dominant factor in the material progress of the world."[16]

Drawing deeply on the racist stereotypes and epithets of segregated America, a number of contemporary authors described the "tawny, dusky, ebony labourers" who made up the great majority of the population of Panama. These "easygoing" "niggers" (and the "sluggish" mestizos) were contrasted with the "lithe . . . clean-skinned, clean-eyed" Americans, who were "really feeling joy in what they are helping to do." Joseph Bishop, a former secretary of the Isthmian Canal Commission, relates that John Stevens, the project's second chief engineer, estimated that the "efficiency" of a West Indian worker was about one-third of that of the "white man." Contrary to notions that "negroes" were immune to "practically everything," Stevens apparently also found that "any white man, under the same conditions will stand the climate on the isthmus very much better than black." And John Foster Fraser, the most blatantly racist chronicler of the project, assured his readers that the Afro-Caribbean workers, who he claimed were all from Barbados, had been able to survive in the Panama environment because "Mosquitos do not like nigger flesh."[17]

Although in the early years of construction some laborers were recruited from the poor of southern Europe, throughout the project those who did the heavy physical work were overwhelmingly workers of African descent recruited from the Caribbean islands. Workers were housed and fed separately and in ways decisively inferior to the amenities enjoyed by white managers and technicians. Socializing across racial and class boundaries—for the most part even with members of the small Panamanian elite—was taboo, and white women who arrived to join their husbands or work as nurses or teachers were carefully watched to guard against inappropriate liaisons. The military personnel stationed in the canal zone were all white, as were the members of the police force, including those in charge of keeping the peace among the laborers. Separate school systems were created for the children of the managerial/technical staff and those of the workers; the former had far superior teacher-student ratios and facilities and covered more grade levels. And though the American authorities provided health care for the laborers building the canal, little or none was extended to Panamanians unconnected to the canal project, and the less vigilant care offered to the workers and their families was reflected in the periodic outbreaks of malaria and yellow fever among the Afro-Caribbean population.[18]

With slight variations, the segregated social structure and American monopolization of technical expertise were reproduced in the countries in the Caribbean and Latin America where the United States intervened militarily, as it did thirty times between 1898 and 1934.[19] This was particularly true of island nations in the Caribbean where the U.S. occupations lasted years or decades, most notably Cuba (1898–1902, 1906–1909, 1912, 1917), the Dominican Republic (1916–1924), and Haiti (1915–1934). In each of these areas, U.S. military and civilian technicians took charge of infrastructural improvements but, in contrast to their approach in the Philippines, made little or no attempt to train the indigenous peoples to plan or oversee such projects. Educational programs were focused on vocational training for men, with some desultory efforts to introduce young women to "domestic sciences." American authorities did little to enhance university education, and in some cases were hostile to higher learning, most openly in Haiti, where they actively worked to undermine the French-inspired, humanistic curriculum that had produced a highly cultivated—but intensely self-serving—creole elite.

Limited educational improvements and reliance on transient foreign technical expertise left the occupied societies with severe shortages of the doctors, engineers, and technicians needed to sustain their own development programs. As in the Philippines—though on a far less ambitious scale—American proconsuls concentrated on public works projects and sanitation schemes while doing little to promote land reform or agricultural assistance for the largely impoverished rural populations. The failure of the underfunded agrarian assistance programs was routinely attributed to the peasants' ignorance, intense conservatism, and at times innate indolence or inability to master new technologies. In combination with U.S. support for local leaders with dictatorial proclivities and American hostility to populist dissent, the U.S. formula for civilizing the Caribbean and Central American "republics" that stretched in a great arc to the north and east of the canal enclave at Panama held little promise for the peasants of Latin America, or for those of Africa and Asia, where resistance to colonial rule was gaining momentum.[20]

As the "titanic work" at Panama neared completion in the fall of 1914, a grand celebration of America's achievement "without parallel in the annals of man" was being prepared near the Golden Gate entryway to San Francisco Harbor. Originally proposed in 1904 to mark the four hundredth anniversary of the "discovery" of the Pacific Ocean by the Spanish explorer Balboa, the Panama-Pacific International Exposition, which opened in February 1915, had in the intervening decade been transformed into a world's fair centered on a large-scale working model of the Panama Canal. The exhibit was equipped with a moving walkway that conveyed visitors around the model of the canal zone, as well as telephone receivers where they could stop and listen to recordings describing the canal's engineering feats. It was housed in the Court of the Sun and Stars, the largest of a complex of exhibition palaces and the structure that linked the Court of Abundance, representing the Orient, and the Court of the Four Seasons, where the achievements of the Occident were on display. Thus the layout of the grounds provided an architectural rendering of the fair's defining theme, "The Land Divided—The World United," which was emblazoned on promotional posters, personified by an "enormous" allegorical statue of a man on horseback holding apart the oceans, and repeated throughout the exposition in sculptures and displays "tell[ing] the story of the unification of the East and the West through the construction of the Panama Canal."[21]

The Panama-Pacific fair was conceived on a scale to rival the Chicago and St. Louis extravaganzas of the turn of the century, and designed to win world acclaim to match that accorded the Crystal Palace exhibition that had opened in London in 1851 and the 1889 Paris exposition featuring Gustave Eiffel's tower. It was also intended to impress the world with San Francisco's resilient recovery in the decade since the earthquake and fire of 1906. But the outbreak of the First World War overshadowed the Americans' celebration of their arrival as the world's foremost technological society and a major global power.

U.S. neutrality in the Great War raging across Eurasia and Africa meant a sudden drop in the nation's importance in international affairs, but it opened the way for America to surpass Great Britain as the world's premier maritime trading nation. The United States moved quickly to dominate markets in Latin America and Asia that the war forced the British, French, and Germans to relinquish, and it became a major provisioner of the Entente partners. At the same time, the British blockade that shut down trade between America and Germany created a situation in which the United States was indirectly at war with the Central Powers long before the spring of 1917. That fact and German contempt for the small and ill-prepared U.S. army go far to explain the German high command's fatal tendency to underestimate the danger of provoking U.S. military involvement in the conflict.[22]

By early 1917, when America entered the war, its trade with Britain and France had increased by 184 percent over peacetime levels, an upturn that helped pull the United States out of a prewar economic slump. During the war the United States went from being the world's greatest debtor to its largest creditor nation, and by the end of hostilities New York had displaced London as the center of the global financial network. Even before it became a combatant, America had lent more than $2 billion to Great Britain and France; by war's end the two allies had accumulated a debt of $12 billion.[23] Their insatiable demand for foodstuffs and raw materials brought about a decade of prosperity, increased mechanization, and renewed expansion of cultivated acreage in America's agrarian heartland. The introduction of a moving assembly line by the Ford Motor Company shortly before the war further enhanced America's considerable lead over the other industrialized powers in manufacturing capacity. In the war years, America shipped over a billion dollars worth of munitions across the Atlantic and became a major supplier of trucks and other vehicles, though it remained dependent on its allies for tanks and fighter aircraft. Little wonder then that by 1916 the British foreign office was cautioning naval commanders who wanted to enforce the blockade against America and other neutral nations that a break with the United States would mean Great Britain's ruin.[24]

America's late entry into the war was critical in shaping a generally positive assessment of the nation's combat experience. The nation's recently acquired naval power made possible rapid and decisive U.S. contributions to the convoy system, which kept Great Britain's oceanic supply lines open despite a daunting escalation of U-boat attacks. The nearly fifty thousand American deaths in just over a year of combat operations paled in comparison to the millions of youths lost by the other major combatant nations.[25] For Americans at least, the notion of a lost generation was a misnomer.

This does not mean that John dos Passos, e. e. cummings, and other young Americans were not traumatized and embittered by their combat experiences, but rather that most of their compatriots viewed American entry into the war as necessary and just and exulted in the nation's victory and sudden preeminence on the world stage.[26] And excepting Dos Passos's *Three Soldiers* (1921), which could have been written by a disillusioned European survivor of the killing fields of the Western Front, no major American writer challenged the assumption that scientific research and technological innovation were fundamentally progressive and beneficent—though this issue was central to European intellectual discourse well into the 1930s.[27] Ernest Hemingway's novel *A Farewell to Arms*, which did much to fix the image of disenchanted American youth after the war, was about Italian soldiers massacred in failed frontal assaults on Austro-German entrenchments. Like the outrage expressed in William Faulkner's *A Fable*, written after the end of a second global war, Hemingway's anger was provoked by the ineptitude of the leaders responsible for the senseless slaughter rather than by the devastation wrought by mechanized killing. Beyond their generic depiction of the degradation of military regimentation, Dos Passos's later works, particularly his sprawling *USA: 1919*, are more notable for their indictment of the smugness, venality, and corruption of corporate society in the United States than for condemnation of the carnage on the Western Front. And though he served at the front in 1918, F. Scott Fitzgerald's postwar novels and stories are almost claustrophobically confined to the malaise and disaffection of America's fashionable elite. Insofar as it occurred, the flight of disenchanted youth from the United States after the war was more often a reaction to the social climbing and hollow materialism personified by Sinclair Lewis's Babbitt than angst concerning the direction in which science and technology were leading humankind.[28]

In large part because the United States entered the war at precisely the stage when innovative weaponry and battlefield tactics were restoring the possibility of maneuver and decisive combat, intervention in the conflict reaffirmed the abiding American confidence that science and technology could be potent antidotes to despotism, militarist aggression, and human miscalculation. By late summer of 1918—when large units of U.S. infantry were first committed to trench warfare—massed air power, tanks (most deployed by the Entente allies), and flame-throwers and "storm troopers" (used to great effect by the Germans) had made possible the great spring offensives that ended the defensive stalemate and appalling war of attrition that had consumed the youth of Europe for over three years.[29]

Not coincidentally, several wartime commanders who would later become influential American proponents of the new technologies were engaged in the allied counteroffensives that broke through the German defenses in the autumn of 1918. Perhaps no officer appreciated the potential of the new weaponry more fully than Billy Mitchell. Placed in charge of the American Expeditionary Force (AEF) air squadrons in August, Mitchell directed the large-scale aerial operations that supported American ground forces in the Saint-Mihiel and Meuse-Argonne offensives, and soon after the war he emerged as the most outspoken American advocate of air power. He shared the conviction of Europeans such as Hugh Trenchard and Guilio Douhet that aerial bombardment could render prolonged wars of attrition obsolete. He held that air bombardment would make it possible to destroy an enemy's factories and communications infrastructure, and thus its ability to sustain military operations. Mitchell and other enthusiasts insisted that if given sufficient resources, national air forces would prove decisive in future wars, and Mitchell was confident that the United States, with its expanding auto industry and its unparalleled capacity for technological innovation, was the best-equipped country to dominate the skies.[30]

Explicitly singling out China, Mitchell averred that nonindustrial societies were not capable of building credible air forces, without which they could not aspire to power on a global scale. But his travels in the Far East had convinced him that war between the United States and a rapidly industrializing Japan was inevitable. In the 1930s Mitchell pushed for a major buildup of American air forces and bases across the Pacific. The Russian pilot and expatriate Alexander De Seversky, who became one of Mitchell's confidants and most fervent supporters, agreed with Mitchell that "American industrial and inventive genius" made air power, not ground forces, the answer to threats of "the teeming hordes of Europe and Asia." Seversky insisted that "because we have more of what it takes in materials, inventive skills, and the natural technological psychology of our people" America could outdo any potential adversary in projecting its power around the globe. Ironically, though few took note of it, he also cautioned that because backward societies, such as China, would offer fewer targets than industrial nations, aerial bombardment was likely be far less effective if directed against them.[31]

The organizational aptitudes and productive capacities Americans displayed in mobilizing for the First World War inspired ever more ambitious formulations of the nation's civilizing mission in Asia and Latin America in the following decades. Despite recurring problems supplying American units advancing into the Saint Mihiel salient and the Meuse-Argonne in late 1918, allied forces were awestruck by the sheer quantity of support for the AEF, as evidenced by the grudging estimate of one British officer that "both in men and materiel . . . they seem to have about five times as much of both as we do."[32] The transcontinental technological systems the United States had been building for a half century provided a powerful matrix for the projection of American technical and material assistance overseas. From the War Industries Board and the Food Administration to federal agencies set up to run railroads or arbitrate disputes between management and labor, government-appointed engineers and entrepreneurs regulated American socioeconomic life to an extent previously inconceivable, an extent that some critics believed was decidedly un-American. And though there was something of a retreat from proactive interventionism overseas under the Republican administrations of the 1920s, the successful management of the war effort fired the aspirations of prominent figures in the newly ascendant social sciences to rationalize human systems and improve social morality through the application of theories informed by analogies to the natural sciences and technology.[33]

Scientism played a major role in the efforts of the proponents of social sciences to establish their authority as social planners and win funding for their endeavors. Claims that they were adapting the procedures of the "pure" sciences were offered to justify the application of their proposed remedies to challenges ranging from the enhancement of labor productivity to the management of natural resources.[34] In a seminal article entitled "The Place of Science in Modern Civilization," published in 1906, the renowned sociologist Thorstein Veblen argued that, in the modern era, "on any large question which is to be disposed of for good and all the final appeal is by common consent taken to the scientist," whose "answer is the only ultimately true one." He averred that the most distinctive quality of "modern civilized peoples" was the "peculiar degree [to which they were] capable of an impersonal dispassionate insight into the material facts with which mankind has to deal," and he judged societies that lacked scientific thought barbaric. Veblen stressed that science and modern machine culture were inextricably linked: "The canons of validity under whose guidance [the scientist] works are imposed by the modern technology." And he insisted on the need for practical application of understandings of the natural world or human behavior arrived at through inductive reasoning and empirical investigation. Veblen concluded that the combination of applied science and the technological breakthroughs (broadly defined to include social organization) it had made possible was the key to understanding the rise of Western civilization to global dominance: "A civilization which is dominated by this matter-of-fact insight must prevail against any cultural scheme that lacks this element. This characteristic of western civilization comes to a head in modern science and it finds its highest material expression in the technology of the machine industry."[35]

Veblen's poorly documented pronouncements about everything from technology as the prime impetus for human cultural advancement to an instinctual urge that "disposes men to look with favor upon productive efficiency" were taken as verifiable truths by leading American social commentators. His imprint was discernible in theorizing as disparate as Walter Lippmann's hostility to tradition and determination to overcome it through technical expertise and John Dewey's conviction that the problem of war could be solved by techniques of social management "comparable to physical engineering devices." The influence of Veblen's vision of social scientists engineering beneficial societal change was evident in the can-do enthusiasm with which engineers and academics had tackled the challenge of organizing American society to fight the First World War.[36] This resounding American affirmation of science and technology as the key to progressive human improvement contrasted sharply with the pessimism then prevalent among Western European—as well as Asian and African—intellectuals. For the latter the appalling destructiveness of the war summoned grave concerns about the directions in which industrialization was leading humanity.[37] The fact that none of the conflict's devastation was felt in the United States itself, and the great gains in the nation's ability to project its productive capacity and military power overseas, went far to explain why American discontent was for the most part focused not on the conduct of the war itself but rather on what was widely denounced as President Wilson's failed peace. And once again Veblen proved astute at capturing the national mood. In his 1922 critique of the botched settlement imposed at Versailles, Veblen prophesied that the failure to arrive at a workable resolution of the war made a second catastrophic conflict inevitable.[38]

In the interwar decades, most American historians who wrote accounts of the Great War also stressed the failed peace rather than the carnage of mechanized combat. They were inclined to view U.S. logistical and military contributions as decisive and to concur with John Maynard Keynes's dismissal of the outcome of the bickering at Versailles as a "Carthaginian" peace. And any anxieties about the future that had been aroused by the war were soon dispelled for most Americans by another wave of technological innovations and a decade of unparalleled, if unevenly distributed, national prosperity. The prewar introduction of assembly-line production and an acceleration in electrification undergirded a consumer boom in the 1920s, driven by mass marketing of such wonders as Model-T Fords, radios, telephones, new home appliances, and motion pictures. By 1930 the suburban sprawl encouraged by the availability of relatively inexpensive automobiles had provided the main impetus for the concentration of nearly half of the nation's population in urban areas.[39]

Even more than in the past, all aspects of American life were suffused with the presence of industrial technologies. Streamlining, allusions to machine tooling, and abstraction were worked into the design of all manner of everyday objects from toasters and clocks to chairs and silverware. The skyscraper boom, which only the Great Depression would interrupt, celebrated the "machine age" on a monumental scale. America's architects incorporated elements of industrial design into the elaborate structures that rose above the skylines of city centers across the nation—perhaps most exquisitely in the Chrysler Building in New York City with its gargoyle-like soaring eagles meant to suggest hood ornaments and its gleaming, streamlined, stainless-steel crest and spire. Artists as diverse as Charles Sheeler, Gerald Murphy, Thomas Hart Benton, and Reginald Marsh used abstraction, allusions to machine parts, and representations of factories in their works. New York, rather than Paris, was becoming the center of avant-garde postwar artistic expression—a trend exemplified by the Atlantic crossings of the French "machine artists" Francis Picabia and Marcel Duchamp.[40]

Charles Lindbergh's solo transatlantic flight to Paris in May 1927 marked the apex of the adulation of the modern as embodied in innovative technologies that swept America in the decade after the war. His feat drew international attention to the human ability to master, rather than be the helpless victim of, industrial technologies that had been called into question by four years of mechanized slaughter. During the conflict, the skill, courage, initiative, and control over machines displayed by the "knights of the air" soaring in the skies above the hapless infantry in the trenches accounted for the paradox that a tiny minority of the war's combatants produced most of its great heroes—including America's Eddie Rickenbacker.[41] The wholesome, youthful Lindbergh and his Ryan monoplane, which he dubbed the "Spirit of St. Louis," seemed an ideal melding of man and machine—a synergy he reinforced by referring to the aircraft and himself as "We." His ecstatic reception by an estimated 150,000 French citizens after his arrival at Le Bourget airport raised doubts about the depth of the disillusionment with science and technology that European intellectuals suggested was pervasive in the aftermath of the war. Lindbergh's triumph implicitly countered the strident rejection by bestselling European writers such as Georges Duhamel of American society as a model for the future of humankind.[42] And the young airman's international celebrity was made possible by yet another round of technological innovations—in this case communications technologies, which included talking films, worldwide radio broadcasts, and machines that transmitted photographs by wire.[43]

Although the number of major technological breakthroughs in the United States dropped off noticeably as the Great Depression took hold, the nation's faith in the progressive, problem-solving potential of science and technology was scarcely shaken. Most Americans blamed bad business practices, not engineers or technicians, for the economic slump. And the New Deal strategies that Franklin Roosevelt's economic planners devised to put the nation back to work gave precedence to large construction projects and infrastructural improvements. In his 1931 speech dedicating the George Washington Bridge, Roosevelt declared that the "story of bridge builders was the story of civilization." Until the Second World War forced a shift in priorities, he pushed vigorously for the building of bridges, dams, and highways across the country in an effort to restore the "heroic optimism" of the machine age and the conviction that "a new and better life was possible through the machine."[44]

One of the cornerstone agencies established under the New Deal was the Tennessee Valley Authority, which targeted a belt of poverty that stretched through the Tennessee river valley across seven states. Anchored by the Muscle Shoals dam and hydroelectric plant in northwest Alabama, which had been built during the First World

War to supply power to factories producing explosives, the project combined the construction of dams and generators to produce electricity with programs of rural electrification, reforestation, soil conservation, and disease eradication. Under David Lilienthal, the first chairman of the TVA, agency bureaucrats were instructed to be attentive to grassroots responses and to draw on local knowledge in planning and implementing these plans. Early TVA administrators stressed the need for decentralization and venues for public participation that would enhance democracy and in effect nurture a new sort of citizenry capable of prospering in the modern world.[45]

In the midst of a second global war, which like the first provided the impetus for the United States to produce its way out of a depression, Lilienthal turned his attention to the TVA as a model for the worldwide reconstruction that would be needed when the conflict ended. In 1944 he reported that representatives from virtually every country in the world had visited the river valley complex, and that the agency had served as a training ground for dozens of technicians from Latin America, China, and the Soviet Union.[46] In part because of Lilienthal's infectious enthusiasm and owing to the real improvements achieved by many of the agency's projects, the large-scale, high-tech TVA model of development assistance would have a profound impact on U.S. foreign policy after the war. But despite its successes, by the early 1940s significant problems in the program and some deleterious side effects were becoming apparent. Although the agency's critics appear to have made little impression on those responsible for shaping policies for assistance to emerging nations in the 1950s and 1960s, they raised serious questions about the appropriateness of the TVA approach for most of the postcolonial world.

Contrary to Lilienthal's confidence that a workable balance between government intervention and local participation could be maintained, in many areas TVA officials and local power brokers had monopolized planning and administrative decisions. The illiterate, often powerless, peasantry of much of the developing world would be even more vulnerable than rural Americans to domination by bureaucrats and local notables. The displacement of people by the flooding caused by TVA dams and power plants would also be more of a problem in countries where peasant or tribal populations were denser and deeply attached to ancestral lands. The construction and operation of hydroelectric complexes profoundly altered regional ecological systems. Strip mining and other extraction schemes led to deforestation and severe pollution of waterways. And other programs, such as rural electrification, foreshadowed the marginalization of women that would characterize U.S. overseas assistance programs in the early decades of the cold war. Although the program increased access to electrical power, the electricity was used overwhelmingly for irrigating fields, lighting farmyards, and running farm machinery. Less than 30 percent was used for household appliances, suggesting the low priority of the domestic chores that women were largely responsible for and necessarily continued to do by hand.[47]

Beyond the TVA, the depression in rural America, which was most dramatically manifested in the Dust Bowl, raised broader questions about the transfer of U.S. models for rural reconstruction to the developing world. The widespread use of tractors and other machinery powered by gas-driven internal-combustion engines, along with a growing dependence on chemical fertilizers, greatly increased the energy needed to sustain U.S. agricultural production. And market-oriented American systems for transporting, processing, packaging, and advertising farm produce also used sophisticated technologies and required significant amounts of energy. As large tracts of marginal, semi-arid land were brought under cultivation, particularly in response to the soaring overseas demand for farm products during the war, additional energy was required to pump water from aquifers and irrigation networks. In the rush to bring new lands into production, the careless cultivation practices employed in clearing the prairie grasses, combined with neglect of basic soil conservation measures, left America's heartland vulnerable to rain and wind erosion, particularly during sustained drought.

If transferred overseas, these practices were certain to contribute to the degradation of the heavily worked lands of Asia—which in many regions had been cultivated for millennia—and the thin, fragile layer of humus characteristic of soils in Africa, Latin America, and other tropical areas. Most underdeveloped postcolonial societies could not begin to afford the mechanized, energy-intensive orientation of American agriculture. In the longer term the polluting effects of gasoline, chemical fertilizers, and herbicides were likely to be even more ecologically pernicious than in rural America. And if the agrarian economists who have calculated that agricultural production over much of the United States requires far greater inputs of energy than the caloric yield of the food it produces are correct, the pattern of large-scale, extensive production that has come to dominate the Great Plains may not be sustainable for the United States, much less for the rest of the world.[48]

The Great Depression brought substantial reductions in the resources that Americans felt they could devote to civilizing projects overseas. Economic constraints also gave added impetus to a shift from missionary societies to philanthropic foundations as the main organizations committed to disseminating the American variant of "modern civilization." The activities of the Rockefeller Foundation in East Asia and Latin America reflected the new alignment in the agents and approaches to development assistance. John D. Rockefeller Jr.'s 1921 visit to China, accompanied by an opulently outfitted entourage, to attend the dedication of the Peking Union Medical College (PUMC) marked a turning point in this transition.[49] The Protestant missionary societies that had been responsible for the introduction of Western medicine and scientific training in China were consulted about what was in effect the consolidation of these enterprises in the impressive new campus that Rockefeller's largesse had made possible. But in the following decades the PUMC would become the epicenter of American efforts to revitalize China through the transfer of Western science and technologies.

In the preceding decades American missionary societies had struggled to upgrade their training of medical missionaries and science instructors and improve their clinical facilities. Some had even sought to provide a patina of social science expertise to lend legitimacy to their proselytizing.[50] Though the Depression made it even more difficult to implement such changes, the missionary offensive had stalled long before 1929. Despite notable successes in Hawaii and the Pacific islands and steady growth in mission activities in India and Africa, the core overseas campaign in East Asia had fallen on hard times. Intensifying nationalism had marginalized American missionaries in Japan. And the ascendancy of the Guomindang in China in the 1920s renewed the anti-Christian hostility that had been so daunting at the turn of the century. The May Fourth movement that arose in 1919 to protest the Versailles settlement (which left the Shandong province a Japanese sphere of influence) soon targeted mission stations, Christian colleges, and YMCAs. These assaults continued through the campaigns in the 1920s and 1930s for secular reforms to strengthen the nation, which often involved efforts to preserve China's venerable cultural heritage.[51]

The PUMC initiative, the most costly and visible of the Rockefeller Foundation's assistance programs in the interwar decades, exhibited both the promise and the shortcomings of more modest projects the foundation funded in Latin America, at times in conjunction with American military occupations. Medicine—whether in the form of education, research, or health service organization—was central to most of its programs. In China the emphasis was on Western medical training; in Latin America most Rockefeller projects were devoted to research and eradication campaigns against tropical afflictions, such as yaws and hookworm, which were of special interest to the foundation's mobile teams. American doctors were usually reluctant to become involved in routine medical treatment for the rural poor, and the medical knowledge yielded by their field research was almost invariably funneled back to research facilities in the United States.[52] Most of the foundation's educators and medical personnel in China also had few opportunities for contact with ordinary Chinese. Their mission was to train *in English* an elite corps of Chinese doctors, thus establishing links between American professionals and Chinese students who were mainly from well-to-do families. The PUMC campus at Beijing, with its elegant buildings in the Chinese style, was designed to be attractive to both first-rate instructors from the United States and the well-educated youths of China's upper classes.[53]

Explorations of the goals and organization of Rockefeller Foundation projects in China or Latin America have often led to critical assessments of the foundation's rationales for assistance to developing societies. Rockefeller disease-eradication programs have been linked to corporate American imperatives for healthy laborers in tropical areas and for the control of diseases in the cities where American businessmen, managers, and other personnel were likely to reside. Critics have argued that Rockefeller policymakers viewed health care initiatives and the transfer of Western scientific knowledge to China as a way of channeling that country's nationalist, revolutionary impulses in gradualist, reformist directions. And at least one commentator in the 1920s argued that programs to improve conditions in China were essential to prevent a flood of Chinese immigration to the United States.[54]

Although corporate concerns were certainly taken into account by the project planners of philanthropic organizations, a mix of motivations, including genuine humanitarian concerns, was clearly at work. National and professional rivalries could also be major factors. In medical research, education, and disease-eradication programs in particular, a quest to demonstrate the superiority of American approaches to those of the French was apparent. Other projects, often associated with military occupations in the Caribbean region, stressed the importance of health care in strengthening regimes or local social groups that were willing to promote American political and economic interests. And Rockefeller support for medical, educational, and rural reconstruction programs in China was frequently explicitly linked to the need to bolster the Guomindang against threats from Japanese military incursions and communist insurgents.[55]

These concerns foreshadowed the cold war contest of superpowers for influence in the postcolonial developing world. The assistance projects they spawned also anticipated some defining features of the development projects after the Second World War. In their emphasis on the training of male doctors or the health of male workers, they displayed an obliviousness to women's issues that would persist well into the 1970s. And though countervailing views were sometimes in evidence in China, there and in Latin America, scientists and technicians working for philanthropic organizations were dismissive of local knowledge, "native" practitioners, and indigenous tools and techniques. They believed that Western science and technology were products of objective empirical investigation and therefore transcended cultural particularities and were transferable to any of the diverse societies categorized as backward, underdeveloped nations. From the failure to build adequate latrines for workers in Brazil, which impeded efforts to eradicate hookworm, to the refusal to push for sweeping social and political reforms in China, which might have made the Guomindang regime more viable, the unwillingness to attend to local concerns, to encourage proactive community participation, and to make use of indigenous knowledge often proved a decisive liability of even the most ambitious and well-funded philanthropic programs.[56]

In the first half of the twentieth century, breakthroughs in applied sciences and technology propelled the United States into the ranks of the global powers and laid the foundations for its dominance of what by the early 1940s Henry Luce would proclaim "The American Century." The medical and engineering feats that were essential for the building of the Panama Canal showed observers around the world that the United States had surpassed its European rivals in capacity for innovation, productivity, and organization. America's role as provisioner of the Entente powers during the First World War and its belated but critical participation in that conflict demonstrated the adaptability of its technicians and the complex technological systems they ran. Wartime mobilization made clear the nation's ability to project its economic and military power overseas. Postwar engineering triumphs and material prosperity enhanced the nation's centuries-old commitment to a civilizing mission that was increasingly envisioned in global terms. Funded mainly by philanthropic foundations, American social scientists, technicians, and social planners pursued ambitious development projects from the Caribbean to East Asia. Their projects were patterned on the large-scale, high-tech approach first fully implemented in the Philippines. Prototypes were provided by the public works projects of the New Deal, from the Verrazano Bridge in New York City to the dams of the TVA. In China, as in the Philippines, the reconfiguration of America's civilizing mission around applied science and advanced technologies was premised on the assumption that at least some nonwestern peoples were capable of mastering the knowledge and skills that would foster nation building on the model of the United States.[57] U.S. funding and technical assistance grew dramatically for projects aimed at bringing nonwestern societies into the modern age in ways compatible with American visions of a viable international order. By the onset of the cold war, development assistance was a fixture of America's global interventionism.

Despite America's rapid ascent from regional to global power, until the Second World War its attempts to exert significant influence overseas often yielded disappointment. Most U.S. interventions—though unilateral and at times military—were rather peripheral when measured against the larger currents of global history. The completion of the Panama Canal did more to solidify the nation's hegemony over the Caribbean and the eastern Pacific than to enhance its influence across the Atlantic or in East Asia, where the Philippines remained a vulnerable outpost of empire. The gains that might have accrued from the Great War were largely frittered away by the failed peace, the nation's postwar demobilization, and its retreat behind the protective barrier of the great oceans. The sudden plunge from the consumerist splurge of the 1920s to deepening depression raised troubling questions about the suitability of the capitalist American model of modernity for the rest of humankind. And the severe economic downturn greatly reduced the funding and expertise that the nation could devote to philanthropic or missionary assistance overseas.

Of all the international setbacks and perils of the 1930s, none matched the turmoil that gripped East Asia in exposing the excessive expectations of the agents of America's civilizing mission, the depth of their misperceptions of nonwestern cultures, and their overestimation of the potential impact of the nation's interventions overseas. As Japan turned in the late 1920s and the 1930s to militarist solutions to its socioeconomic dilemmas that were translated into its own expansionist mission, it became plausible to those who thought beyond the xenophobia that was preparing the way for the Pacific war that Perry's forcible, abrupt "opening" of the island nation might have been ill-advised. It also became evident that the U.S. tutorial had failed and the sophisticated culture of the Japanese could not be remade according to American specifications. If Japan posed an unpalatable alternative to modernism American-style, China offered the prospect of a descent into chaos that the agents of U.S. intervention appeared to be incapable of preventing. But reluctantly, and contrary to the policy of Cordell Hull and other policymakers who sought to preserve a semblance of American neutrality in the face of the full-scale war that was developing between China and Japan,[58] the U.S. government became a major source of financial and technical assistance and eventually military support for the beleaguered regime of Chiang Kai-shek. The second, and equally unacceptable, East Asian alternative path to national development represented by Chiang's communist adversaries ensured that U.S. aid would continue long after Japanese militarism had been crushed. But the Guomindang's inability to check the Maoist revolutionary surge or cope with the Japanese invasion of Manchuria and China proper underscored the limits of America's interventions. However great its technological edge, in the coming decades the United States would find that the nation-building side of its civilizing mission often provoked resentment and resistance rather than grateful compliance. And a persisting tendency to underestimate the ingenuity and resilience of those in the nonwestern world who opposed America's hegemonic agenda would lead to a massive increase in the nation's costly military commitments across the globe and at times to sobering military debacles.

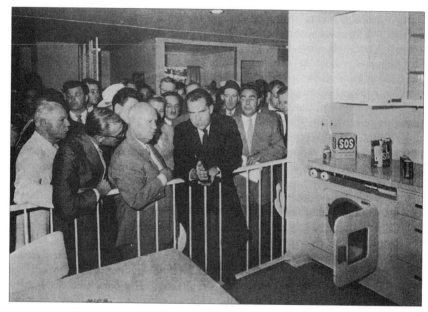

In the unlikely setting of a model American kitchen on display in Moscow at a high point of superpower rivalry in 1959, Vice President Nixon and Premier Khrushchev debate about the merits of capitalist consumerism versus the communist goal of a classless utopia. Associated Press, Valeo Clearance License 3.5721.4626108-94346.

We must embark on a bold new program for making the
benefits of our scientific advances and industrial progress
available for the improvement and growth of underdeveloped
areas.

—HARRY S. TRUMAN, 1949

5

IMPOSING MODERNITY

By late 1944 there was good reason to believe that
the apocalypse had arrived, at least in Eurasia.
From Great Britain to Japan, large swaths of cities and thousands of
towns in what had long been regarded as the core areas of the civilized world lay in ruins. Between 60 and 85 million people, the great
majority of whom were civilians, had been killed and countless others maimed, in what one U.S. congressman characterized as a "War
of Armageddon." The war had also made refugees of some 25 mil-

lion noncombatants, whose numbers were nearly matched by the 23 million people deported or forcibly resettled. In Asia, hundreds of millions faced starvation as drought and disruptions of transport and production led to widespread famine in some of the most densely populated areas on earth. In some regions, such as British India, the government's inability to prevent widespread starvation accelerated the process of decolonization. In others, such as China and Vietnam, the ability of communist-led revolutionary parties to provide relief to tens of millions of desperate peasants proved critical to their efforts to mobilize support for their ultimately victorious bids for power.[1] As the allied armies advanced into the crumbling empires of the Axis powers in the spring and summer of 1945, the systematic brutality and mass murder of the prison camp networks across Eurasia were revealed. In early August, the near-instantaneous obliteration of the Japanese cities of Hiroshima and Nagasaki convinced some of the scientists and military personnel involved in developing and deploying atomic weaponry that the final cataclysm was under way and the end of days imminent.[2]

Alone of the major combatants, the United States was spared the unprecedented levels of destruction and suffering made possible by over a century of industrialization and innovations in warfare. American combat deaths, which by some estimates were as high as 400,000, were undoubtedly sobering. But they were not of the same magnitude as those of the Chinese, Germans, Japanese, or Russians, which numbered in the millions. And although much of America's island empire in the Pacific fell to early Japanese offensives, no part of the continental United States was occupied by enemy forces. Nor did the American civilian population have to endure massed aerial assaults or the depredations of invading armies. The sanctuary that two vast oceans still provided, despite interwar advances in air power, meant that total American losses were even more disproportionate relative to those of most other major combatants. The contrast with the Soviet Union, the only other great power to emerge from the war with its political influence and military strength greatly enhanced, is particularly telling. American deaths were but 2.5 percent of the Soviet military and civilian total. Even much smaller nations suffered much higher *absolute* losses than the United States. Poland, for example, absorbed 6 million dead, or 17.5 percent of its prewar population; Yugoslavia lost between 1.5 and 2 million, which exceeded 12 percent of its total population before the German invasion.[3]

By late 1944 the transportation and communications infrastructures of all great industrial powers but the United States, as well as much of the colonized world, had been severely disrupted or utterly destroyed. Within the United States, both networks had been expanded and upgraded during the war. In the postwar years the American merchant navy made up nearly two-thirds of the world's total, including 60 percent of oil tankers, and the United States possessed the only sizeable fleet of transcontinental aircraft.[4] The rapid shift of the American economy into production geared for waging a global war had brought an end to the persisting depression, which a proliferation of New Deal programs had been unable to surmount decisively. In stark contrast, the industrial and agricultural sectors of the other major combatants—ally and enemy alike—had been laid waste or pushed to the point of exhaustion.[5] The immunity from direct assaults that North America enjoyed, and the U.S. economy's seemingly limitless capacity for industrial and agricultural expansion, ensured that the United States and its allies would emerge victorious from the "gross national product war"[6] that Axis aggression had inflicted upon much of Eurasia.

An economy that, despite the Great Depression, had produced 42 percent of the world's capital goods in 1939 (compared with 15 percent for Germany and 14 percent for Great Britain) expanded substantially in virtually all sectors during the war.[7] Within a year and a half of the Japanese attack on Pearl Harbor, the United States had achieved full employment and transformed its military into the best supplied and by far the most mechanized in the world. In addition, the United States was the only nation that was able to mobilize the diverse resources required to develop atomic weapons. In fact, the U.S. government found the funding and scientific expertise needed to underwrite several different programs to achieve that end—a strategy deemed necessary because no one was certain which approach would succeed. Informing the American people that an atomic bomb had been dropped on Hiroshima in early August 1945, President Harry Truman did not stint on hyperbole in proclaiming America's victory in the "battle of the laboratories":

> But the greatest marvel is not the size of this enterprise, its
> secrecy, nor its cost, but the achievement of scientific brains
> putting together infinitely complex pieces of knowledge
> held by many men in different fields of science into a work-
> able plan. And hardly less marvelous has been the capacity
> of industry to design, and of labor to operate, the machines
> and methods to do things never done before so that the
> brain child of many minds came forth in physical shape and
> performed as it was supposed to do . . . What has been done
> is the greatest achievement of organized science in history.[8]

Even while the United States was building the "the largest armed force in the history of the world," many sectors of the civilian economy were expanding as well, and despite some shortages and rationing, the standard of living for most Americans actually improved during the war years. American agricultural production grew by 33 percent and the nation's extraction of coal by 32 percent, crude oil by 40 percent, and iron ore by a remarkable 400 percent. In 1941, while the shift to a full wartime economy was still under way, the United States produced more steel, aluminum, oil, and motorized vehicles than all the other adversary powers combined. Between 1941 and 1945 America's manufacturing output doubled, and its GNP rose from $88.6 to $198.7 billion. By contrast, Great Britain, the former global hegemon, was bankrupt by 1940 and the world's largest debtor by 1945. In the same half-decade, the GNP of Western Europe as a whole fell by over 25 percent. At war's end, the United States possessed by far the largest, most innovative, and most dynamic economy in the world. Americans, who made up only 7 percent of the world's population, controlled more wealth than all of the rest of humanity.[9]

A strong consensus emerged in the United States after the war that American productivity and military prowess had been essential to the allied victory.[10] Franklin Roosevelt's pledge in 1940 that America would transform itself into the "arsenal of democracy" had been fulfilled. Even before formal U.S. entry into the war, the nation's productive capacity was essential to sustaining the struggle against the Axis powers in both Europe and the Pacific. On the western periphery of Eurasia, the lend-lease program provided Great Britain with ships, food, and war materials that made it possible for the British to fight on alone against Nazi Germany after the rout of their continental allies. Far to the east, assistance under the original lend-lease agreements negotiated with the Guomindang regime contributed significantly to the frustration of Japanese designs in China. Despite the need to supply its own armed forces, America's material and financial support for its allies increased substantially and steadily after it entered the war in late 1941. Under another lend-lease arrangement, for example, between 1942 and 1945 the United States supplied the Soviet Union with over a half a million trucks, jeeps, and other vehicles, which proved to be "the backbone of the Soviet motorised supply system."[11] During the same years, the United States produced approximately two-thirds of the military equipment (including half of the weapons) that carried the allies from near defeat to total victory. How fitting then (if in retrospect also remarkable) was Stalin's impromptu toast, at Tehran in late 1943, to American war production, "without which," he acknowledged, "our victory [over the Nazi invaders] would have been impossible."[12]

The U.S. capacity to supply its allies with such quantities of essential materials surprised even the most knowledgeable observers in view of the fact that before the war the American military establishment—particularly its army and air components—had been undermanned, poorly funded, and ill-trained relative to those of the other major combatants. By the early 1930s the armed forces the nation had assembled to fight the Great War had been all but dismantled by successive demobilizations and steady cuts in funding. The anemic condition of America's military was readily revealed by its ranking in terms of size—eighteenth—among the armies of the world. In 1940, with the Japanese in control of the most urbanized and developed areas of China and Nazi armies poised for the conquest of the Low Countries and France, military expenditures amounted to just 2 percent of America's GNP. As late as the spring of 1942, even as victories in the Battles of the Coral Sea and Midway ensured eventual allied triumph in the Pacific theater, the Japanese enjoyed marked superiority over U.S. forces in key aspects of aircraft design and firepower, and in the numbers of warships and planes.[13]

As in earlier wars, sheer economic productivity, advanced technology, and technical proficiency—all of which contributed to an overwhelming edge in firepower—were critical to U.S. success. Despite the economic distress and widespread disillusionment with industrial capitalism brought on by the Great Depression, as the war approached America remained a culture awash in mechanical gadgetry and devoted to technological advance. Donald Nelson, who headed a bewildering succession of boards appointed by Roosevelt to oversee production and procurement during the war, captured the longstanding national affinity for machines in his postwar memoir:

> In men who understood machines, how to tinker with them
> and run them, we were the foremost nation of the world.
> We had more first-rate engineers and mechanics, more sci-
> entists and graduates of scientific and engineering schools,
> more professional and amateur inventors, more boys who
> liked to take watches and motors apart . . . For generations,
> boys had been driving the family car and exploring the mys-
> teries of the internal-combustion engine. Some elementary
> knowledge of mechanics had almost become second nature
> with at least two generations. The country, from the days of
> the Yankee pedlars, had always been gadget-crazy. It would
> not be difficult to initiate this yeomanry into the intricacies
> of tanks and planes, ships and radar.[14]

American familiarity with and access to diverse technologies com-plemented a national public education system, sports culture, and corporate ethos that inculcated *both* individual initiative and com-petition *and* team work and group cooperation. These qualities ren-dered the 12 million men and women mobilized into the U.S. armed forces by 1944 the best prepared of all the combatants to cope with the highly mechanized, unprecedentedly mobile, and rapidly shift-ing conflict in which they were engaged. Whether or not the com-bat proficiency of American soldiers was a match for their Japanese or German adversaries, the overwhelming material superiority of the United States, and the technical and cognitive aptitudes that en-abled U.S. forces to make effective use of it, proved critical to allied victories on all fronts.

In the Pacific theater, where Americans spearheaded the allied as-sault on the empire that the Japanese had conquered with such re-markable speed in late 1941 and early 1942, U.S. advantages in weaponry, material sustenance, and logistical support were obvious from the outset and grew more pronounced as the conflict moved closer to the Japanese home islands. Even before America's wartime recovery from the Depression, its lead over Japan in level of mechanization was dramatic: in the late 1930s, for example, for every 1,000 Americans there were more than 200 automobiles and trucks; for every 1,000 Japanese there was less than one. By late 1942 the material discrepancy between the two adversaries was demoralizing: for every American serving in the Pacific theater, there were 4 *tons* of supplies; for every Japanese soldier, barely 2 *pounds*.[15]

Although the material advantages of U.S. forces over German units were less pronounced, the combination of logistical support and levels of firepower and mechanization possessed by American-supplied allied armies proved daunting to even the most accomplished German commanders. After the North Africa campaigns, Field Marshall Erwin Rommel complained that Anglo-American air power had reduced his crack Panzer divisions to "savage" fighters locked in unequal combat with the most modern of enemies. And one German officer commented about the Normandy invasion in 1944: "If I did not see it with my own eyes, I would say it is impossible to give this kind of support to front-line troops so far from their bases."[16]

The economic and military power the United States was able to project on a global scale in a war that was so devastating to its industrial rivals served to bolster Americans' long-held belief in their nation's exceptionalism. But it also intensified the paradox that had muddled exceptionalist thinking from the time of the earliest European settlement in North America. Although Americans deemed their nation's historical development both unprecedented and unique, they also saw American institutions, ideas, and modes of organization as models for all societies. The experience of the Second World War heightened the moral and millenarian dimensions of this teleology, which had long been grounded in Americans' conviction that they were a people of unprecedented inventiveness and technical aptitude. As Hans Morgenthau, the distinguished proponent of "realism" in international politics, observed shortly after the end of the war, technological prowess had come to be conflated with national virtue—with Americans' age-old sense of "moral superiority"—and both were seen to confirm America's mission as "the lawgiver and arbiter of mankind."[17]

In assessing the significance of their material contributions to the victory over the Axis powers, Americans came to view the struggle as a crusade of the righteous to rid the world of the palpable evil unleashed by Nazism and Japanese militarism. American political leaders and social commentators insisted that the proliferation of concentration and prison camps, the genocidal Holocaust, Japanese atrocities in China, and the draconian regimes of occupation in Europe and Asia demonstrated that the Axis nations had aimed to build a global order based on violence, oppression, and tyranny.[18] This tendency to see the war in terms of Manichaean absolutes, as a struggle without quarter between the forces of light and darkness, goes far to explain the allies' insistence on the unconditional surrender of their adversaries. It also bolstered the victors' determination to purge the German and Japanese political and sociocultural systems of all traces of Nazism and *bushido* militarism. The international tribunals that the Americans and their allies conducted after the war were intended to document the crimes of the vanquished regimes and to dramatize the punishment meted out to the principal agents of the evil empires, west and east.[19]

Given the American conviction that the Second World War was a crusade against the evil embodied in the Axis regimes, it is not a little ironic that even before the war ended the locus of evil was shifting in key policymakers' thinking to America's erstwhile communist allies. Over the first decade of the cold war, the steadily expanding global rivalry between the United States and the Soviet Union eroded the ebullient confidence that U.S. leaders had initially exuded as they plunged into the task of rebuilding a world order shattered by decades of depression, autarky, totalitarian tyranny, and armed conflict. Those who shaped postwar U.S. policy viewed the Soviets' extension of the communist realm in Eastern Europe as a direct challenge to the global hegemony that the United States had earned by virtue of its unparalleled economic, technical, and military feats during the war. American wealth, high-tech weaponry, and prodigious capacity for providing material assistance proved critical in checking the communist advance in Europe. The retention of a substantial U.S. military presence in Western Europe, billions of dollars in financial and technical assistance and technology transfers, and a steadily growing arsenal of nuclear weapons effectively stalemated—despite periodic crises—the superpowers' contest for dominance in Europe for the rest of the cold war.[20]

As is often the case with those who believe themselves the bearers of a special dispensation, American cold warriors and their Soviet counterparts, and by the early 1950s the Maoists in China, conjured up and greatly inflated threats posed by competing powers or rival ideological systems. Stalinist, McCarthyite, and Maoist purges at home were coupled with often violent interventions abroad aimed at blocking the spread of what were viewed as heretical systems, forcing apostates back into the camp of the righteous, or opening up new realms for the spread of the supposedly irreconcilable approaches to developmentalism championed by the cold war rivals.[21]

In its new incarnation the Manichaean dichotomy constructed by the rival superpowers was, as least until the Sino-Soviet split became apparent in the mid-1960s, less complicated by multiple polarities and more encompassing than its Second World War predecessor. The global struggle between what American politicians, journalists, and public intellectuals routinely termed totalitarian communism and capitalist democracy revived many of the binary conceptualizations of international relations that had dominated foreign policy in the 1930s and 1940s. Free enterprise and free trade were pitted against the stifling state controls and autarkic ambitions of the communist bloc. The rule of law and human rights were contrasted with the Soviet show trials, the gulags, and state censorship. An innovative, "open," and competitive American or "free world" system, conducive to scientific and technological innovation and economic growth, was juxtaposed to the claustrophobia of life behind the iron and bamboo curtains, where stolid bureaucrats repressed individual excellence, critical inquiry, and dissent.[22] The myopia, paranoia, and systematic brutality of communist regimes were confirmed by widely publicized exposés written by defectors or persecuted intellectuals who had been consigned to Siberia for speaking out against the political corruption or miserable living conditions within the workers' utopias.[23]

The tropes of despotic repression and enslaved peoples that came to dominate American impressions of the Soviet Union and Eastern Europe were informed by longstanding perceptions of the Russians as not fully European but rather a polyglot, semi-oriental people. Russian émigrés and American academics had built distinguished careers around debates over the legacy of centuries of "Tartar" control of the Russian heartlands. And despite a rather disconcerting absence of centralized irrigation networks in Russian history, the political philosopher Karl Wittfogel repeatedly cited examples from tsarist and Soviet Russia to support his controversial theories linking hydraulic systems to "oriental despotism." His frequent use of the machinations of Lenin and Stalin to illustrate the precepts and behavior of despotic rulers was echoed, perhaps more subtly but often

every bit as purposefully, in biographical studies that explored the
Asiatic lineages of these leaders.[24]

Whether commentaries on oriental contributions to Russian
character stressed racial factors or cultural influences, the conclusions were similar: the first peoples to embrace the Marxist variant
of communism as their national ideology had historically demonstrated a deeply rooted, Asiatic proclivity for authoritarian rulers and
regimentation. As early as the 1930s, American journalists in the Soviet Union had argued that both Stalinist brutalities and peasant tolerance for these abuses were rooted in the Russian "national character." Some even attempted to justify Stalin's harsh methods with
colonial-style diatribes against the innately passive, fatalistic, and reactionary disposition of the Russian peasantry.[25] During the cold
war, the linkage of supposed flaws in Russian character with oriental
influences figured prominently in scathing critiques of, rather than
apologies for, Soviet tyranny.[26] The hybrid nature of the peoples and
culture that had launched the communist experiment in the USSR
was seen to render the project suspect as a model for the future of
human societies. Semi-oriental peoples were no more worthy than
the fully Asiatic Chinese, whose country would soon emerge as a
rival of both the superpowers, to take up the challenge of the worldwide civilizing mission that Americans had come to see as synonymous with the spread of modernity.

The totalitarian control and institutionalized repression that many
Americans believed inevitable under communist regimes not only
appeared to rule out accommodation, they were seen as critical
sources of the persisting material backwardness and lamentable
living standards of peoples trapped behind the iron curtain. Eyewitness accounts of life in the Soviet Union and Eastern Europe returned again and again to the contrast between American prosperity
and the bleakness of everyday life under communism. Pundits and
editorialists celebrated the "people's capitalism" that had liberated

Americans from "grinding poverty" and thereby transformed the United States into "an oasis of prosperity in a world desert of human misery." They touted the worldwide appeal of America's "bloodless revolution," which had succeeded where the communists in both the Soviet Union and China had failed so miserably, in "creating almost overnight the most abundant and most truly classless society in history." The task of diffusing this revolution to the "underdeveloped" world was seen to be as "great a challenge to Americans as [was the settlement of] the virgin prairies," a global project that would yield "equal miracles of human cooperation and achievement."[27] Images of Soviet, and later mainland Chinese, economic failures and material backwardness compared with the capitalist West—which implicitly came to include Japan—were quite consciously employed to demonstrate that the American path to modernity was the obvious choice for the peoples of "emerging" postcolonial nations.

From streamlined automobiles to cosmetics, the relatively low cost and widespread availability of consumer goods in America—and increasingly Western Europe and Japan—were compared with the shoddy quality, high prices, and scarcity of even basic material amenities in communist societies. American periodicals of the 1950s chronicled the elaborate stratagems Soviet citizens devised to secure the most commonplace foods, furnishings, or appliances. Charles Thayer, a former head of the Voice of America in Russia, reported that items most Americans regarded as staples, including "matches, salt, soap, thread, needles, buttons, kerosene, toothpaste, razors and razor blades," were "often unobtainable outside Moscow and the other big cities."[28] American visitors to the Soviet Union bemoaned the "perpetual shortages of materials and supplies," which, along with poor workmanship, they blamed for the dilapidated state of even recently constructed buildings and the "general inadequacy" of highways. Even Soviet consumer goods that Americans deemed of reasonably high quality, such as radios, were reported to be much too expensive for anyone but members of the elite to purchase. And television sets, not to mention automobiles, were said to be luxuries that the average Russian could not even imagine owning. Reasonably inexpensive amusements that were available to the masses, such as motion pictures, were reported to be shown in dreary, dirty, and uncomfortable theaters. And to partake of such dubious entertainments, ordinary Russians were forced to endure the red tape and delays of the Soviet bureaucracy to purchase admission tickets that had to be arranged weeks in advance.[29]

American observers also lamented the shortages of staple foods, including meat and butter. Thayer pointedly reminded the readers of *Look* magazine that these had been major export commodities before the Bolshevik revolution but were being imported in large quantities by the 1950s. Thayer and others linked these everyday deprivations to the failure of Soviet agriculture. Hamstrung by forced collectivization and the absence of profitable market outlets, they argued, agricultural production and food distribution had deteriorated in the Stalinist era. Livestock herds had dwindled, production quotas were mere fantasies, and schemes to bring "virgin lands" in Central Asia under cultivation were turning Russia's steppe regions into "gigantic dust bowls." Although some observers conceded that similar mistakes had been made in settling the western prairies in the United States, they underscored the current contrast between the low productivity of Soviet farmers and the market-oriented, highly mechanized U.S. agricultural sector, which not only fed Americans better than any people on earth but made the nation the world's leading exporter of foods and other farm products.[30] The British journalist Edward Crankshaw reported that the Soviets' abysmal failure to build a productive agrarian sector and deliver consumer goods had sparked widespread resistance in the countryside and growing disenchantment among urban industrial workers, disaffection that was becoming a serious threat to Stalin's communist regime.[31]

In the early cold war decades, the very layout of popular magazines such as *Look* and *Life* suggested the stark contrasts in the levels of consumer comforts provided by capitalist and communist systems. Articles on shortages behind the iron curtain were routinely flanked by ads for the latest models of American cars, TV sets, dishwashers and other household appliances, and variety packs of breakfast cereals. Even Soviet advertising was dismissed as a pathetic imitation of its western counterpart: heavy-handed Russian marketing techniques were depicted as crude attempts to use capitalist devices to invigorate faltering state-controlled economies.[32] A different sort of juxtaposition contrasted the restoration of West German cities and the growing availability of automobiles, clothing, and even bathroom fixtures in the allied zones with the grimness of life in East Germany and other Warsaw Pact countries.[33]

Similar material contrasts were also deployed to demonstrate the more genuine and fulfilling nature of women's liberation in the capitalist world. Press accounts of international exhibitions, such as those at Brussels in 1958 and Moscow in 1959, devoted extensive coverage to fairgoers' enthusiastic responses to dazzling displays of American household appliances. In contrast to the heavy machinery featured in nineteenth-century showcases of modernity, U.S. exhibits in the early cold war were organized around well-appointed kitchens, lavish beauty parlors, and sleek automobiles. Exhibition crowds were treated to fashion shows and cosmetic clinics that stressed the casual, leisure-oriented quality of American women's lives. It was no coincidence that the famous 1959 "kitchen debate" between U.S. Vice President Richard Nixon and Soviet Premier Nikita Khrushchev took place in the ranch-style model home that was the focal point of the American National Exhibition in Moscow. Appropriately, given the venue, household technologies and contrasting approaches to the gendering of modernity were the focus of the contentious exchange between the two leaders. In his election-minded 1962 memoir *Six Crises,* Nixon recounted pointing out to Khrushchev that the washing machine on display was a superb example of the way affordable consumer goods had improved the lives of American women. In a somewhat oblique rebuttal, Khrushchev declared Nixon's equation of consumerism and societal advance to be symptomatic of a "capitalist attitude" and dismissed the household appliances as inconsequential "gadgets."[34]

While projecting images of American women enjoying stylish lives as both homemakers and ladies of leisure, the U.S. media also conveyed bleak impressions of the fate of women under communism. Rather remarkably, the trope of the squaw drudge, so influential in shaping attitudes toward Amerindians in earlier centuries, was reincarnated in accounts of the dreary lives of wives and mothers in the Soviet Union and Eastern Europe. American commentators often conceded that women were strongly represented in high-status professions, such as medicine and higher education, but they noted that these pursuits did not enjoy the prestige or the high pay accorded them in the West. And far more of the media coverage of women in communist societies focused on their roles as workers. Photographs and provocative captions portrayed women engaged in physically demanding tasks, such as harvesting grain or pouring concrete. American journalists also stressed that after working full shifts as laborers, Soviet or Polish women returned home to clean, cook, and raise their children, tasks their menfolk were said to eschew. Amenities such as nylons and trips to the beauty parlor were reported to be in short supply or beyond the means of most women. And much was made in the press of how eagerly these simple pleasures were sought by women like the "model Commie girl" whose flight from East to West Germany was chronicled in a 1951 *Life* article illustrated with pictures of the attractive young woman in a stylish swimsuit and shopping in a well-stocked West Berlin market. At a more explicitly political level, the dumpy, ill-coiffed, and frumpily dressed wives of communist premiers and commissars were routinely disparaged in the American media and popular banter.[35]

American representations of the contrasts between capitalist and communist societies were, not surprisingly, almost completely reversed by Soviet analyses and images of the West. Communism was celebrated as a system that promoted social solidarity, the material well-being of all citizens, and equal access to health care, education, and job security. The United States, meanwhile, was denigrated as the epitome of capitalist exploitation and social inequity. Unemployment, racial and ethnic discrimination, homelessness, and urban decay were depicted as pervasive and decried as the inevitable outcome of a capitalist system that denied the most basic human rights to its citizenry. Drawing on themes from contemporary American bestsellers such as *The Lonely Crowd* and *The Organization Man,* Soviet commentators averred that cutthroat competition, unbridled individualism, and a political system that mainly served the interests of corporate and military cliques had produced a society riddled with anomie, loneliness, crime, and disaffection. Social divisions and political dissent, from the civil rights movement to the urban riots and antiwar protests of the late 1960s, were extensively reported in the Soviet press (and later by state-sponsored television). These disruptions were interpreted as symptomatic of the contradictions that would bring about the demise of the capitalist order predicted by Marx and Lenin. Soviet—and at least until the early 1960s, Chinese—interventions in other countries were characterized as anti-imperialist, motivated by an altruistic determination to forestall capitalist exploitation of colonized and recently liberated peoples. The United States and its allies were caricatured as aggressive, expansionist, and desperate to avert the collapse of their doomed system through the forcible neocolonial control of foreign markets, investment outlets, and raw materials.[36]

Even at the height of the cold war, some Soviet academic analysts and *Pravda* editorialists tempered sweeping criticisms of America with positive assessments of U.S. science and technology. Although they frequently prefaced their comparisons with claims of superiority in key sectors, Soviet observers conceded that the USSR lagged behind the United States in overall scientific and technological advancement. Despite the adversarial relationship between the two powers and strict state censorship regarding such strategically vital issues, Soviet commentators were often effusive in their praise for America's innovativeness, well-funded and sophisticated research and development, and highly skilled and productive workers. Some went so far as to hold up the United States as a model—albeit a deeply flawed and transient one—for the industrial development of the Soviet Union and other socialist states.[37]

Seen in this frame of reference, it is likely that Nikita Khrushchev's infamous taunt in a speech at the United Nations in the early 1960s—that the USSR would bury the United States and the capitalist world—was not the bellicose bravado that was widely reported in the American press, but rather a prediction of the inevitable eclipse of the capitalist countries by the rapidly industrializing communist bloc.[38] On the one hand, Khrushchev's boast was a routine iteration of orthodox Marxist doctrine that capitalism was doomed by its fundamental internal contradictions. On the other, it was an expression of confidence that the Soviet Union would soon outstrip the West in industrial productivity and in the living standards of its citizenry. That at least some Americans understood the Soviet leader's challenge to be economic—and specifically technological–rather than military was reflected in ads in popular periodicals that called for increased spending on the U.S. rail system and featured a photo of a combative Khrushchev exclaiming: "Thank you America for what you have done to your railroads!"

Soviet representations of America were routinely dismissed in the United States as propaganda that projected crude caricatures grounded in Marxist ideology rather than contemporary reality. Through the first decade of the cold war, most Americans who wrote for popular consumption concurred with Dwight Eisenhower's assessment in 1951 that "in total wealth, material strength, technical scientific achievement, productive capacity, and in rapid access to most of the raw materials of the world, we, the free nations, are vastly superior to the Communist bloc."[39] Media pundits, and indeed the majority of U.S. policymakers, were also confident that the capitalist path to development would be the inevitable choice of the peoples of the nations then emerging from colonial rule. But as the Soviets developed nuclear weaponry in the late 1940s and early 1950s, and as they repeatedly demonstrated the scientific and technological prowess to match or surpass a steady American arms buildup, a different source of anxiety surfaced in comparisons of the productive capacity of market capitalist and command communist societies.[40]

Despite Eisenhower's sanguine estimate of the global balance of power, Americans' complacency regarding the superior capacity of their open and competitive system to produce *both* consumer goods and advanced armaments was periodically unsettled by reports of the growing sophistication of Soviet weaponry or the sheer size of the Soviet arsenal. Prominent columnists began to ask whether the United States could afford both its high levels of civilian consumption and the rising costs of the arms race. Richard Wilson pithily summarized these alarmist concerns in *Look* magazine in early 1951: "Russia built a radar network—we splurged on television." Fierce

debates within successive administrations over U.S. military preparedness surfaced in the popular press in articles such as Wilson's,
which asked "Can Our A-Bombers Get Through?" and in editorials
warning that the pursuit of material comforts had begun to "Undermine Our Military Posture." Election campaigns featured highly
charged arguments about alleged "missile gaps," the nation's vulnerability to surprise attacks, and reports that the outnumbered and outgunned NATO forces in central Europe would be overrun by a determined Soviet offensive. Democrats charged that the Eisenhower
administration had underfunded key military projects while Americans heedlessly pursued consumer satisfaction. In the 1960 presidential debates John F. Kennedy quipped that Nixon had conceded this
perilous tradeoff in the "kitchen debate" by telling Khrushchev:
"You may be ahead of us in rocket thrust, but we're ahead of you in
color television."[41]

The Soviet Union's launching of an unmanned satellite, *Sputnik
I*, into outer space on October 4, 1957, had turned the arms-versus-
butter debate a good deal more vitriolic and moved it to center
stage in American politics. A *Life* editorial in mid-October mocked
the scientists and (unnamed) Pentagon planners who had underestimated Soviet scientific and technological advances because they
"confuse[d] scientific progress with freezer and lipstick output." A
month later, after the Soviets launched *Sputnik II*, the prominent
scientist George R. Price, who, *Life*'s editors emphasized, had
worked on the development of the atomic bomb, declared that
Americans must make "fundamental changes in [their] values and
purposes in life," that would force them to "set a higher value on
liberty than upon luxury." Ironically, these concerns regarding the
citizenry's addiction to consumerism were underscored by the appearance of space toys and designer clothing featuring space motifs
within weeks of the Sputnik launchings.[42]

The Sputnik challenge gave added impetus to the expansion of

America's military-industrial complex, which had begun to accelerate in response to perceived threats from communist rivals after the
end of the Second World War. The coalescence of government, corporate, military, scientific, and academic interests that makes up the
military-industrial complex can be traced back to the end of the
nineteenth century. But the post-1945 decades saw a pronounced
intensification of the links among these sectors, which was facilitated by the interconnected technological systems that had multi-
plied and grown more sophisticated over the first half of the twentieth century. The sharp decline in military spending that came with
postwar demobilization and conversion to civilian production had
slowed, then reversed, in the face of a succession of challenges from
the Soviet Union's development of an atomic bomb to the outbreak
of the Korean war. By early 1950 Eisenhower's security advisors were
sufficiently alarmed to urge a massive shift in resources from consumer production to weapons development to contain the spread of
communism. In the following decades government funding for re-
armament rose steadily—often at the expense of social welfare programs and research and development in the civilian sector. Though
much of the military spending was invested in missile systems and
high-tech weaponry to match the Soviets' buildup, from the late
1950s the upgrading and expansion of conventional forces and technology transfers to meet rapidly proliferating commitments in developing nations became a major focus of government contracts for
both corporations and academic institutions.[43]

The possibility that Russia's space triumphs signaled America's
eclipse as the global leader in science and technology was seen as a
major setback in the cold war competition to convince developing
nations to adopt the market capitalist rather than the command
communist model of social and economic development. Soviet celebrations of the satellite launchings, which coincided with the 40th
anniversary of the Bolshevik revolution, were reported around the

world. Elaborate displays of Russian pride, and even taunts against the United States (such as a circus clown carrying a balloon labeled "the American Sputnik"), were predictably featured at these state-sponsored festivities. But the presence in Moscow of prominent leaders from the developing world, including Chairman Mao Zedong, dramatized the propaganda value of the satellite launchings. Many critics of the Eisenhower administration's defense policies concurred with Senator Scoop Jackson's blunt assessment that the USSR had delivered "a devastating blow to the prestige of the United States as the leader of the scientific and technical world." CIA Director Allen Dulles warned that the Soviets were using Sputnik and other technological breakthroughs to gain "maximum leverage in the Middle East and, more generally, to demonstrate the effectiveness of the Communist system to the underdeveloped countries." Undersecretary of State Christian Herter found overseas reactions to the Soviet breakthroughs "pretty somber," and a joint report by the State Department and the CIA warned that if the United States did not match the Soviet achievements in rocketry and satellite technology, America's influence over its allies and capacity to promote "neutrality" among the nonaligned nations would be seriously compromised.[44]

Mao Zedong's presence in Moscow's Red Square during the 40th anniversary celebrations, and his speech exhorting the communist nations to work together, provided an indication of just how attractive the Soviet formula for development had become in much of the postcolonial world. But Mao's public enthusiasm for the Soviets' technological feats also represented another strand in a multivalent, and increasingly potent, communist challenge to the U.S. model of modernity as the template for all of humanity. From the first years of the cold war, many American policymakers and academic specialists had seen the communists' victory in the decades-long struggle to rule mainland China as a significant victory for the Soviet rivals.

Mao's proclamation of the People's Republic of China in Beijing's Tiananmen Square on October 1, 1949, added a vast swath of Asia to, and more than doubled, the population in the "communist camp." In the acrimonious aftermath of the Chinese "defection," American politicians and commentators warned that the Chinese communists' seizure of power imperiled the global supremacy of the United States and its "free world" allies and threatened friendly or nonaligned regimes in the decolonizing regions of South and Southeast Asia.

The fundamental historical trajectories that the People's Republic of China shared with virtually all the "underdeveloped" or "emerging" nations of Asia and Africa, which were routinely and inappropriately lumped together as the "Third World," meant that China had much more potential than the Soviet Union to pioneer a path to development that might serve as a model for emerging nations. Although China had not been formally annexed by the western imperialist powers, it had been as economically exploited, politically humiliated, and culturally disoriented as the colonized regions of Africa and Asia. In fact, because no western nation formally governed—and thus had to take responsibility for—China, the impact of colonialism was a good deal more devastating there than in most areas ruled by a single imperial power. The Japanese invasion and brutal occupation of large portions of northeast and coastal China in the 1930s and early 1940s brought socioeconomic disintegration and human suffering to almost unimaginable levels. Like other emerging nations, China had been colonized and pillaged. Like them, China had been left after the colonizers' departure with massive poverty, a largely illiterate peasant base, a paltry and backward technological endowment, and rates of population growth that far outpaced its ability to expand productivity or even raise the living standards of the mass of the people through forced redistribution.

By the mid-1950s the Maoists' commitment to the spread of a revolutionary struggle for economic growth and sweeping social reform had become a much greater threat to America's global dominance than China's increasingly shaky alliance with the USSR.[45] The convergence of the challenges posed by Russian scientific and technological advances—epitomized by the Sputnik launchings and Soviet participation in high-profile development projects like the Aswan Dam in Egypt—and the appeal of Maoist revolutionary formulas in much of the postcolonial world, rendered it imperative that Americans come up with convincing counters to the development options championed by their communist rivals.

Apprehension over reverses in Asia intensified efforts to overhaul U.S. approaches to diplomacy and the projection of military and economic power overseas. The systematic gathering of information about foreign peoples and cultures, which had been officially pursued in a transitory and site-specific manner by successive Philippine Commissions at the turn of the century and expanded by Woodrow Wilson's inquiry team at the Paris Peace Conference in 1919, was fully institutionalized and globalized in the 1950s by a proliferation of area studies programs, think tanks, and government-funded research. These initiatives were perhaps most fully integrated at the Center for International Studies (CENIS) at the Massachusetts Institute of Technology. By the mid-1950s CENIS had brought together a diverse team of distinguished social scientists and policy specialists who quickly established the program as a major force in the emerging field of "development studies." CENIS conferences, publications, and collaborative research projects played critical roles in the gestation of what came to be called modernization theory and its establishment as the dominant American ideological and policy response to both the threat of communist expansion and the instability spawned by decolonization. In modernization theory, American social scientists and political leaders found a way to resuscitate the technological triumphalism that had been so pronounced at the end of the Second World War. Modernization theorists would elevate generalized claims regarding America's superior technical prowess and material culture into central elements in the nation's cold war crusade, particularly in the contested zone that made up the "Third World."[46]

Although it has been argued that modernization theory was a product of a specific, peculiarly American, moment in the history of the cold war,[47] the intellectual underpinnings of the modernization discourse can be traced at least as far back as the Enlightenment era. And virtually all the basic tenets of the cold war modernizing ideology can be found, if often in inchoate form, in nineteenth-century European and American expansionist rhetoric. Champions of modernization theory shared with the *philosophes*, earlier western colonizers, and Soviet ideologues the conviction that technological innovation was a critical source of national power, and that it was, with applied science, essential for achieving the high and sustained levels of economic growth associated with modernity. For example, in each of the five phases that W. W. Rostow identified in his influential 1960 book *The Stages of Economic Growth*, which traced the trajectory from the backwardness and relative poverty of a "traditional" society through the "take-off" to full modernization, breakthroughs in science and technology played pivotal roles. Rostow insisted that it was "an essential condition for a successful transition that investment be increased and—even more important—that the hitherto unexploited back-log of innovations be brought to bear on a society's land and other natural resources." And he declared that in Britain, the first industrial nation, and in the United States, the one that had advanced the furthest, "the proximate stimulus for take-off was mainly (but not wholly) technological."[48] Rostow's conviction that applied science and industrial technology were core prerequisites for modernization was widely shared, at times implicitly, by most of the social scientists studying modernization in the 1950s and early 1960s. But some modernization theorists stressed specific technologies, such as transport and communications networks, while others sought to discern particular social configurations or world views that had produced the spurts of discovery and innovation that accounted for scientific and industrial transformations.[49]

Rostow divided human societies according to a "pre-Newtonian" and "post-Newtonian" dichotomy—a variation on the distinctions between the traditional and the modern that preoccupied many modernization theorists. Like American proconsuls in the Philippines and medical missionaries in China, Rostow deplored the passivity and fatalism of those who lived in societies that had not yet entered the Newtonian epoch. And he celebrated the take-charge, problem-solving propensities of those who had come to "believe that the external world was subject to a few knowable laws, and was systematically capable of productive manipulation." Rostow averred that risk-taking entrepreneurs, and polities and societies that were able to accommodate or advance their ambitious schemes for increasing productivity and profits, were critical to the take-off into sustained investment and economic growth that he considered a necessary step toward modernity. In his view, post-Newtonian societies were committed to "endless change." And his version of the modernization process as a predictable sequence of stages of progressive development captured one of the persisting themes in the exceptionalist vision of the American historical experience.[50]

Modernization theorists treated the emerging postcolonial nations as a single, undifferentiated entity. In keeping with the fondness for binary opposites exhibited by earlier European colonizers, they routinely lumped these so-called underdeveloped societies together as "traditional." Rostow and other American modernizers were certainly cognizant of major differences in the vast region they conceptualized as the "Third World"—differences in religious and ethnic makeup, geography, and relations with the adversarial cold war blocs.[51] But the social science paradigms and formulas for development they proposed under the rubric of modernization obscured elements of diversity through essentialized analyses of a timeless and unified tradition, which they viewed as the source of major impediments to the achievement of an equally essentialized state of modernity. As the stark bifurcation between tradition and modernity suggests, American developmentalist thought in the 1950s and 1960s was often organized around dichotomous and antithetical ideal types. Modernity was celebrated for its emphasis on rationality, empiricism, scientific investigation, activism, technological innovation, continuous change, and the domination of nature. Tradition was equated with superstition, intuition, a veneration for untested custom, indiscipline, fatalism, material backwardness, stagnation, and helplessness in the face of the "whims" of nature.[52]

This essentialized vision tended to discourage serious investigation into the history and current situations of specific postcolonial societies, often leaving policymakers with only superficial impressions or generalizations about areas they deemed vital for demonstrating the superiority of American approaches to development over communist alternatives. As had been the case in the Philippines, China, and elsewhere, abstract theories and social engineering projects designed to accord with American historical precedents were favored as responses to the challenges presented by the poverty, backwardness, and instability of postcolonial nations. For modernization theorists, the preconditions for achieving the take-off to sustained growth were uniform and universal rather than society-specific. And because they assumed that the United States had vastly excelled all other nations in levels of industrial development and societal well-being, they were confident that America could best serve as a model for the rest of humankind as well as marshal the resources that would be needed to modernize the emerging nations.[53]

In comparison with what they regarded as the "normal" path to modernity traversed by Britain and the United States, Rostow and most other modernizers considered communism a malformed, corrupt, and brutal system that had failed to provide the basis for genuine take-off in any of its permutations. Rostow likened communism to a disease that spread readily in the unstable period of transition from tradition to modernity. He frequently contrasted India's post-independence economic growth with China's stagnation. He noted India's ability to sustain the world's largest democracy, despite obstacles of overpopulation and poverty of the same magnitude as those faced by totalitarian China. Echoing many commentators of the early decades of the cold war, he concluded that oppression and sacrifice under communist regimes had resulted in societies that were "drab and hungry," a record that did little to recommend Marxist-inspired alternatives to the rest of the underdeveloped world.[54] In view of the much more rapid pace of modernization in the societies of "free Asia," Rostow was convinced that "history" was on the side of "free men"—that the United States and its allies would prevail in the cold war.[55] In modernization theory, he and other American policymakers believed they had found a persuasive response to the challenges posed by appeals of command communism for the development-minded societies of the "Third World." And though their formulations rarely made explicit connections to America's expansionist past, they had found a way to reconfigure the ideology of the nation's civilizing mission to address the contingencies of the global rivalries of the cold war.

Despite an obsessive preoccupation—both during the cold war and in much of the subsequent historical literature on that era—with the ideological differences and global rivalries that separated the superpowers, American policymakers and their Soviet counterparts had a good deal in common. They shared the conviction that their political and social system was the best model for the development of human societies. And though each of the rivals was determined to propagate its own version of the gospel of development, their ostensibly irreconcilable approaches to this process shared as much as, perhaps even more than, they differed in terms of historical origins and fundamental presuppositions. The advocates of market capitalism and command communism deployed remarkably similar strategies for providing development assistance to emerging nations. Both viewed the mastery of western science and industrialization as essential to the improvement of humankind.[56] And, at least until the early 1970s, and then only marginally in the Soviet bloc and communist China, neither side gave much thought to the environmental consequences of its formula for achieving sustained economic growth and social stability in the developing world. This convergence of the rival systems provides a hitherto neglected explanation for the persisting accommodation that made it possible to avert a third global war in the last half of the twentieth century.

There were, of course, major differences between the systems. At the onset of the cold war, the United States enjoyed a seemingly decisive advantage in the level of resources it could direct toward development projects in emerging nations. With the USSR devastated by the Nazi invasions of the Second World War, the Soviets were compelled to concentrate on rebuilding at home and providing assistance to the satellite states they had occupied in Eastern Europe. But in the competition for influence in postcolonial nations the Soviets had an advantage as well: their commitment (though often more rhetorical than realized) to full employment, state-supported education at all levels, universal health care, and affordable housing addressed the immediate needs of developing societies much more directly than did the American emphasis (again often rhetorical) on market-based alternatives and democracy.

Despite global depression and world war, the Soviets had within a generation transformed a society of poor, uneducated peasants into a major industrial and military power. For the leaders of emerging states in the decolonizing world, who had rather abruptly assumed responsibility for similar agrarian, backward, and unstable societies, the Soviet model of centralized control and five-year plans had great appeal, particularly if they managed to overlook the sobering human costs of Stalin's brutal forced march to modernity. And the Soviet emphasis on a strong state and top-down control was compelling for elites who sought to hold together newly created nations with artificial boundaries and invented pasts, which were often deeply fragmented by ethnic and religious rivalries, class divisions, and periodic eruptions of communal strife. The critical roles played by autocratic rulers and centralized bureaucracies in "late" industrializing countries, such as Germany and Russia, had impressed many nationalist leaders with the need for similar political structures to oversee development in their own nations.[57]

The modernization option promoted by the United States assumed that, with the exception of primary education, the quality-of-life concerns of the "masses" would be satisfied not by state programs but by the mechanisms of a free market, individual enterprise, and competition. The benefits emphasized by American developmentalists, such as freedom of the press, assembly, and speech and opportunities for self-improvement, often had little resonance in societies that lacked historical or cultural prerequisites for representative democracy and were just emerging from the autocratic rule of colonial proconsuls. Democratic processes, particularly elections, often intensified ethnoreligious differences and provoked communal strife. Reforms, even in established democratic societies, tended to be contested and slowly implemented, but the leaders of developing nations usually required swifter remedies for the urgent challenges of rapid population increase, pervasive poverty, and social unrest. In any case, America's ideological commitment to human rights and democracy was often at odds with its propensity to support "Third World" dictators and overlook their human rights violations, especially when their appeals for U.S. assistance stressed the threat that communist subversives posed for their regimes.[58] This convergence of U.S. policy with the Soviets' tolerance for top-down, coercive leadership underscores the need to explore the similarities between two systems that have routinely been treated as implacably opposed.

Like Marx's dialectical version of history and Rostow's stage-sequence counter to it, both modernization theory and command communism were grounded in assumptions of the Enlightenment era. The development theories, policies, and assistance projects devised by both sides were premised on the conviction that there were manifold possibilities for the rational manipulation of both human societies and the natural world. Theorists in each camp insisted upon the importance of empirical inquiry and quantifiable data. With the exception of Maoist hardliners in the 1950s and 1960s, they sanctified what were regarded in the West as scientific approaches to the study of the natural world, which had increasingly been applied to resource extraction and manufacturing, social organization, and even political and military theory. Their scientific fetishism helps to account for the ever greater influence accorded to hyperspecialization and cultures of expertise. Although proponents of both ideological alternatives deplored the exploitation and excesses of European colonialism, both communist and capitalist developmentalism were, like the civilizing mission ideology of the colonialists, profoundly ethnocentric and patterned on the historical experience of western European societies.[59]

American and Soviet ideologues alike sought to propagate scientific epistemologies, procedures, and research systems as culturally transcendent and universally valid. They shared the diffusionist vision of this process held by their nineteenth-century counterparts.

And they constructed comparable scales of evolutionary development, supplanting the primitive-barbarian-civilized triad of the colonialists with variations on the phases of development delineated
by Marx or the modernizers. Developmentalism in both its modernization and communist permutations assumed the necessity for
continuous economic growth and, often implicitly, embraced a
progressivist vision of human evolution premised on the possibility
of unlimited scientific and technological advance. Ideologues on
both sides of the cold war divide were "drunk on speed, efficiency,
[and] the magic of machines."[60] They viewed invention and recurring innovation as essential to achieving the high levels of resource
extraction, agricultural output, and manufacturing productivity that
they considered essential components of successful development
strategies.

In the middle decades of the cold war—roughly the mid-1960s to
the late 1970s—when competition between developmentalist ideologies was a focal point of contests between the superpowers, "almost
no one thought about the connection between development and the
natural environment." The myriad government reports and social
science publications that addressed the challenges of political stability, economic growth, and social welfare in emerging nations rarely
even mentioned the environmental consequences of development
projects. Rostow, for example, did not raise environmental factors as
serious considerations for development planners until 1976.[61] Both
modernizers and communist planners concentrated on deciphering
indexes of growth, transferring technologies, and devising formulas
to increase productivity and raise levels of savings and investment.
The "Newtonian" (or more aptly, the Baconian[62]) world view, to
which both camps fully subscribed, prompted theorists and project
managers to treat nature as a constant in their equations for development. They were convinced that societies with modern science and
technology could always find ways to extract sufficient natural re-

sources and, if necessary, could reconfigure whole ecosystems to satisfy aspirations for material improvement and higher standards of living. This confidence was buttressed by the superpowers' possession of vast, resource-rich frontier regions in the American West and Soviet Siberia. Not only had these endowments proved critical to each nation's rise to global power, but as secure power bases, they undergirded structures of accommodation designed to avert open conflict between the cold war adversaries.

From the time of the colonization of the Philippines, American proconsuls and technical consultants had foreshadowed the modernizers' determination to inculcate "backward" nonwestern cultures with "the basic tenets implied by science—that man is the master of his own fate, that his environment is controllable."[63] Nurtured through centuries of U.S. expansion into seemingly limitless frontier lands, these assumptions contributed to the profligate exploitation of the natural resources and the severe degradation of the air, land, and water in much of North America. As Gifford Pinchot, a luminary of the early American environmentalist movement, admitted at the turn of the nineteenth century, conservation "stands for development . . . of the resources with which this country is so abundantly blessed." And the exploitation of that bounty was deemed essential to sustain the ever-rising levels of consumption that most of the American citizenry had come to regard as an entitlement. Pinchot's conservationism and the more aesthetically oriented preservationist approach to environmentalism were supported by progressive politicians in both major parties from the end of the nineteenth century through the Great Depression. But neither strand of environmentalist concerns was very well suited to the rather different resource, social, and demographic challenges that confronted American consultants and indigenous planners working in the developing world before and after 1945.[64]

Many emerging nations began the quest for development with large and dense populations, and in all postcolonial societies the rate of population increase was high and accelerating. Most citizens of newly decolonized societies were poor, illiterate, and rural, struggling to eke out a living on lands severely depleted by millennia of deforestation, mineral extraction, and intensive cultivation. In these circumstances conserving natural resources that were vital to sustaining the population was difficult, if not impossible. And often these resources were the only products that developing nations could export to industrial nations in order to pay for the machinery, consultants, and manufactured goods they needed to industrialize or to achieve sustained economic growth.

American-style preservationism was even less appropriate than conservationism for most emerging nations. Reserves set aside exclusively for wildlife were misplaced in societies where forests and savannas had long been home to hunting and gathering peoples, shifting agriculturalists, and pastoralists. Not only were these groups dependent for their survival on food and other raw materials drawn from potential reserve areas, but efforts to protect feral animal species intensified those animals' competition for pasture lands with the domesticated herds of farmers and pastoralists. When appeals to establish animal reserves were coupled, from the 1970s onward, with requests to refrain from developing wilderness areas, especially rain forests, to avoid exacerbating global environmental problems, leaders and activists in the emerging nations came to regard the American ecological agenda as myopic and blatantly self-referential, if not hypocritical.[65]

In the decades between the Progressive Era and the Second World War, severe depression and global conflict focused American national attention on employment issues, more efficient methods of resource extraction, and ways of raising productivity in agriculture and manufacturing. Despite a number of landmark legal and legislative victories won by preservationists in the 1950s and 1960s,[66] the full ascent to Rostow's stage of "high mass consumption," coupled with the cold war arms buildup, effectively marginalized environmental issues in public discourse. Rachel Carson's exposé *Silent Spring* (1962) proved the catalyst for a more broadly conceived and widely supported environmentalism than the progressives had inspired. But her book appeared just as modernization theory was exerting its greatest influence, and widespread environmental activism and popular awareness did not fully surface until the end of the decade, when they coalesced around the demonstrations, teach-ins, and clean-ups associated with the first Earth Day in April 1970.[67] Although the connections between modernization theorists and the policymakers who fashioned the debacle in Vietnam were perhaps more critical, it was not coincidental that as environmentalist consciousness mounted in the late 1960s, modernization theory came increasingly under critical scrutiny and gradually fell from favor.

Even after pollution and sustainability became national concerns, the "ecology of affluence" that preoccupied American environmentalists often had little more relevance for developing nations than preservationism. As John Kenneth Galbraith trenchantly observed in 1958, the conservationists' concern with managing nature to ensure that Americans would be able to sate their appetite for raw materials begged the more fundamental question: "What of the appetite itself . . . if it continues on its geometric course, will it not one day have to be restrained?"[68] Quality-of-life and aesthetic concerns, epitomized by Lady Bird Johnson's highway beautification campaigns, were luxury issues for most of the rest of humanity. Efforts to curb pollution held little appeal for planners in developing countries whose main priority was boosting production to keep up with alarming population growth. Environmental activists in Nigeria, India, or Mexico were concerned with finding ways to pay for the treatment of sewage, which was dumped raw into rivers and seas over much of the postcolonial world, or to reduce the emission of highly toxic chemicals by multinational factories that would not have begun to meet environmental standards in the West or Japan. But most of the antipollution technologies generated by the "new" environmentalism were far too costly to be deployed in financially strapped developing nations.

If the United States was not a suitable environmental model for emerging nations, development strategies modeled on those of command communist regimes were formulas for ecological disaster.[69] In pursuit of what Trotsky proclaimed the "proper goal of communism . . . the domination of nature by technology, and the domination of technology by planning," the Warsaw Pact nations pillaged their agricultural sectors and extracted natural resources with abandon in order to build the massive industrial complexes that were showpieces for communist cold war propagandists.[70] Although in the early 1920s prominent Soviet scientists were able to voice concern about these trends, Stalin's rise to power, and the consequent demise of any remaining openings for dissent within the Bolshevik hierarchy, silenced those who sought to buffer the environmental impact of successive five-year plans. Anxious to curry favor with Stalin and other power brokers, quota-conscious commissars ruthlessly prodded the workers and peasants to complete grandiose state-sponsored projects, regardless of their effects on the Soviet Union's air, land, and water.

As in the U.S. version of developmentalism, science was enlisted to validate these initiatives, which were pursued at breakneck speed. But the science invoked was often the bogus, regime-pandering strain made infamous by the geneticist Trofim Lysenko. Under Stalin, scientists were reduced to serfs of the state, whose duty was to facilitate material progress, the all-encompassing measure of success for the command communist approach to development. And the statistics cited to verify the heroic victories of the Soviet people in their struggle to build a modern industrial society were as likely to be fabricated, in critical instances by Stalin himself,[71] as the lengthy honor roll of inventors the regime credited with all manner of modern technological breakthroughs. Stalin and his successors passed off economic failures and environmental disasters, including Khrushchev's ill-advised schemes to extend grain production into the arid southern steppes, as triumphs of socialist planning. In reality, the Soviet path to modernity was clogged with slag heaps and industrial waste, blighted by chemical pollutants, and imperiled by slipshod construction of everything from high-rise housing to nuclear reactors. The persisting toll of decades of state-directed bludgeoning of the environment has been recently manifested in the myriad health problems and declining life expectancy of peoples throughout Eastern Europe and the former Soviet Union.

The communist regime that took power in mainland China in 1949 initially relied heavily on Russian advisors and Soviet-style planning in its quest for economic growth. But as Sino-Soviet relations deteriorated in the early 1960s, the Chinese strove to pioneer an alternative approach to development, which they proclaimed was best suited to other postcolonial societies. Mao Zedong and his supporters fashioned a modified version of command communist development that turned out to be even more environmentally devastating than the one the Soviets had pursued. Because China lacked the abundant natural resources of the United States and the USSR, and because it began the drive to "take-off" with a much larger population that was growing at alarming rates, the Maoist model was far more labor intensive and conversely less dependent on capital inputs, both monetary and machine. But both within China and when exported to other countries, it shared the Soviet and American contempt for indigenous methods of production and low-tech solutions as well as the mistaken assumption that local ecologies were both highly malleable and interchangeable.

Soon after they captured power, the Maoists imposed a succession of calamitous development experiments on Chinese society. Deliberately flouting constraints that were deeply embedded in ancient Chinese philosophical systems, particularly Taoism and Buddhism, communist planners exhorted the workers and peasants to subdue the natural world and refashion it according to wildly unrealistic development agendas. Mao's invocation of the example of the legendary Monkey King, who defied the laws of nature, became a mantra for the masses waving their "little red books." Draconian measures were taken to force the pace of industrialization, both to hasten the arrival of the socialist utopia and to strengthen China against external enemies, real and imagined. From the Great Leap Forward of the late 1950s through the Cultural Revolution of the mid 1960s, Maoist development projects devastated China's already depleted forests, fouled its river systems, exacerbated already high levels of pollution in urban areas, and so disrupted agrarian production that at the end of the 1950s tens of millions of peasants perished in famines that may have been the deadliest in all human history.[72]

❖   ❖   ❖

In the early cold war decades both the command communist and market capitalist camps tended to favor development assistance that promoted industrialization and large-scale infrastructural projects, including hydroelectric dams and transportation systems.[73] Both sides accorded a much lower priority to more modest and diffuse improvements in the agrarian sector, despite the fact that the overwhelming majority of the population of developing societies lived in rural areas. Some development experts, including Rostow, sought to correct for this imbalance by pointing out the many ways in which the success of industrialization depended on resources and markets located in rural sectors.[74] But even when assistance was earmarked for agriculture, it was often used to promote extensive, mechanized crop production, though programs sponsored by China were often organized around the mass mobilization of rural laborers. In agrarian programs backed by communist donors, collectivization and state management predominated; in those underwritten by the United States or its Japanese and European allies, production tended to be oriented toward cash rather than subsistence crops, and corporate interests often figured prominently.[75]

This urban, industrial bias of American, Soviet, and—after 1949—Maoist developmentalism was linked to a mistrust of, even an antipathy to, the peasantry, which was seen as a repository of tradition and a potential obstacle to the ascent to modernity or the realization of the socialist utopia. Many of the qualities associated with the peasantry by theorists and planners in both cold war camps echoed earlier colonial (and, in the case of Marx, anticolonial) discourses.[76] Peasants were dismissed as lazy, fatalistic, passive, and hopelessly provincial. They were seen to be submissive to the forces of nature rather than striving to dominate them, suspicious of new technologies or modes of organization, hostile to influences from outside their communities, and highly resistant to change.[77]

With rare exceptions, American modernization theorists viewed the transformation of tradition-bound, subsistence-oriented peasants into enterprising farmers, producing primarily for the market, as one of the central goals of the development process. Despite a half-century of governance in largely peasant societies in the Philippines and the Caribbean, U.S. policymakers continued to assume that, given the proper incentives, the interests and responses of peasants in developing societies would (or ought to) resemble those of farmers in the United States. And American farmers were idealized in ways that rendered them the very antithesis of the submissive, reactionary stereotype of the peasantry. At times in defiance of actual American experience, farmers were celebrated for their fierce spirit of independence, profit- and market-orientation, and pursuit of material and social advancement. The modernizers' determination to turn peasants in developing societies into farmers, and their conviction that this metamorphosis was not only a critical precondition for take-off but inevitable once infrastructural improvements had been introduced, would prove as misguided as later romanticizations of an imagined, proactive revolutionary peasantry that came into vogue in the decade of America's interventions in Vietnam.

Soviet planners, in keeping with orthodox Marxist thinking, also viewed peasants as tradition-bound and reactionary. But as Stalin's genocidal assaults on the kulaks of Ukraine and enterprising landholders elsewhere made clear, peasants were not to be transmuted into market-oriented farmers. From the first campaigns to collectivize agricultural production in the late 1920s, Soviet bureaucrats treated the peasantry, and the rural sector of the Soviet empire more generally, as little more than a reservoir of cheap labor and surplus agrarian production. Both were considered essential for rapid industrialization, the all-consuming priority of the Stalinist regime. The brutal incorporation of the peasantry into the Soviet variant of command communism in the 1930s completed the revolutionary process that had begun with the Bolshevik seizure of power in 1917. Through much of the decade after that coup, effective communist control was limited mainly to the cities and the working classes, particularly those of central Russia. Over much of the rest of the former tsarist empire, the peasants had made their own revolution, forcibly throwing off the control of the landed classes and dividing up their estates. Lenin's grudging introduction of the New Economic Policy in the early 1920s, which in effect conceded economic control in rural areas to market-minded peasants like the kulaks, left most of the peasantry uncaptured and much of the Bolsheviks' Marxist agenda unrealized.[78]

Despite persisting peasant resistance in the USSR, in the 1920s and 1930s the Comintern's propaganda mills played up the revolutionary potential of the agrarian classes in colonized areas.[79] But in the early decades of the cold war, Soviet developmentalists, like their American counterparts, pushed for dams and steel mills in emerging societies and factored the rural sector into their equations mainly as a source of labor, food, and capital for urban industrialization. Nowhere was this emphasis more apparent than in the People's Republic of China during the 1950s, when five-year plans and mega-projects were de rigueur and Soviet experts ubiquitous, and when the drive for industrialization eclipsed the regime's revolutionary emphasis on the "Yan'an Way" to the reconstruction of peasant society.[80] The development rhetoric the Maoists promulgated to win allies and client states in postcolonial nations stressed popular participation and labor-intensive projects. But much of the development assistance they actually proffered in the 1950s and 1960s was directed toward large-scale, highly centralized, state-directed projects resembling those the Maoist true believers were imposing within China. Although some Chinese aid to Africa was devoted to processing and pharmaceutical plants, much of it went to mega-projects, such as dams and steel mills. The largest foreign development project of the Maoist era, the Tanzania-Zambia Railway, consumed more than 20 percent of China's aid to Africa and 10 percent of its total assistance to noncommunist nations.[81]

In Tanzania, Ghana, and other aspiring socialist states of sub-Saharan Africa, the Chinese model was eagerly emulated, often in defiance of local conditions that rendered it even less productive and more environmentally harmful than it was in China itself. Maoist-style collectivization in Tanzania, for example, led to the disintegration and often the physical destruction of rural communities that had endured for generations. They were replaced by ill-designed, sterile "planned" villages, where uprooted populations were forcibly settled. Chinese-inspired communal farming operations, imposed on such villages by zealous Tanzanian bureaucrats, failed to meet unrealistic production quotas. Like their Soviet and Maoist counterparts, African functionaries cooked the totals they reported to the government and paid little heed to the environmental degradation caused by ill-considered schemes to bring marginal lands into cultivation or build showcase dams.[82]

Although the rhetoric of command communist regimes and the legal codes of newly independent African and Asian nations claimed women's liberation as a major priority, communist and modernization ideologies shared a strong bias toward men as the agents and main beneficiaries of development. Development experts, whether communist or capitalist, were, with such rare exceptions as Barbara Ward, male, and they overwhelmingly lumped women in postcolonial societies with the peasantry as repositories of tradition and potential obstacles to their transformative schemes. Although the competing paradigms of development were gendered in rather different ways, both subordinated women's issues and improvement to those of men.

Nowhere was the gender imbalance more pronounced than in modernization theory and development programs promoting competitive market capitalism. The social scientists who churned out position papers on the modernizing process in the 1950s and 1960s were virtually all American men. And though the gauges of modernity they privileged varied somewhat, most of the influential works in this field stressed the capacity of individual *men* to master innovative industrial technologies and adopt a progressive world view. As its title made clear, for example, the much-cited *Industrialism and Industrial Man* (1960), by a team of male scholars led by Clark Kerr, focused almost wholly on male managers and factory workers, mentioning women only in a couple of generalized asides. Two years earlier Daniel Lerner had launched his academic career with a study of the impact of communications systems on the process of modernization in six Islamic countries. His ostensibly representative sample of informants at different stages along the path to modernity, as measured by the subjects' exposure and receptivity to "modern" media, consisted only of men. In fact, Lerner concluded that the degree to which male informants had moved beyond the home and the family was critical to their ability to navigate the passage from tradition to modernity. A half-decade later Alex Inkeles also sought to identify attributes that were shared by "modern men" in developing countries. Among these, he included their belief in the efficacy of "modern" (or western) science and medicine, and their active commitment to technological advance, which they believed would raise their standards of living.[83]

Drawing heavily on the presuppositions that had informed the civilizing initiatives of European colonizers and American social engineers in the Philippines, East Asia, and the Caribbean, modernization theorists contrasted male and female attributes in terms of elaborate dichotomies that were meant to affirm the superior receptivity of males to various aspects of the process of modernization. Because female informants were rarely—and almost never systematically—canvassed in field research, women were inevitably essentialized on the basis of vague impressions that frequently recycled colonial stereotypes. Researchers who saw themselves as employing the latest social scientific methodologies to establish measures of modernity and definitive parameters for the modernizing process routinely ignored the impact of that process on the lives of women. Those who bothered to address women's issues settled for generalized pronouncements about the benefits women received, despite their backwardness and resistance, from the societal improvements generated by modernization.[84]

The very skewed notions of male and female attributes and capacities arrived at by such research were naturalized and explicitly linked to physiological differences. Qualities that were gendered male were correlated with high receptivity to modernizing. These attributes included a strong sense of individual identity, as reflected in clear recognition of the distinctions between the male respondent and his family and between the self and the natural world. Modern men, or those in transition, were found to be proactive, and to have an affinity for machines, a proclivity for empirical investigation, and an appreciation of the virtues of clock time, discipline, and punctuality. Women, when they were discussed at all, were seen as repositories of "tradition"—modernity's polar opposite—and hence as instinctively conservative and resistant to development initiatives. Modernization theorists assumed (often implicitly) that women were enmeshed in and subject to the forces of nature, including the cyclical rhythms of the seasons and crop production that rendered them oblivious to clock time. Women's capacity for critical inquiry and scientific investigation was thought much more likely than men's to be stunted by religious beliefs, folk customs, and superstitions. Women's responses were assumed to be emotion-ridden and intuitive rather than carefully reasoned. And women's concerns and loyalties were deemed to be centered on the family and local community with little or no interest in or identification with the state, much less the broader international community.[85]

Like those of their colonial predecessors, the modernizers' essentialized notions regarding the orientations and aptitudes of men and women were translated into development policies that privileged males in virtually all respects.[86] In the early 1960s Rostow unabashedly affirmed that modernizing projects were "typically" initiated and directed by "urban coalitions of professional men, soldiers, politicians, civil servants" from within developing societies. These "urban men," he averred, were "reacting against the traditional rural society—or removed from its old orbit—motivated strongly by the desire to see their nations assume a dignified modern stance on the world scene."[87] Rostow might have added that nearly all the American, European, and increasingly Japanese planners and technical experts who went out to postcolonial nations to serve as advisors were men. In most instances, a predisposition to male-directed modernization initiatives was strongly reinforced once they were in the field by both the underlying presuppositions of developmentalism and the patriarchal mind-set of the national and local leaders with whom they collaborated.

None of the competing economic formulas associated with modernization factored household labor or even women's contributions to subsistence farming into computations of productive assets or indexes of economic growth. This failure to count a very substantial proportion of women's work meant that women's skills and productive capacity were greatly undervalued.[88] This tendency is illustrated by William Lederer and Eugene Burdick's novel *The Ugly American*, ironically one of the most scathing cold war critiques of American approaches to development assistance programs. In one semifictionalized vignette, an American woman living in Southeast Asia notices that all the old local women are painfully disfigured by stooped postures. Surmising that their short-handled brooms are responsible for their chronic discomfort, she introduces American-style brooms, which she is confident will straighten their backs. Rather than faulty brooms, however, it is likely that the women's back ailments actually resulted from decades of bending over to plant rice seedlings in nursery beds and later to transplant the young shoots. This misperception on the part of authors notable for their sensitivity to the needs and cultural predispositions of local communities underscores the propensity of American development specialists to relegate women to the domestic sphere (not coincidently in accord with middle-class ideals in the United States in the 1950s and

1960s) and to ignore their contributions to agrarian production, marketing, and wage labor.[89]

Lederer and Burdick's misreading of local conditions suggests some of the ways in which even observers critical of modernization theory reinforced patriarchal systems that were the norm over most of the developing world. Modern men, after all, were expected to engage the world beyond the home. But they also remained in firm control of domestic hierarchies that continued to be ordered by age and gender markers. Development schemes informed by modernization theory, which generally promoted cash-crop and export-oriented production, often diminished opportunities for women outside the home, particularly in societies where they had dominated local marketing or been crucial participants in subsistence agriculture. While men all but monopolized positions in emerging factory workforces, particularly when skilled labor was involved, women were left to compete for poorly paid and physically grueling work in sweatshops or on labor-intensive local construction projects. As in the colonial era, development planners overwhelmingly favored men in the competition for access to education, particularly advanced training in science or technology. Modernization agendas also slighted women in programs as vital as health and nutrition, housing, and legal rights, including property ownership. The frustration, even despair, evoked by this peripheralization of women was captured by a Zambian woman interviewed in the mid-1970s. Exasperated that her efforts to establish self-help networks among her fellow villagers had failed because of a lack of resources, she declared that it "was time the planners paid some attention to the women's needs."[90]

In principle at least, Marxist-inspired approaches to development espoused a fundamental commitment to liberating women from the oppression and exploitation they had suffered in the patriarchal households of traditional, or in Marxist parlance, feudal societies.

From the Bolsheviks' revamping of the institution of marriage to Mao's dictum "Women hold up half the sky," official decrees and legal codes in newly established communist regimes suggested that enhancing women's reproductive rights and integrating them fully into the wage-labor force would be major priorities of state-directed drives to attain modernity. But long before Marxist parties seized power in Russia or China, tensions were evident between measures that would empower women and those that would advance the interests of the working classes.[91] And in postcolonial societies where revolution-minded elites imposed command communist approaches to development, many of the tensions between ideological commitments to women's liberation and class struggle were readily apparent.

Like peasants, women were likely to be reduced by the schemes of revolutionary ideologues to regimented objects rather than proactive agents working for their liberation. Although some women gained access to political power in communist regimes, they seldom rose above middle-range positions and remained a small minority of state bureaucrats. Once drawn into the wage-labor force, they generally received lower wages for performing the same tasks as men. And when they were able to establish themselves in traditionally esteemed professions, such as medicine in the Soviet Union, both the pay and status of these pursuits declined markedly. As in noncommunist developing societies, women's work in the home was not included in growth indexes or in projections relating to material condition and social status. This mode of accounting tended to obscure the double burden that most women bore under command communist regimes: full-time employment plus continuing responsibility for childrearing and housekeeping. And despite high-minded state condemnations of what were characterized as male-chauvinist attitudes and activities deeply rooted in tradition, women continued to be subjected to patriarchal household arrangements that left them under the control of their fathers or spouses with regard to critical life decisions from reproduction to career choice.[92]

❖   ❖   ❖

The subordination or neglect of peasant and women's issues meant that local knowledge systems and techniques of production and extraction of natural resources were, with rare exceptions, ignored by lending agencies and field consultants from both sides of the cold war divide. These dismissive attitudes also characterized the modernizers' and the commissars' views of nonwestern epistemologies— including religious and philosophical systems, which were usually treated as obstacles to development—as well as the time-tested methods of problem solving and even the cultural values of postcolonial societies. American modernizers were determined to overcome the attachment of peoples in targeted societies to what they saw as antiquated techniques and deeply embedded superstitions. From the modernizers' perspective these were simply barriers to the transfer of superior American technologies and scientific approaches to farming.[93] Soviet commissars, who had paid little heed to the local wisdom of their own peoples, were highly unlikely to seriously consider those of African or South Asian peasants or pastoralists in designing projects for developing societies. This contempt for local knowledge owed much to colonial precedents, but agents of both modernizing and communist developmentalism were confident not only that their technologies and techniques were incomparably superior to those of postcolonial peoples but that they could implement on a global scale the ascent to modernity that had eluded their colonial predecessors.[94]

Ironically, what may have been the most beneficial of all the cold war development programs, the "Green Revolution," provides some of the best examples of the developmentalists' marginalization of grassroots concerns, local knowledge, and groups such as women and poor peasants. Spanning decades and genuinely global, the quest to develop high-yielding varieties (HYVs) of key staple food crops has been by far the largest assistance program targeting the agrarian sectors of emerging societies. Initiated in Mexico in late 1944 and financed in its early stages by the Rockefeller Foundation, the Green Revolution was also driven by its proponents' sense of the urgent need to expand food production to match the unprecedented and accelerating population growth in developing nations. However problematic this reading of the causes and magnitude of the Malthusian challenges facing humanity in the early cold war decades, when there were much-publicized fears of imminent global famine, a clear priority for Norman Borlaug and others who pioneered research on HYVs was to alleviate hunger and avert starvation for the developing world's population.[95]

Measured in terms of increases in staple grains productivity, particularly over much of Asia and Latin America, the Green Revolution was a great success. India, Indonesia, and other densely populated developing countries, which had been heavily dependent upon imported grain, declared self-sufficiency within decades of the introduction of the HYVs, and more recently China has made great strides toward this end. These successes won substantial U.S. government funding, garnered international recognition for Borlaug and others associated with the project, and turned research on and dissemination of HYVs into a high-profile initiative of the United Nations through the 1970s and early 1980s. The Green Revolution was hailed as a panacea for everything from the West's need to counter the Sputnik challenge to the search for an "epochal [technological] innovation" that would "kick start" the process of modernization in developing economies.[96]

In the years when HYV cultivation techniques were being introduced, much-hyped gains in agricultural productivity masked serious social and environmental drawbacks. To thrive and enhance yields, the hybridized plants required large inputs of fertilizer, water, and pesticides throughout the cropping cycle. The resulting dependence on petroleum-based chemicals proved highly disruptive to the economies of many developing nations when OPEC raised oil prices after the 1973 Yom Kippur war in the Middle East. The cost of fertilizers and pesticides, as well as the irrigation systems needed to supply adequate water, meant that in most areas only the more prosperous farmers could afford to cultivate the new varieties. Substantial landholders and early innovators also were able to take advantage of large-scale production, connections to grain merchants and transport companies, and contacts with government bureaucrats in charge of research stations and credit agencies. In many regions the Green Revolution undermined small landholders who lacked the resources to innovate and compete, many of whom ended up losing their land. It also led to the dissolution of longstanding patron-client bonds between entrepreneurial farmers and tenant families, as the latter were replaced by cheaper landless laborers. And, as has often been the case with the introduction of new farming technologies and techniques, men dominated the cultivation of the new crop varieties, and women's roles in agricultural production were consequently diminished.[97]

Because, particularly in its early stages, HYV research was focused overwhelmingly on increasing the output of marketable rather than subsistence crops, well-to-do and middle-income consumers were often the major beneficiaries of its potential for nutritional improvement. The considerable fossil fuel and technical inputs the new seed varieties required also made recipient countries and regions more dependent on multinational petrochemical and agro-industrial corporations as well as more susceptible to global price fluctuations. Furthermore, chemical fertilizers and pesticides seeped into soils and polluted irrigation and river systems, and it was not clear what long-term effects the new modes of production would have on agricultural lands. As the Green Revolution gained momentum, whole regions that had formerly been patchworks of fields planted in many varieties of rice, wheat, or corn were devoted to a single hybrid variety for the market. The spread of these monocultures, which defy the evolutionary principle that diversity enhances the odds in favor of adaptation, left staple agricultural production in many developing countries vulnerable to decimation by plant diseases. This threat was often countered by excessive use of pesticides, which resulted in still higher costs of production and energy inputs, as well as further damage to the environment.[98]

The environmental degradation that has resulted from the introduction of HYVs has raised troubling questions about the sustainability of this high-tech, fossil-fuel-dependent approach to staple food production. There has also been growing recognition that much of the environmental fallout and social dislocation brought on by the Green Revolution might have been alleviated if its proponents had made serious attempts to take into account local knowledge and customary farming practices.[99] And from the late 1960s these critical issues were compounded by concerns of a rather different order. By then substantial USAID funding had been added to that provided by the Rockefeller and Ford Foundations to support the elaborate research institutes, epitomized by the ultra-modern complex at Los Baños in the Philippines, where the hybrid grain varieties were developed. It was soon clear that U.S. government funding came at a high price. American-backed autocrats, most egregiously the Filipino president Ferdinand Marcos, manipulated the "miraculous" advances of the HYV programs to legitimize their corrupt regimes. In addition, the United States used dwarf rice strains to enhance its leverage in Indochina. In the late 1960s they were introduced into South Vietnam in the hope that they would resuscitate a moribund development campaign. Concurrently, the possibility that

HYVs would be made available to the communist government in the North was raised during negotiations aimed at extricating the United States from a losing war.[100]

The shared assumptions of American, Soviet, and Chinese developmentalists were at best misplaced, and often potentially devastating, for emerging nations, many of which had dense and rapidly growing populations of which the vast majority were impoverished, illiterate peasants. Even the better endowed of the "underdeveloped" nations, such as Brazil and the Congo, were unpromising sites for large-scale, fossil-fuel-guzzling development schemes. Developing nations had only limited domestic markets for whatever industrial products they produced, and they had to compete for export outlets with established industrial areas. Because the prices of primary products declined steadily relative to those of manufactured goods in most of the last half of the twentieth century, most postcolonial societies were compelled to pay a higher price for their development initiatives than nations that had industrialized earlier. Developing societies sold foodstuffs and minerals at bargain—and notoriously unstable—prices to the industrial economies of the West and the Soviet bloc. At the same time, they paid ever higher prices for manufactured goods and technology transfers. The fluctuating prices of primary products also made their economic planning uncertain at best. And very often the resources sold at bargain rates were nonrenewable assets that emerging nations would need for their own growth in the longer term.[101]

Despite the formidable obstacles to the successful transfer of approaches to development worked out in the West and the Soviet Union, the leaders of the newly independent nations of Africa and Asia displayed a decided preference for steel mills and large dams over handicraft revivals and grassroots initiatives. Their commitment to western-style industrialization had been initially inculcated by colonialists who extolled European science and technology as the keys to western global dominance and as empirically verifiable proof of "Aryan" or "Caucasian" superiority.[102] But in contrast to the resistance of African and Asian nationalists to the colonizers' self-appointed mission as civilizers and improvers, the leaders and planners of postcolonial states were drawn to both the capitalist and the communist development ideologies in part because they were explicitly anti-racist. The indigenous elites of the new nations shared the confidence exuded in at least the public pronouncements of American and Soviet diplomats and technical advisors that their citizenry would be able to master the scientific methods and technical skills needed for industrialization. And most assumed that with the ample transfer of scientific knowledge and advanced technologies from industrial nations and their dissemination through the growth of indigenous technical education, development could be accomplished within one or two generations. Few leaders or intellectuals in postcolonial nations would have disputed Rostow's claim that development was essential to achieving national dignity, which could be fully realized only when their societies were on the way to becoming fully modern. And modernity nearly always involved attaining high levels of mechanization and heavy investment in state-sponsored public works projects conceived on a grand scale. Almost without exception, postcolonial African and Asian leaders and planners were determined "to 'catch up with the West' by reproducing Western science-based, industrial economies in their countries."[103]

Jawaharlal Nehru, for example, ruminating in the early 1940s on the high price in terms of state violence and "wholesale regimentation" that the peoples of the Soviet Union had paid to launch their society on the trajectory to modernity, deeply regretted the human suffering involved, but concluded that it had been essential to the breakthrough to "a new order based on peace and co-operation and real freedom for the masses." Although Nehru was convinced that by adopting a milder form of revolutionary socialism, such as that theorized by the British Fabians, India could avoid much of the misery associated with the Stalinist strategy for industrialization, he "was impressed by the reports of the great progress made by the backward regions of Central Asia under the Soviet Regime." He also acknowledged that the "scientific outlook of Marxism" had guided the USSR's drive for modernity, and judged that "in the balance . . . the presence and example of the Soviets was a bright and heartening phenomenon in a dark and dismal world."[104]

The majority of indigenous planners, managers, and engineers who oversaw development projects in the "Third World" were as oblivious to environmental concerns and as dismissive of the potential of traditional modes of production and local knowledge as aid specialists from industrial countries.[105] In postcolonial societies where there was substantial support for alternative approaches to achieving economic sufficiency and social equity, these options were marginalized, and in some instances forcibly suppressed. And without question, the most comprehensive and influential alternative to highly centralized, large-scale, industrially oriented development projects was devised by the single most influential leader in the worldwide struggles for decolonization. Beginning in the years before the First World War, Mohandas Gandhi and his followers not only fashioned a form of civil disobedience that would prove the most effective mode of anticolonial protest over the next half century, but in their ashrams and model villages they also worked out a radical alternative to models of development based on the experience of the West.

Although Gandhi readily acknowledged his debt to the Russian utopian Leo Tolstoy and the American naturalist Henry Thoreau, his prescription for improving living conditions in impoverished and technologically backward societies was deeply rooted in Indian thought and rural community life. Drawing on age-old philosophical precepts and folk knowledge, the Gandhian approach to development blended social, economic, and moral imperatives. Inverting the priorities of both Marxist and modernizing options, it focused on local conditions, grassroots issues, and renewable sources of energy. Gandhi stressed decentralized planning, community input into development decisions, and local consumption of the bulk of agrarian and handicraft production. And despite critics who parodied his alleged aversion to modern innovations, he and his followers envisioned vital roles for constructive technologies ranging from railroads to Singer sewing machines. They also conceded that some political centralization would be required to ensure regional cooperation in emerging nations, and they found a place in their development strategies for selected large-scale industrial facilities, particularly those that would supply electrical power. At the same time, the Gandhian approach favored manual over machine production as appropriate for economies with high unemployment rates and meager capital endowments. It relied on small-scale workshops, handicraft techniques, and time-tested tools wherever possible. Gandhi himself put great value on diversified tasks for laborers (and time to pursue creative pursuits) rather than on the highly specialized division of labor associated with industrialization.[106]

In the decades of decolonization, Gandhi's radical thought represented just one strand of a larger critique of industrialization and mega-project developmentalism. But in contrast to the probing inquiries of such intellectuals as Jacques Ellul and Herbert Marcuse,[107] Gandhi's writings and community revival schemes exemplified activist responses to these hegemonic paradigms and provided detailed and plausible alternatives. Consequently, his localized vision of development inspired grassroots activists across the decolonizing world. It also influenced those who sought to promote

environmental priorities and moderate consumerism in the industrial West. The impact of Gandhi's thinking is evident, for example,
on E. F. Schumacher's *Small Is Beautiful*, which is widely regarded
as the classic exposition of the precepts of the alternative technology
movement. Gandhi's thought, as Schumacher acknowledges, in-
forms the German economist's advocacy of what he calls intermediate machines, localized and labor-intensive production, environmental checks on human artifice, and sustainability.[108] By the mid-
1970s such ideas were occasionally factored into policy debates
about foreign assistance. More rarely they became part of efforts to
stimulate public awareness of the importance of these issues—perhaps most notably in the didactic tales in Lederer and Burdick's
Ugly American, which were meant to demonstrate that hands-on, localized projects and simple technologies could transform the lives of
the peasant majority of the underdeveloped world.[109]

Just as the lessons Lederer and Burdick were intent on populariz-
ing were lost on U.S. policymakers, the grassroots projects advocated
by alternative technology movements were at best marginalized, and
often actively opposed, by American, Soviet, and Maoist developmentalists. Even in India, alternative approaches to development
were shunted aside and denied funding that was often allocated instead to Soviet-inspired five-year plans, big dam projects, and a high-
tech scientific establishment intent on catapulting the country into
the nuclear age. And this quite deliberate reduction of state support
for the Gandhian vision of development to little more than handicraft production and model villages for the tourist trade was overseen by one of Gandhi's closest disciples, Jawaharlal Nehru, and
other leaders whom Gandhi had done much to bring to power.[110]

This is not to suggest that there was something sacrosanct about
Gandhi's proposals. In fact, as his critics have pointed out, Gandhi's
views on economic growth and social change were problematic in
many ways, ranging from his dubious, essentialist views about hu-

man needs to the levels of integration required for viable national states in an age of industrialization and globalization. Given the diversity of the political and sociocultural systems undergoing transformation, neither Gandhi's nor any other single approach to development was or is likely to prove a formula for universal success. But in its respect for local knowledge, serious attention to grassroots concerns, insistence on appropriate technologies, and focus on poverty, the rural sector, and community cohesion, Gandhian thinking has much to contribute to what must inevitably be composite and site-specific project designs. It is also critical to stress that appropriate technology should not be equated with simple or primitive tools. Though fuel-efficient and low-maintenance machines are often best suited to local communities with limited resources and technical training, sophisticated technologies, particularly personal computers, cell phones, and web-based communications networks have been used to great advantage in myriad grassroots projects with aims as diverse as upgrading irrigation systems, tracking market trends for agricultural produce, and promoting women's education and career development.

It is not likely that the Gandhian or other alternative approaches to development would have yielded higher rates of economic growth, at least at the aggregate national level, than the modernization or command communist options that held sway in the early cold war decades. But alternative strategies certainly would have done far less damage to the environment of the developing world, and they would very likely have distributed income increases and improvements in living standards more equitably. Often compounding misguided initiatives from the colonial era, the large-scale, high-tech development schemes favored during the cold war misfired in innumerable ways. Many disrupted—or, in the case of large dams, utterly eradicated—local communities and ecological systems. Others, such as those in Sudanic Africa, led to the breakdown of centuries-old symbiotic interactions between sedentary agriculturalists and nomadic pastoralists. This process resulted in a steep decline in the production of subsistence and fodder crops, which contributed in turn to an acceleration of desertification and ultimately severe droughts and famines. In locales as disparate as sub-Saharan Africa, Indonesia, and the Amazon basin, the introduction of industrial machinery to clear forests or deep-plow tropical farmlands was responsible for devastating flooding and severe soil erosion, leaching, and lateralization over large areas. In many developing countries, peasants were pressured or enticed with consumer rewards to substitute the production of export crops for the cultivation of time-tested subsistence staples. In the short term at least, the shift from subsistence to market production brought higher living standards, especially for landed peasants and mercantile groups. But often the skewed distribution of rising incomes and subsequent market downturns engendered chronic indebtedness, growing landlessness, and widespread poverty amid degraded environments that neither scientific nor technological fixes could begin to restore.[111]

The underlying similarities in the developmentalist ideologies and policies advanced by the rival superpowers in the cold war meant that the options available to leaders and planners in postcolonial societies were much more limited than currently prevailing interpretations of the era, which stress superpower differences and competition, would suggest. Development assistance undoubtedly contributed to infrastructural improvements, economic growth, and higher living standards for some—especially urban and elite—social groups in emerging nations. But the development formulas promoted by both capitalists and communists neglected—or espoused transformations that were contrary to—the interests of most of the people in these nations. The pattern that emerged first under the progressive, improving U.S. colonial regime in the Philippines, and shortly thereafter in the Caribbean and Central America, was generalized in the cold war decades to much of the developing world under the rubric of modernization theory. High-tech mega-projects were substituted for basic social reforms and redistribution of economic assets. At times inadvertently but often quite consciously, diplomats and developmentalists from both sides of the cold war divide colluded with indigenous politicians, experts, bureaucrats, and local power brokers to obfuscate the failure to address endemic inequities.[112] The worsening plight of much of the peasantry and the rapidly growing ranks of the urban poor (many of whom were migrants from depleted rural areas) has been repeatedly confirmed by the widening gap in living standards and mortality rates between these groups and affluent minorities, including middle-class urban dwellers, prosperous landowning households in the countryside, and expatriate minority groups with privileged access to the global economy.

Modernization theorists sought to dispute these trends or minimize them as predictable and short-term consequences of the inevitable lag caused by traditional obstacles to change. But even the more upbeat spokespersons for leading international development agencies, such as the United Nations and the World Bank, were forced to admit that in most emerging nations, infrastructural improvements and the gains of the landowning and middle classes had been more than offset by the meager rise, or actual decline, in the standard of living of substantial majorities of the population.[113] By the mid-1970s it was clear that the accelerated economic growth since the early 1950s had greatly expanded, rather than reduced, gaps in income and standard of living between industrialized and developing nations and between the haves and have-nots within postcolonial societies. According to statistics gathered by the United

Nations, the World Bank, and other international organizations, the per capita income of the richest one-fifth of humanity had increased during that period 67 times more than the meager gains of the poorest one-fifth. By the 1970s the average annual income of the poorest half-billion people on the planet—most of whom lived in developing nations—was roughly $87, which represented a gain of just 73¢ since the beginning of the cold war. Even after the formation of OPEC and the oil embargos of the early 1970s, in 1976 the most prosperous fifth of the world's population, most of them citizens of developed countries, purchased more than 80 percent of its marketed resources. At the same time, the richest 20 percent consumed 17 times more resources per capita than the poorest 20 percent, and per capita energy use in the United States was nearly 450 times that of such poor but populous nations as Bangladesh. Between 1960 and 1980 the per capita wealth ratio between the richest and poorest fifths of humanity rose from 20:1 to 46:1. Within the developing nations themselves, World Bank estimates indicated that between the 1950s and the mid-1970s, half of the aggregate income was controlled by the wealthiest 20 percent of the population, while the poorest 40 percent received only 15 percent of the total. In India, by 1968 the richest 5 percent of the population (25 million) spent more on consumption annually than the 30 percent (150 million) at the low end of the income scale.[114]

Such statistics emphatically demonstrated that the nations that had already achieved high levels of mass consumption, as well as elite minorities in developing societies that had attained comparable levels of affluence, grew steadily richer in a postcolonial order structured overwhelmingly to the advantage of the industrial nations. And despite dramatic growth in the extraction of nonrenewable natural resources and an alarming acceleration of environmental degradation, for the large majority of humanity that was concentrated in the developing world, the "take-off" to the good life was at least as remote in the early 1970s as it had been in 1945. Neither consumer capitalist nor command communist development formulas had provided appropriate or effective approaches to the "great ascent" to earthly utopia envisioned by the ideologues of both the rival systems. And while human misery increased exponentially over much of the globe, the superpowers profligately expended time, vast resources, and much of their scientific and technical expertise on nuclear weaponry and proxy wars to determine which of their systems truly offered (to paraphrase Abraham Lincoln) the last, best hope for humankind.

U.S. Army UH-1 helicopters flying in tight formation near Phuoc Vinh, north of Saigon, South Vietnam, 1965. Associated Press, Valeo Clearance License 3.5721.4626108-94602.

A field of elephant grass weighted with wind, bowing under the stir of a helicopter's blades. The grass dark and servile, bending low, but then rising straight again when the chopper went away.

—TIM O'BRIEN, *The Things They Carried*, 1990

6

MACHINES IN THE VIETNAM QUAGMIRE

The "battle" at the Mekong Delta village of Ap Bac, forty miles southwest of Saigon, encapsulated the mistaken assumptions and arrogant miscalculations that persistently shaped U.S. interventions in Vietnam.[1] Like most of the combat in the Vietnamese wars of liberation from French colonial rule and then from American occupation, the fire fight at Ap Bac developed quite rapidly, lasted only hours rather than days or weeks, and ended abruptly as the guerrillas faded into the countryside at nightfall. The clash was precipitated by intelligence reports that the National Liberation Front (NLF) was making radio transmissions from Tan Thoi, a hamlet just north of Bac. Having determined that Vietcong forces were concentrated in the area, the Army of the Republic of South Vietnam (ARVN) launched an assault on the two hamlets in the early morning hours of January 2, 1963. The original objective of the operation was modest—to capture the NLF transmitter and kill or capture any guerrillas found in the hamlets. But when the first clashes made it clear that the NLF had substantial numbers of soldiers (in fact, between 350 and 400) entrenched in the area, the ARVN commanders, and especially their American advisors, welcomed the rare opportunity to engage the guerrillas in a set-piece battle. Over the course of the day, more than 1,400 ARVN infantrymen, supported by helicopters and fighter-bombers, were ordered into the fight.

Since the late 1950s the outgunned and usually outnumbered NLF forces had relied on ambush-and-withdraw tactics in the civil war for control of South Vietnam. But at Than Thoi and Ap Bac the guerrillas had taken up fixed positions. The American advisors attached to the ARVN units saw this misguided deployment as a chance to destroy the guerrilla units, which would be caught in positions surrounded by open fields that offered little chance of escape. The Americans' anticipation of victory was bolstered by their knowledge that thirteen M-113 armored personnel carriers had been called up to support the infantry. They considered these vehicles precisely the sort of advanced technology that would give ARVN forces the firepower and mobility to fix and destroy rebel forces. The M-113 had been specifically designed for pursuit of guerrillas in a landscape crisscrossed by dikes and irrigation canals and dominated by flooded paddy fields fringed by palm groves and mangrove swamps. The tracked M-113s were engineered to move with equal facility on dirt roads, along soggy embankments, or through mud and water.

Despite the overwhelming superiority of the ARVN forces in numbers and firepower, from the first clashes the guerrillas outfought the government troops. The South Vietnamese infantry displayed little inclination to engage the enemy. Those in the main assault force approaching Ap Bac over the mud-clogged fields to the west took heavy losses in the opening minutes, then crouched for much of the day behind the low dikes near the helicopter landing area. The bulky helicopters that had brought them into battle were mauled by machine-gun, mortar, and automatic weapons fire from the entrenched NLF forces along the tree line. Within minutes five helicopters had crashed into the paddies and most of the others had been hit. Because aerial observers, including the ranking American advisor on the scene, John Paul Vann, had difficulty locating the well-camouflaged NLF positions, early attacks by fighter bombers and Huey attack helicopters destroyed peasant houses, granaries, and livestock but had little effect on the guerrillas. The NLF positions also proved largely impervious to long-distance ARVN artillery fire, which was poorly targeted and, despite repeated calls by American advisors for corrections, overshot the tree-lined embankments through most of the day.

As the infantry literally bogged down in the paddies near the hamlets, Vann and the other Americans counted on the arrival of the M-113s to turn the stalemate into a rout of the NLF forces. But the indecision and politicking of the ARVN commanders and the difficulty of crossing large irrigation canals delayed their entry into the battle until early afternoon. Though furious at the delays, Vann was consoled by the thought that there were still seven hours before the guerrillas could retreat under cover of darkness. Deploying the armored personnel carriers like tanks, with infantry in support, several ARVN commanders, with steady prodding from American advisors, charged toward the NLF lines on the embankment. But the guerrillas, with weaponry inadequate for a fire fight with the armored vehi-

cles, held their ground and inflicted further losses on the soon retreating M-113s and infantry. Subsequent air strikes could not dislodge
the guerrillas, and a much-delayed paratroop assault came only as
the sun was setting.

With darkness, the opportunity to engage and destroy a major
concentration of NLF insurgents was lost. The NLF commanders
oversaw a perfectly executed withdrawal, including most of their
dead and wounded. At day's end, the ARVN had suffered its most
humiliating defeat in more than half a decade of fighting. The NLF
officers and fighters were justly celebrated as heroes in propaganda
posters that quickly appeared throughout the Delta. Adhering to
what had become routine practice, Paul Harkins, the general who
headed the U.S. Military Assistance Command Vietnam (MACV),
declared the loss a victory because the "Vietcong," as the NLF
fighters were known, had abandoned their defensive positions at the
end of the battle.

The "battle" at Ap Bac would have been little more than a skirmish
in the conventional wars that had dominated French and American
military history. But it came at a critical juncture in the U.S. effort to
build a viable noncommunist nation in South Vietnam. By early
1963 it was becoming clear that most of the initiatives the United
States had taken since the mid-1950s to secure the permanent partition of Vietnam were failing. The government headed by America's
hand-picked dictator, Ngo Dinh Diem, was on the verge of collapse
in the face of guerrilla assaults from without and enemies, including
its own military officers, from within. U.S. counterinsurgency strategies to crush the NLF and win the "hearts and minds" of the South
Vietnamese had mostly backfired. In many instances mishandling of
programs for rural reconstruction and economic growth had only in-

creased hostility to the regime. Perhaps even more critically, Diem and his entourage remained unresponsive to American exhortations to undertake political reforms and launch economic projects that would improve the lives of the peasantry and the urban poor. Although U.S. involvement was substantial by early 1963, with more than 11,000 military and secret service personnel assigned to train, monitor, and advise the Saigon regime, Americans had not yet taken direct charge of either counterinsurgency operations or development programs, and the United States was not yet irrevocably committed to participating in the war. Meanwhile, the situation inside Vietnam was complemented by international factors that made for an atmosphere conducive to what was perhaps "the last real opportunity for a negotiated settlement."[2]

In private conversations and meetings, but not in public, President John F. Kennedy had expressed serious reservations about expanding America's commitment to Diem's corrupt and inept government. His misgivings were magnified by reports from CIA operatives, disillusioned American journalists, and a handful of military advisors such as John Paul Vann. Lt. Col. Vann put a distinguished military career at risk after the Ap Bac fiasco by insisting not only that ARVN had lost the battle but that the Saigon regime was losing the war. His refusal to be silenced by his military superiors cost him what had once seemed an inevitable promotion to general and prompted his resignation from the army in the summer of 1963. By the early fall of that year, Kennedy's inner circle was so divided and the news from Saigon so grim that Robert Kennedy wondered aloud whether South Vietnam might be the wrong place to do battle to stop the advance of communism. But precisely because President Kennedy and most of his policymaking team had come to see the struggle for Vietnam as a pivotal confrontation in the cold war, warnings about a debacle in the making were overridden by concerns, both global and domestic, that had little to do with Vietnam.[3]

By early 1963 many of the key trajectories of the cold war contest between the superpowers had converged in Vietnam. The underlying causes and specific decisions that brought Vietnam to the epicenter of the cold war have been subjects of much debate. Nonetheless, the general contours of the escalating American intervention have been delineated in scholarly, popular, and polemical works. Well before the end of the Second World War, the fate of French Indochina had begun to figure (if marginally) in American plans for the postwar global order. Risking the ire of his British and French allies, Franklin Roosevelt deplored the impoverished conditions in the region as evidence of the bankruptcy of old-style European colonialism and the need for the United States to promote economic development and social reform. Faced with the threat of the spread of communism in Europe after the war, Roosevelt's successor, Harry Truman, yielded to warnings from the French that the loss of their empire would increase their vulnerability to communist takeover and approved American backing for French efforts to recolonize Vietnam and the rest of Indochina.[4]

In the following years the French grew steadily more dependent on American funding and military hardware in their efforts to suppress or co-opt the forces of national liberation in Indochina. Favoring weapons transfers over the introduction of manpower, American policymakers resurrected, on a much reduced scale, what amounted to a give-away version of the lend-lease formula. But as the decisive military confrontation between the French and the Vietnamese unfolded at Dien Bien Phu in late 1953 and early 1954, American strategists weighed the costs of sending U.S. troops (or even using tactical nuclear weapons) to bail out the French. Having so recently reconciled themselves to a stalemate in the costly conflict in Korea, Eisenhower and his advisors balked at committing the country to another land war in Asia. Chagrined by the American refusal of rescue, decisively defeated at Dien Bien Phu, and deeply divided at home, the French signed away their Asian empire in Geneva on July 20, 1954.[5]

Although the Americans were ambivalent at best about the French retreat from Indochina, their policy in the region was increasingly driven by cold war calculations. Despite Ho Chi Minh's attempts to enlist U.S. backing for his struggles, and his linking of the Indochinese war for independence with ideas espoused by rebellious colonials in the American Revolution,[6] U.S. policymakers perceived his Viet Minh–dominated coalition of resistance forces as a communist organization. When this perception was confirmed by the Viet Minh's explicitly communist social and economic programs and their efforts to eliminate noncommunist allies, American strategists regarded what had become the mainstream nationalist movement in Vietnam as a hostile and tyrannical force that must be denied power. American support for French efforts in Indochina was the entering wedge of ever more ambitious U.S. interventions, first to defeat the Viet Minh and later to contain the new Democratic Republic of (North) Vietnam that was formally recognized at Geneva. An ominous sign of American determination to block the advance of communism in Southeast Asia was the U.S. refusal to recognize the 1954 peace accords. Even more critical were American schemes to ensure the permanent partition of Vietnam at the 17th parallel, even though that division was explicitly intended by the signatories at Geneva to be temporary. In direct defiance of the provision in the peace accords for free elections within two years for an all-Vietnam government, in the mid- and late 1950s the Americans scrambled to establish a separate, noncommunist nation in South Vietnam.[7]

By the time President Kennedy took office the threat that communist Pathet Lao insurgents would overthrow the monarchical government in Laos was more of a concern for U.S. policymakers than the guerrilla operations in South Vietnam. Post-Korean projections of massed communist forces, whether Laotians supported by the Viet

Minh or Chinese or both, advancing against what the United States recognized as the legitimate government in Vientiane were unacceptable to the cold warriors of Kennedy's inner circle. Until it was (officially at least) neutralized in 1961, Laos was the flashpoint of the superpower standoff in Southeast Asia. In dealing with the crisis in Laos, Kennedy's advisors displayed both the ignorance of local conditions and the overweening faith in technological solutions that would drive U.S. interventions in Vietnam. Proposals were drawn up for the creation of an American-style army to bolster the regime in Vientiane, and some thought was given to development projects intended to immunize the Laotian peasantry against the revolutionary propaganda of the Pathet Lao.

Above all, Kennedy's policymakers, particularly Robert McNamara, counted on sheer military might in the form of massive aerial assaults against the Pathet Lao forces to decide the issue. But tense meetings of the National Security Council (NSC) revealed that none of Kennedy's team had given much thought to what the North Vietnamese or the Chinese might do if the United States suddenly bombed Laos. As in 1954, when American air power had been touted as the salvation of the French at Dien Bien Phu, no one had considered what to do if the bombing failed, though the use of nuclear weapons was again proposed should the Chinese intervene as they had in Korea. Alternatives were not seriously debated because Kennedy and his advisors were certain that American technological superiority would ensure that Laos would not be drawn into the communist camp.[8]

As a "neutralized" Laos was downgraded to a peripheral, but nonetheless casualty-ridden, standoff in the cold war contest, Vietnam emerged as the focus of American determination to halt the advance of communism. And if one did not pay too much attention to its complex political situation or to its ancient history and culture, Vietnam seemed like an ideal, even inevitable, place to draw the line. With its peasantry left impoverished by French colonial rule and its resources drained by an economy controlled largely by French *colons* and Chinese merchants, the southern part of Vietnam, which had been denied to the Viet Minh at Geneva, was an irresistible target for communist insurgency and takeover. As such, it came to be seen by U.S. policymakers as a superb arena in which to test American approaches to development and to demonstrate the advantages of capitalist democracy over communist alternatives. No other contested site in the "Third World" seemed better suited to impress upon the global community America's technological superiority over its communist rivals and thus its incomparable capacity to deliver economic and technical assistance to developing countries.

On the military side of the cold war rivalry, the conflict between the Diem regime and the NLF offered an opportunity for the United States to test strategies of counterinsurgency, which the Viet Minh's success against the French and the proliferation of guerrilla-based peasant wars elsewhere in the postcolonial world had elevated to a major priority. But few American policymakers could have imagined that U.S. involvement in the civil war in Vietnam would grow into a massive commitment to a prolonged and lethal conflict. No other locale in the "Third World" provoked such massive interventions by the major cold war adversaries. For nearly three decades the United States, the Soviet Union, and China backed rival political regimes or rebel forces that were touted as standardbearers for one or another developmentalist ideology as well as exemplars of or counters to guerrilla insurgencies.

The months between the ARVN's loss at Ap Bac in January 1963 and the overthrow of the Diem regime in November proved to be one in a succession of transitional interludes when U.S. options in Vietnam were intensely debated. Although the weight policymakers gave to different factors and rationales shifted over time, each decision for deeper involvement from the late 1940s until the mid-1960s was informed by the calculus of gain or loss in both U.S. domestic politics and cold war competition with the Soviet Union and China. These concerns and the disparity in power between the United States and the Vietnamese communists meant that, at least until the mid-1960s, the reasons for escalating American interventions in Vietnam had more to do with the miscalculations, personal anxieties, and global ambitions of America's leaders than with the impact of French colonialism, the civil war in southern Vietnam, or even the responses of Vietnamese leaders, North or South. Through the gradual buildup to an Americanized and full-scale war in 1965, growing U.S. commitments in Vietnam provided much of the impetus for ever more lethal conflict. This asymmetrical dynamic goes a long way to account for the fact that most of the historical analysis of the causes for the American phase of the Vietnamese wars for liberation has been devoted to decisions reached in Washington and the policymaking processes that shaped them.[9]

However heatedly policymakers debated decisions to prop up floundering regimes in Saigon or to punish the North Vietnamese and the NLF, decisions under five presidents to intensify U.S. engagement in an ever-widening conflict were premised on the conviction that America's unmatched wealth and technological development guaranteed eventual success. As Colonel Harry Summers, who served in the war, observed, no American leader dreamed that a "10th rate" nation like Vietnam could defeat the United States.[10] And even in the mid-1960s, when the NLF's refusal to be cowed by the escalating application of U.S. economic and military power prompted pessimism about the outcome of the war, a can-do confidence in technological fixes served, at least in the short term, to assuage doubts and to counter arguments in favor of a scaling back of American participation.

Among U.S. policymakers at all levels the consensus regarding the decisive disparity between America's power and that of its North Vietnamese and NLF adversaries was so thoroughly internalized that it was rarely factored explicitly into deliberations. Instead, the obvious asymmetry was invoked at critical moments to underscore the reasonableness of further escalation. When, for example, after months of fractious meetings, Lyndon Johnson announced to the NSC in July 1965 that he had decided on major increases in the number of American troops in Vietnam, he assured his advisors: "We can bring the enemy to his knees by using our Strategic Air Command and other air forces—blowing him out of the water tonight. [But] I don't think our citizens would want us to do it." Several years later, as the newly elected President Nixon and his national security advisor, Henry Kissinger, sought to find a way to extricate the country from Vietnam, Kissinger said unilateral withdrawal was unthinkable because the United States could not let itself be defeated by a "third-class Communist peasant state." At another point, Kissinger expressed confidence that a "fourth-rate power like North Vietnam" had a "breaking point" that could be reached if America applied sufficient force. But such blatant expressions of techno-hubris were relatively rare in policymaking circles. They were rendered superfluous by the implicit conviction that the next injection of development assistance, or more typically military might, would stabilize the regime in Saigon or force Hanoi to negotiate an acceptable settlement.[11]

Only at moments of severe crisis or during transitional interludes like the one after Ap Bac did questions about how much manpower and weaponry it would take to achieve American objectives in Vietnam figure prominently in policy deliberations. Even then, the issue was never whether or not the United States *could* reach its goals if it applied its gargantuan resources and power in an open-ended way that might result in the utter devastation of Vietnam.[12] The talisman of technological superiority was ever present. The questions addressed were how much additional U.S. commitment would be needed to force Hanoi to the bargaining table and how much damage each escalation would do to America's international standing, its crusade against communism, or its president's ranking in public opinion polls. By early 1965 the disappointing results of successive and substantial escalations of U.S. involvement in the war had pushed these questions to the top of policymakers' agendas, where they stayed until all pretense of victory in Vietnam was abandoned in the early 1970s.

In the summer of 1965, after months of bombing and troop increases had achieved little, George Ball and (less confrontationally) Clark Clifford disputed the projections of McNamara, Walt Rostow, and others that further increments of high-tech firepower would force the communists to the bargaining table. And they dared to ask if the war was winnable, or at least to question whether the cost of winning would be too high. Ball expressed "serious doubt that an army of westerners [could] fight orientals in Asian jungles and succeed" (presumably whatever their advantages in weaponry) and mused that the greatest humiliation would come from the perception of allies and adversaries that "the mightiest power in the world is unable to defeat a handful of guerrillas." Ball and Clifford's pessimistic assessments were countered by spirited rebuttals from Johnson's other advisors, who stressed the necessity of making the enemy fight on ground of America's choosing in order for its undeniable technological advantages to be fully deployed. If the North Vietnamese responded to U.S. escalation with large injections of troops, General Earl Wheeler asserted, their "greater bodies of men [committed to conventional warfare]" would allow American forces "to cream them." Although Wheeler also insisted that with sufficient forces the United States could defeat the enemy even in the jungles, Henry Cabot Lodge urged that sea and air power were the keys to victory: "We have great seaports in Vietnam. We don't need to fight on the roads. We have the sea. Visualize our meeting [the] VC on our own terms. We don't have to spend all of our time in the jungles." McNamara, Rostow, and others weighed in with more measured support for escalation, each urging that Johnson approve the major increases in American combat forces requested by General William Westmoreland, the U.S. commander in Vietnam. Defenders of escalation provided an array of statistical data to support their conclusion that the troop buildup, in conjunction with steady increments in air strikes against North Vietnam, would *in time* force the enemy to yield to American demands.[13]

The misgivings of the minority of advisors who opposed further escalation were given a lengthy hearing. But in July 1965, as in earlier policy debates under both Kennedy and Johnson, dissenting views stood little chance of significantly affecting policy toward Vietnam because they clashed with the recommendations of McNamara and his systems analysts, who until that juncture were allied with Rostow, Dean Rusk, McGeorge Bundy, and the military chiefs. Ball's assessments were tempered by long experience in foreign affairs, but they were subjective, decidedly unquantifiable, and charged with emotion; thus, from the viewpoint of McNamara and his technoanalysts, they were little more than informed hunches. To counter such largely impressionistic warnings, McNamara, Rostow, and their aides churned out situation reports and position papers based on the reels and reels of computerized data that had been processed on their mainframes. Within months after McNamara and his "whiz kids" moved from restructuring the Ford Motor Company to "rationalizing" the operations of the defense department bureaucracy, they had established an imposing mystique for decisive "scientific management."

Drawing on vast amounts of information, McNamara's newly created Office of Systems Analysis (OSA) performed cost-benefit analyses for tasks as diverse as weapons procurement, streamlining the defense bureaucracy, and responding to the volatile situation in Vietnam. When they argued for widening the war, they prided themselves on using scientific procedures and verifiable (that is, wherever possible, statistical) data, which they believed made their decisions far more objective than the recommendations of the critics of escalation. Only when McNamara began to have serious doubts (which mounted steadily through 1966 and 1967 but which were consistently at odds with his public assessments) about the war did he recognize the limitations inherent in policymaking that was so heavily dependent on quantification and the narrow range of responses the American planners deemed rational.[14]

With the "facts"—usually in the form of a flood of statistics—apparently confirming their take on the situation in Vietnam, the systems analysts overrode the doubts and alternatives offered by even the closest confidants of successive presidents for most of a decade. When lower-level officials, such as Desmond FitzGerald of the CIA, dared to question the certitude of predicting outcomes on the basis of numbers alone, they were frozen out of the policymaking team.[15] In the critical deliberations of July 1965, McNamara was ready with a surfeit of statistics and elaborate computer projections to deflect Ball and Clifford's questions as to whether or not a military victory was possible and what "winning" would mean in view of the fragility of the Saigon regime and the undiminished determination of its enemies to destroy it.[16] To those who championed systems analysis, the possibility of losing in Vietnam, given America's verifiably overwhelming military assets, was counterfactual.

Ensconced within the blinkered realm they themselves had created, the advocates of escalation evinced scant interest in the history, social structure, and cultural proclivities of the peoples of Indochina. Alain Enthoven, recruited by McNamara from the Rand Corporation to head the OSA, could not conceive of a meaningful way to factor historical experience (which after all could not be quantified) into planning for high-tech military forces.[17] Despite Rostow's admonitions to be attentive to the diversity of postcolonial societies, he and other presidential advisors consistently regarded Vietnam as a "laboratory" for testing counterinsurgency strategies, development projects, and innovative weaponry that could, if successful, be deployed elsewhere in the "Third World."[18] McNamara repeatedly made the sort of controlled tours of Vietnam, featuring scripted briefings and staged visits to strategic hamlets, that William Lederer and Eugene Burdick had mocked in *The Ugly American* as exercises in disinformation. And after each progress through the countryside, the secretary of defense departed Saigon confident that he had gained a "good feel" for the situation on the ground. The journalist David Halberstam, who witnessed a number of these charades, characterized McNamara's fact-finding trips: "He epitomized booming American technological success, he scurried around Vietnam, looking for what he wanted to see; and he never saw nor smelled nor felt what was really there, right in front of him. He was so much a prisoner of his own background, so unable, as indeed was the country that sponsored him, to adapt his values and his terms to Vietnamese realities."[19]

In view of the "whiz kids'" much-hyped insistence on "hard" data, McNamara's whirlwind visits to Vietnam provide a revealing contrast to the laborious gathering of information in the Philippines just over a half-century earlier undertaken by the commissions that launched America's foray into direct colonial rule. In the buildup to what in effect was the recolonization of Vietnam in the mid-1960s, none of the many government agencies involved undertook systematic investigations of local social systems, the Saigon regime, or even the history of the Vietnamese civil war. Some CIA operatives did gather information on rural conditions and political rivalries, and by the early 1960s American social scientists had begun to engage in extensive fieldwork in relatively secure areas. But there is little indication that policymakers in Washington or Saigon took their findings seriously into account. On the contrary, the recommendations of CIA agents with in-country experience in Vietnam were consistently contravened by desk-bound systems analysts relying on very different sorts of data from the Kennedy through the Nixon phases of the war.[20]

As the sorry history of American interventions in Vietnam would make clear, social science paradigms and cost-benefit analyses could not compensate for the policymakers' woeful ignorance of the history of the refined and deeply rooted societies and cultures of Indochina. Quantification and abstract modeling obviated serious consideration of historical precedents and enduring socio-cultural patterns that, if they been taken into account, might have significantly altered decisions about U.S. involvement. McNamara admitted decades later: "We clearly lacked the understanding of Vietnamese history and culture that would have prevented us from believing that they would reverse course as a function of being 'punished' by U.S. power."[21] The near obsession of McNamara and other key presidential advisors with computerized data and their consequent tendency to disregard the historical dimensions of Vietnamese responses were quite consistent with the antipathy to "tradition" that was pervasive among the modernization theorists in the Kennedy administration. These predilections produced policymakers who could not meaningfully factor Vietnamese society, culture, or history into their momentous decisions. As Daniel Ellsberg observed in the early 1970s, when America was in full retreat from Vietnam:

> It is fair to say that Americans in office read very few books, and none in French; and that there has never been an of-

ficial of Deputy Assistant Secretary rank or higher (including myself) who could have passed in office a midterm freshman exam in modern Vietnamese history, if such a course existed in this country. (Until recently, there were two tenured professors in America who spoke Vietnamese; now, I believe, there are three.) Is this failure to translate or learn from the French to be seen as the root or the symptom of American officials' near-total ignorance of the "First Indochina War"—or one more phenomenon of their "need not to know"?[22]

American ignorance of Vietnamese history and society was matched by insensitivity or outright hostility to Vietnamese culture, which was apparent at all levels from war councils in the White House to U.S. infantrymen confronting peasants in the countryside of South Vietnam.[23] In the context of the civil war that was so inextricably bound up with the recent past of Vietnamese society and culture, this insensitivity was a fundamental source of the massive debacle for combatants on all sides that expanded with each escalation of U.S. involvement. An appreciation of the historical pattern of fierce Vietnamese resistance to foreign invaders might have resulted in more realistic projections of the human and monetary costs of ever expanding military interventions. In their recurring expectation that the massive resources and firepower of the United States would force the leaders of tiny, poor, underdeveloped North Vietnam to the bargaining table, American policymakers failed to take into account the Vietnamese people's sense of themselves as slayers of powerful dragons, which was as ancient as their resistance against Chinese invasions in the last centuries B.C.E. And in their disregard for history and tradition, U.S. military planners did not heed the insistence of Pham Van Dong and other North Vietnamese leaders that among the great assets the Vietnamese possessed in their unequal struggle against the United States were patience and a capacity to endure hardships when in pursuit of a righteous cause, such as independence from foreign domination.[24]

Recognition of the long history of Vietnamese resistance to Chinese, and later French, political domination might also have revealed the folly of seeing the Viet Minh bid for power as simply an extension of a monolithic communist drive for global dominance, directed from Moscow or Beijing. American planners consistently underestimated the strength of Vietnamese nationalism and the appeal of Ho Chi Minh and the Viet Minh as champions of national liberation.[25] And the failure to comprehend the major role that Vietnamese women had played in resistance to foreign domination since the earliest risings against the Chinese conquerors meant that U.S. policymakers consistently underestimated the contributions women would make in the last phase of the twentieth-century wars of liberation.[26]

Wedded to development strategies dependent on market expansion, U.S. policymakers also overlooked the extent to which most Vietnamese associated capitalism with colonial exploitation. And they did not seriously consider the possibility that by taking the side of the corrupt, inept Saigon regime the United States would become the target of the accumulated hostility engendered by decades of French misrule.[27] In part because the modernizers viewed Buddhism and other non-Christian religions as repositories of stultifying tradition, which modernization programs were intended to supplant, monks and their lay followers were largely excluded from U.S.-backed efforts to fabricate a nation in South Vietnam. As the clashes between government forces and the Buddhists that did so much to bring about the overthrow of Diem indicated,[28] his regime's narrow base of Catholic and landlord support had alienated the great majority of South Vietnamese, who were Buddhists and tenants or landless laborers. And though the NLF often matched

Diem's functionaries in their capacity for brutal reprisals, at least they offered the promise of genuine social reform and improvements in the living conditions of peasants and workers.

McNamara's insistence, in his late 1990s retrospectives, that information on these critical dimensions of the Vietnamese situation was not available in the early 1960s,[29] does not square with the substantial corpus of work that appeared in the decade and a half after the Second World War. Apparently American policymakers assumed there was little useful to be learned from accounts by French officials, scholars, and journalists long associated with Vietnam, such as Paul Mus and Bernard Fall. Works of obvious relevance, such as Mus's *Viet Nam: sociologie d'une guerre* (1952) or the English translation of Fall's *The Viet Minh Regime* (1954), were in part discredited because they were written by Frenchmen, whose country had ignominiously "lost" Vietnam. U.S. policymakers approached the challenge of combating the communist insurgency in Vietnam in much the same frame of mind as that in which their predecessors had tackled the challenge of building the Panama Canal. Drawing upon its superior resources, deploying its vaunted technological capabilities, and proceeding with its can-do determination and a "moral authority" that the French colonizers had lacked, the United States would surely once again succeed where the French had failed.[30] But those formulating U.S. policy appear to have ignored not only French but also British and American academics and journalists with expertise in Southeast Asia and Vietnam.[31] At the very least, these writers' insights into the complex forces at work in Indochina might have complicated the bipolar presuppositions that drove American interventions.

The failure to give serious consideration to Vietnamese history and culture was compounded by the fact that the statistics the systems analysts fed into their computers were often inaccurate and at times utter fabrications. Most notorious were estimates of enemy troop levels, which were consistently too low, and body counts of NLF and later PAVN (People's Army of [North] Vietnam) troops killed in action, which were routinely inflated. Body counts in particular, and the "kill ratios" of US-ARVN to NLF-PAVN casualties extrapolated from them, became the main gauge of progress (or setbacks) in the war for observers as disparate as field commanders in the Mekong Delta, members of the NSC in Washington, journalists in-country, and the American public back home, which tracked the first televised war through the kill ratios reported on the nightly news. Because promotions in both the U.S. and the ARVN military depended heavily on estimates of enemy losses, dead Vietnamese, whether they were in fact guerrilla fighters or civilians caught in the crossfire, were regularly counted among the NLF casualties. Framed by an ethos of "corrupt competitiveness" and "compulsive falsification," body counts were inflated as they passed up the bureaucratic ladder from provincial killing grounds to the data processors in Saigon and Washington. As one American soldier who had been present during the My Lai massacre admitted, if a Vietnamese person was dead "it *had* to be a VC. And of course the corpse couldn't defend itself anyhow." Philip Caputo, who was in charge of counting the enemy dead, maimed, and captured at a series of base camps, recalled that "almost every unarmed male Vietnamese found in enemy-controlled areas . . . ninety percent of [whom] turned out to be innocent civilians," was labeled a Vietcong suspect and interrogated.[32] Although ARVN commanders were usually cautious about admitting to such blatant falsifications, one general made it clear in the early 1960s that Diem's functionaries were quite willing to cook up "statistiques" about everything from NLF casualties to numbers of strategic hamlets to satisfy the Americans' incessant demand for quantifiable data.[33]

As the defeat at Ap Bac amply demonstrated, unreliable statistics and faulty intelligence reports could prove lethal. Although radio intercepts had pinpointed the NLF forces at Ap Bac, their numbers had been underestimated and American planners knew little about the hamlets they occupied or the surrounding countryside. By contrast, thanks to their own intercepts of radio messages, the guerrilla fighters knew that a large-scale ARVN assault was coming, though they did not know its precise target or timing. Of all the misinformation associated with Ap Bac, the greatest harm in the longer term was done by General Harkins's implausible declaration of victory, which maintained the posture of can-do optimism that suffused the progress reports of MACV while Harkins was in charge between 1962 and 1964. The credence Harkins gave to the exercises in obfuscation that passed for statistics among Saigon's bureaucrats, combined with the reluctance of the American ambassador, Frederick Nolting, to speak ill of Diem or his cronies, meant that U.S. policymakers had to rely on dissenting generals, such as John Paul Vann, or on reporters to begin to fathom the disarray in South Vietnam.[34] Although some sources of information improved markedly as the United States steadily took over the war in the mid-1960s, Washington analysts would continue to rely upon inflated statistics and wildly optimistic assessments long after they had abandoned any hope of "winning" the war.[35]

Communist fighters proved adept at circumventing the most sophisticated American surveillance devices, and the strength of their military forces repeatedly confounded estimates of the damage done by U.S. bombing raids or infantry assaults. Perhaps the most embarrassing failure of U.S. intelligence involved the air force's $5 billion Operation Igloo White, designed to detect and destroy enemy supply columns that moved back and forth between North and South Vietnam along the network of jungle pathways known as the Ho Chi Minh Trail. From 1967 to 1972 computers in an imposing command center in northeast Thailand were connected to thousands of miniaturized sensors scattered along the trail where it cut through southern Laos. Disguised as dung or tropical plants, the sensors could detect various traces of human presence, from the smell of urine or truck exhaust to the sounds and motion of convoys. Once the analysts at the command center plotted the location and direction of enemy forces, the coordinates were transmitted to fighter jets to guide saturation bombing of the area identified.

According to official estimates, "cutting edge" surveillance and coordinated strike operations destroyed as much as 90 percent of the supplies transported along the Ho Chi Minh Trail during the years Igloo White was targeting them. But in reality, though some supplies were destroyed, the communists quickly adapted by triggering the sensors with bogus signs of activity and then resuming transport operations after the jets finished dropping their bombs on empty patches of rainforest. Thus the exorbitantly expensive Operation Igloo White had only a negligible effect on the flow of supplies from North Vietnam to the NLF and PAVN forces in the South. It has been estimated that as much as $100,000 was expended for each truck destroyed on the Ho Chi Minh Trail, and that more than 300 aircraft were lost monitoring or attacking enemy convoys. But the highest price was paid by the ARVN and American soldiers taken by surprise in the major offensive, supported by tanks and heavy artillery, launched by the North Vietnamese and NLF in South Vietnam in 1972 after nearly five years of Igloo White operations.[36] Many American infantrymen were embittered by misinformation and lapses in intelligence that exposed their units to ambushes or clashes with unaccountably superior enemy forces. Bad statistics and faulty computer projections had long been lampooned and deeply resented by the grunts who struggled to survive the war the systems analysts had steadily expanded in the confidence that it could not be lost.[37]

Through much of the decade after the 1954 Geneva accords, U.S. efforts in Vietnam were often focused not on military matters but on development projects and economic assistance. The first specialists assigned to the Diem regime were mainly army engineers. The pretext suggested by Maxwell Taylor, who urged the initial introduction of military advisors in 1961, was flood relief.[38] Before the military buildup of the Johnson years, Rostow and like-minded presidential advisors emphasized the importance of Vietnam as a testing ground for modernization programs. For Rostow American-style development was more than just a way to inoculate emerging societies against the "disease" of communism. It was also the key to the American mission of fostering industrialization and democracy in developing nations without disrupting global financial institutions and trade networks. If allowed to fester, he warned, deprivation could give rise to communist solutions hostile to the capitalist world order, and even direct threats to the security of the affluent West and Japan.[39]

Precedents for the mix of economic aid and military support that Rostow and other Kennedy advisors advocated were in place soon after the Second World War. As early as 1947, economic and technical assistance were established as vital components of a broader counterinsurgency strategy devised to suppress a communist guerrilla movement in Greece.[40] The strategy worked out for Greece, which was soon dubbed the Truman Doctrine, was enshrined as a pillar of U.S. cold war policy in Harry Truman's inaugural address in 1949. In "Point Four" of the major ways he proposed to contain the spread of communism, Truman envisioned a mix of economic aid, food relief, commercial expansion, and above all the sharing of scientific and technological expertise with developing nations. He declared that "the old imperialism—exploitation for foreign profit—has no place in [America's] plans," which he was confident would "greatly increase the industrial activity in other nations and . . . raise substantially their standards of living." For Truman, as later for Rostow, America's development mission transcended cold war rivalries. Truman viewed the campaign to eradicate hunger and poverty as both a moral imperative and a security measure for wealthy, technologically advanced societies like the United States. Like Rostow, he believed that the fortunate minority of humanity who enjoyed the relative abundance of life in the industrialized nations could not long be secure in a world peopled mainly by disgruntled "have-nots."[41]

Truman's resuscitation of America's civilizing mission seemed to many observers to be at odds with his informal approval a year later of the NSC's recommendations for a massive arms buildup and a more aggressive posture in the "peripheral" regions of Asia, Africa, and Latin America. And there was good reason to expect that his emphasis on material assistance as a counter to communist expansion would be superseded by the militarized and confrontational approach advocated by the NSC.[42] But, despite the rapid escalation of the nuclear arms race during the Eisenhower administration, there were significant increases in technical assistance and development projects slated for emerging nations. Even the outbreak of the Korean war in 1950 was gradually offset by a growing interest, particularly among academics, in modernization theory as an ideological and programmatic response to underdevelopment and the spread of communism.

Those who championed modernization as the solution to South Vietnam's ills could point to development programs initiated in South Korea beginning in the late 1940s, which in the early 1960s were finally yielding substantial returns in the form of economic growth and improved living standards. But as the ARVN's reverses at Ap Bac made painfully apparent, South Korean analogies for the situation in Vietnam were misleading at best. In South Korea, where assaults by conventional North Korean and Chinese forces had been repulsed by an American-led, U.N.-sanctioned alliance, there had never been a significant popular-based insurgency to match the NLF in South Vietnam. Despite widespread corruption, the dictatorial regime of Syngman Rhee had put large portions of U.S. development assistance to productive use. The Korean "economic miracle" was just getting under way, but foreign investors had begun to recognize the region's potential. By contrast, a decade of Diem's paternalistic, myopic, and family-aggrandizing rule had done little to improve the condition of the South Vietnamese peasantry. If anything, the great majority of the rural population, landless and impoverished, were more exploited by corrupt bureaucrats and rapacious landlords than they had been under French rule. Diem's longstanding opposition to communism and refusal to serve in the French colonial bureaucracy, which his American sponsors saw as sufficient credentials to certify him as a nation builder, meant less to his peasant subjects than his Catholic roots and his failure to tackle meaningful land reform. The alienation of the rural population from his inept, often brutal, functionaries was critical to the spread of the communist insurgency.[43]

In an attempt to manufacture support for Diem, his American backers pumped ever increasing resources into rural development schemes that coalesced into the Strategic Hamlet Program, which Diem and his brother Ngo Dinh Nhu were confident would win back the countryside from the NLF. Resettling much of the peasant population in government-controlled, fortified villages would, its advocates hoped, gradually deprive the insurgents of food, housing, intelligence, recruits, and sanctuary.[44] Supporting rural development programs had the added advantage of lending credibility to claims that U.S. intervention in Vietnam was constructive rather than simply preventive: that the objective was to put South Vietnam securely on the path to modernity. Those making a case for the development side of the U.S. mission in Vietnam resurrected the well-worn colonial imagery of fatalistic, passive peasants who needed to be activated by American-supplied consumer incentives, education, and the promise of a better future.[45]

Despite glowing press reports in its early stages, the Strategic Hamlet Program was patterned after dubious precedents from earlier counterinsurgency campaigns in Southeast Asia and jerry-built on rotting foundations.[46] In many ways, the fortified hamlets of 1962 and 1963 were more ambitious versions of the *agrovilles* that Diem and Nhu had sought to establish in rural locales beginning in 1959. Although the agrovilles were more obviously intended to carve out rural fiefdoms for Diem and especially Nhu, they shared the later program's emphasis on surveillance and control rather than development. Both schemes involved massive, and highly unpopular, displacements of peasants and their resettlement in sterile fortified villages, surrounded by trenches and barbed wire and regulated by often abusive troops. Advocates of the Strategic Hamlet Program argued that only these extreme measures would make it possible to provide security for a rural population that over much of the Mekong Delta lived in small hamlets in which their dwellings were strung out along waterways or ridges.[47]

But many peasants resisted resettlement and had to be forcibly moved while their home villages were burned so they could not return. Payments promised for those who cooperated were rarely made, often having been siphoned off by corrupt officials and local notables. For much of the relocated population, the loss of their communities meant abandoning ancestral graves and village shrines that were vital sources of familial continuity and personal identity and the locus of religious rituals and seasonal celebrations. From the outset, the paucity of young men among the peasants herded into the strategic hamlets was a worrisome indicator. Whether they were already active supporters of the NLF or displaced persons who would soon make willing recruits, these youths regularly infiltrated the fortified settlements to visit their families and often served as the base of support for NLF operations within areas that the Saigon regime designated as "secure."[48]

Living conditions in the fortified villages, which their occupants were more likely to compare to prisons than to sanctuaries, ensured that the program would garner far more rural support for the insurgents than for the Diem regime. Elaborate press controls and public relations campaigns were launched to obfuscate the often miserable conditions endured by peasants crowded into the treeless villages, which one *New York Times* reporter compared to concentration camps. Government troops sought to regulate the residents' movement to and from the fields they cultivated, to enforce strict curfews, and to pry into the most private aspects of their lives. Villagers were coerced into providing poorly paid or unremunerated labor—reminiscent of that demanded by the French colonizers—for the construction of their hamlet's fortifications. Few of the schools, clinics, and other improvements that the program's designers believed essential to winning the loyalty of uprooted communities ever materialized. And when they did they were usually understaffed and poorly maintained. Much of the funding that was supposed to be spent on improving living standards was pilfered by government functionaries and local military commanders. Very often supplies for village projects ended up for sale at inflated prices on the local black market. And the resettlement scheme's material and moral bankruptcy was indicated by the fact that diet foods containing cyclamates, which had been banned in America as cancer-causing, were included among the provisions supplied to the resettled peasants.[49]

By the autumn of 1963 the Strategic Hamlet Program was so obviously failing that the peasants who were being relocated to a settlement named after Ambassador Nolting were not allowed to attend the festivities to inaugurate the village *for security reasons.* As the ease with which NLF guerrillas had moved into Ap Bac made clear, the fortified hamlets had secured little of the strategically vital Mekong Delta. Despite the stress on surveillance in the planning of the strategic hamlets, radio intercepts rather than local informants were responsible for the detection of the NLF forces in the Ap Bac area. Widespread local support for the insurgents was demonstrated by the impunity with which they were able to operate in the countryside and by their superbly executed retreat at nightfall. Their soon to be legendary escape, in which all the wounded were evacuated, did much to bolster the morale of the outgunned guerrillas. And because they managed to carry off all but three of their dead, the NLF units denied the ARVN and the Americans the body counts that had become the critical measure of success in the war.

With the fall of Diem and the ensuing political turmoil in Saigon, the Strategic Hamlet Program fell into disarray, and it was effectively abandoned by the end of 1963. Peasants who had been forced to relocate to the fortified villages burned many of them to the ground before dispersing into the countryside. Others were overrun by guerrilla forces. In the years that followed, American interventions in defense of a succession of autocratic regimes in Saigon were concentrated not on modernization but on military campaigns against the communist insurgency in the South and increasingly its North Vietnamese supporters.

From the first days of Kennedy's presidency, finding ways to defeat peasant-based guerrilla movements in the developing world was a major priority. Maoist guerrillas in China had been victorious in the late 1940s, despite massive American assistance to the Guomindang. American overestimation of the role of guerrilla warfare in Fidel Castro's rout of the Batista regime in Cuba in the late 1950s, and Khrushchev's declaration in January 1961 that the liberation struggles of the Third World had opened up a new front in the cold war, lent special urgency to the development of strategies for coping with nonnuclear, nonconventional military challenges. Notwithstanding his willingness to win votes in the 1960 election by stirring up fears of a missile gap that made the United States vulnerable to a Soviet nuclear assault, Kennedy was determined to scuttle the defense policy of the Eisenhower years, which was centered upon nuclear retaliation. And no confrontation made this shift more urgent than the Cuban missile crisis of October 1962. The specter of nuclear conflagration raised by the face-off in the Caribbean gave added impetus to measures already undertaken by leaders on both sides to reduce the possibility of direct confrontations that might lead to a global conflict.

Both superpowers remained deeply ambivalent about the scientific and technological capabilities of their rivals and suspicious about the uses to which these were likely to be put. But Soviet achievements had established a global stalemate in strategic terms, ensuring mutual destruction should the superpowers be drawn into a major conflict. Until the early 1960s, global strategic planning and the buildup of nuclear arsenals had been given first priority by both sides. Nuclear stockpiles in both camps continued to grow alarmingly in the following decades, but both superpowers and the People's Republic of China, now asserting itself as an independent force, placed increasing emphasis on regional contests to advance their influence globally. Each deployed a mix of development assistance, military support for client governments or dissident groups, and (more rarely) direct interventions. But often lavish American financial and military support for compliant regimes could not stem the tide of communist insurgent successes in Asia, which had become the epicenter of the cold war rivalries. American policymakers believed that approaches to counterinsurgency worked out in South Vietnam, if successful, could be adapted to defeat communist-led peasant movements elsewhere. When linked to fears of falling dominoes and communist challenges to U.S. credibility, this expectation ensured that Indochina would become the main arena in which modernizers and communists vied—very often violently—to win converts to their competing ideologies.

As John Paul Vann's initial confidence at Ap Bac suggests, the assumption of Washington planners that technological supremacy would prove decisive in Vietnam was shared by U.S. commanders in the field. Vann and other advisors to the ARVN forces had few illusions about the combat effectiveness of the ARVN. Most ARVN soldiers had evinced little sense of loyalty, much less ideological commitment, to the Diem regime. In part because of uncertainty about the reliability of their soldiers, most ARVN officers had developed a strong aversion to combat with guerrilla forces, which was shared by Diem himself. And their reluctance to engage the enemy was intensified by American criticisms of ARVN commanders of units with high casualty rates, which were seen as jeopardizing their chances of promotion.[50] Thus Vann's optimism before Ap Bac had much less to do with the numerical advantage of the government forces than with the impressive weaponry that they were about to deploy against the hitherto elusive guerrillas. Vann went into the battle assuming that the vastly superior firepower of American-supplied artillery, helicopters, and M-113 armored personnel carriers would compensate for the ARVN units' inferiority to the NLF guerrillas in terms of motivation and morale. Vann and other advisors displayed chagrin, and at times outrage, during the battle when ARVN soldiers did not make effective use of this sophisticated weaponry. This reaction suggests that they had expected the ARVN forces to become bolder and more willing to fight once they recognized the advantage in killing power conferred on them by American technology.

Vann's faith in high-tech weaponry was shared in varying degrees by most of the Americans who served in Vietnam. Their respect for the courage, adaptability, and tenacity of NLF guerrillas and North Vietnamese regulars increased substantially as the war expanded, thereby dispelling earlier American notions that Vietnamese men (largely because of their comparatively small size and gentle de-meanor) were lacking in virility and physical courage.[51] But even af-ter the stunning Tet offensive in January 1968, American command-ers and policymakers remained steadfast in the conviction that with sufficient injections of sophisticated weaponry and U.S. forces to wield it, the communist adversaries, however brave or tenacious, could be beaten into submission.

The Americans' can-do confidence in their own technological capabilities was complemented by highly essentialized, and very often racially charged, assessments of their Vietnamese adversar-ies, which often evoked longstanding U.S. images of Asians more generally. As Douglas MacArthur told William Westmoreland when Westmoreland was about to assume command in Vietnam in Janu-ary 1964: "The Oriental is terrified of artillery; if used properly it is the best way to bleed his morale." Westmoreland's predecessor, Harkins, infamously countered a reporter's question about the use of napalm on Vietnamese villages with the quip: "It really puts the fear of God into the Viet Cong . . . And that is what counts." When Da-vid Halberstam, in a *New York Times* article, contested Harkin's mantra that the war was going well by giving a detailed account of the deterioration of ARVN forces and the increase in mobile warfare by NLF units, one of Harkin's aides curtly dismissed the notion that the guerrillas could be mobile: "Mobility means vehicles and air-craft. You have seen the way our Vietnamese units are armed—fifty radios, thirty or forty vehicles and mortars and airplanes. The Viet-cong have no vehicles and no airplanes. How can they be mobile?" Halberstam's rejoinder captured the folly of the military's blind faith in mechanized solutions: "To be sure, the Vietcong did not have mobility by Western technological standards. But they had an Asian ability to filter quietly through the countryside unobserved, to move twenty-five miles a night on foot, or in sampans with excellent local guides, to gather and strike quickly and then disperse before the

Government could retaliate. That was the only kind of mobility they had, but it was a far better kind than the Government's, and too many American generals never understood it."[52]

In his retrospective defense of his badly flawed strategy for winning the war, Westmoreland professed to be a staunch exponent of innovation, who would "exploit any technology that scientists might develop." But his enthusiasm for the use of tanks in the "rice paddies [which] constituted muddy morasses" and the "impenetrable mountainous jungle" of Vietnam belied the visionary image he sought to project. His vastly inflated estimate of the efficacy of tanks was not only at odds with his admission that the main effect of "mechanized cavalry" in these environments would be psychological, it also indicated that he had little understanding of the nature of the insurgency he had been sent to put down. His confidence that massive firepower would terrify the enemy made it apparent that he shared MacArthur's stereotypical view of "the Oriental."[53]

This unyielding reliance on high-tech weaponry made it all but impossible for American commanders like Westmoreland, and their counterparts in the ARVN, to take advantage of opportunities to seriously address the challenges posed by guerrilla warfare. Without question Kennedy and his advisors, having pledged to develop America's counterinsurgency capability, enthusiastically set about the task of mastering the fundamentals of insurgency, as propagated in the works of Mao Zedong and Vo Nguyen Giap and analyzed by American experts on the Vietcong.[54] And yet the United States fought a conventional war against both the NLF and the North Vietnamese. Though modified in some respects in response to the situation on the ground, U.S. tactics and overall strategy were consistent with a longstanding "American way of war"—that is, oriented to offensive operations designed to draw enemy forces into set-piece battles, in which advanced U.S. weaponry was expected to result in decisive victories. American forces in Vietnam enjoyed a staggering

advantage over their communist adversaries in weaponry and material backup, an edge that was made possible by the unprecedented
size and complexity of the logistical support system that equipped
U.S. servicemen (and women) in Vietnam at levels exceeding anything seen in the previous history of warfare.[55]

In the absence of a clear strategy for fighting the war, the application of ever more destructive technologies became the mainstay
of American responses both to guerrilla operations in the South and
to conventional support for the NLF from the North. Before direct
participation by U.S. combat troops became substantial, American
planners emphasized training for conventional warfare and the provision of up-to-date weaponry for the armed forces of South Vietnam. Tactical air support for ARVN forces and helicopters to enhance their mobility were deployed extensively but often not very
effectively. As the number of American soldiers engaged in the conflict rose dramatically in the mid-1960s, air power and ever more sophisticated technologies became the dominant components of U.S.
operations. And American soldiers were deployed in conventional
ways, as epitomized by the "search and destroy" operations that
Westmoreland insisted would destroy the NLF infrastructure. This
very traditional approach to combating the guerrillas' highly innovative tactics, which were continuously reworked by Giap and other
communist military planners, drew American forces into a bloody
war of attrition that favored the deeply committed and patient Vietnamese insurgents.[56]

For over two decades U.S. military strategists failed to understand, or
to admit, that their high-tech, massed firepower conceded the initiative to the enemy and allowed the insurgents to choose the sites
where combat would take place. The Americans' ignorance of Viet-

namese society and geography—and the flaws in U.S. and South Vietnamese intelligence or its inadequate transmission to the soldiers on
the ground[57]—provided abundant opportunities for NLF and North
Vietnamese forces to offset their adversaries' technological advantages. The communist forces' capacity for adaptation was critical to
their success in turning the conflict into a prolonged war of evasion
and small-unit actions. And war on the enemy's terms proved more
and more exasperating to American planners, both military and civilian, who were impatient for clear-cut victories and under increasing pressure to extricate their forces from the Vietnam quagmire.

The very environment in which the NLF forces concentrated
their operations greatly reduced the effectiveness of much U.S. technology. It also rendered American war-machines, such as helicopters
and M-113 armored personnel carriers, highly vulnerable to even
crude weaponry. The rainforests and wetlands provided myriad hiding places for NLF (and later PAVN) encampments, hospitals, ammunition depots, and command centers. Insurgent units in these
areas could be supplied and coordinated from base camps in officially neutral Cambodia, which—until the spring of 1969 when
President Nixon approved secret bombings—was largely untouched
by the devastating technowar the Americans were inflicting on peasant society across the Mekong.[58] Rainforests and swamplands also
served as superb cover for ambush tactics, which the NLF excelled
at executing.

Ensconced within the verdant or rugged terrain of South Vietnam, insurgents could shoot down the slow-moving helicopters that
carried American and ARVN units into battle, supported them in
fire fights, and evacuated the wounded and dead. The noise and
bulk of the machines that were so essential to U.S. operations, and
the less than stealthy movement of American or ARVN infantry patrols, gave guerrillas warning of impending attacks. U.S. and ARVN
troop activities were also monitored by civilian informants, from

peasant cultivators and itinerant merchants, many of them women, to young boys herding water buffalos, to "fun girls" and servants working near or within U.S. military bases. These surveillance networks repeatedly denied U.S. or ARVN attack forces the element of surprise.[59]

Surprise, initiative, and the choice of combat sites almost invariably belonged to the NLF. In the small clashes that accounted for most of the military casualties in the war, NLF or PAVN forces usually fired first and from defensive positions they had prepared in advance. U.S. or ARVN infantrymen normally went into the field with only vague estimates of the strength of the enemy forces and little sense of how they were deployed. The absence of any semblance of a front line, along with the constant movement of guerrilla fighters, made it difficult to bring to bear the superior firepower of U.S. air and artillery support units. It also meant high levels of American and ARVN casualties caused by friendly fire. As a veteran U.S. artilleryman observed bitterly: "You had all this firepower and didn't really have anything to release it on, and you couldn't really cut loose on someone. We were just firing in the dark too much, just shooting off our guns. We didn't know where to shoot or where to go. We had the punch but never really could throw it."[60]

The elusiveness of the enemy was one of the greatest sources of frustration for American commanders and infantrymen. Whenever possible, insurgent units practiced "shoot and scoot" tactics, withdrawing into the forest or countryside before devastating U.S. firepower could be directed against them. As American soldiers were painfully aware, the positions they "captured" from retreating insurgents were theirs only as long as they physically occupied them. Very often they would have to return again and again to recapture patches of territory where fellow soldiers had been maimed or killed in earlier assaults. U.S. and South Vietnamese forces had to defend exposed areas, such as cities and base camps, as well as the lines of communication that were vital to their massive logistical systems, while enemy fighters could ambush them at any instant and from any direction. The search-and-destroy missions favored by Westmoreland simply gave the enemy more opportunities for successful guerrilla attacks. These operations involved sending often inexperienced American units into rainforest terrain that the insurgents knew well and had often supplied with booby traps, underground communications networks, pillboxes and shelters, and avenues for retreat.[61]

Particularly in the more heavily populated areas, NLF fighters blended into the peasant population, on which they depended for shelter, food, intelligence, and the storage of supplies. As Michael Herr observes in *Dispatches*, the guerrillas controlled the night and everything under the ground; the Americans and their ARVN protégés struggled, often in vain, to maintain a semblance of authority in the daylight hours and above ground.[62] Raids into the network of underground tunnels the NLF had established throughout South Vietnam were the most resented and terrifying assignment for U.S. infantrymen. These hand-dug shelters and communications channels extended over thousands of miles, and the most elaborate housed storage depots for food and ammunition as well as hospitals, kitchens, shops for printing and weapons manufacture, and assembly rooms. Like the rainforest, the tunnels were familiar territory for the guerrillas but nightmarish, snake- and spider-infested labyrinths for American grunts. They were well provided with lairs for concealing enemy fighters, exits and dead-end passages, and ingenious booby traps. Aside from carpet bombing by B-52s, which was not usually considered an option in heavily populated areas, the massive counterinsurgency arsenal was rendered impotent by the underground sanctuaries. Once in the tunnels, more rudimentary weapons—pistols, bayonets, hunting knives, and even flare guns— were the key to an infantryman's survival. Cunning, physical agility, or brute strength determined the outcome of hand-to-hand combat in the dark and confined spaces where high-tech weaponry could be more a liability than an asset.[63]

Besides finding ways, as they had in their resistance against the Japanese and then the French, to reduce the advantage conferred by their enemies' technological superiority, the guerrillas kept U.S. and ARVN forces off guard by using ancient Vietnamese defenses, such as sharpened stakes planted in camouflaged pits—some large enough to impale helicopters, others designed to penetrate combat boots and inflict foot injuries. The insurgents also recycled American weaponry, such as the casings of unexploded shells and the rubber treads or armor of gutted M-113s, into mortars, booby traps, pulleys and drive belts for rudimentary machines, and pillbox fortifications. And downing helicopters and ambushing U.S. and ARVN units provided NLF forces with a substantial arsenal of captured weaponry.[64]

However seriously American planners underestimated the NLF's capacity to withstand high-tech assaults, their approach in South Vietnam betrayed an even more basic misconception of the nature of guerrilla warfare. The writings of Mao and Giap may have impressed the Kennedy team with the importance of development programs to win the hearts and minds of the Vietnamese peasantry. But somehow Kennedy's circle, and most of their intelligence analysts, failed to understand that a counterinsurgency strategy relying on massive firepower—delivered mainly by remote helicopters, fighter-bombers, and artillery—was fundamentally counterproductive. The Maoist dictum that guerrilla fighters were as dependent on the peasantry as fish were on water meant that to have a chance at being effective, counterinsurgency had to be oriented to grassroots programs for rural development and peasant security. This realization did provide the rationale for the ill-conceived Strategic Hamlet Program, which achieved neither goal. But it was never meaningfully factored into the military assault against the NLF. As the destructive power of American weaponry mounted with each new influx of U.S. soldiers, counterinsurgency operations often devolved into indiscriminate and brutal assaults on the South Vietnamese peasantry. The "malignant obsession" with body counts of dead Vietnamese provided a strong incentive for the slaughter of civilians. This "grotesque technicalization" not only exposed the American attempt to portray the war as a moral crusade as cynical rhetorical exercise, it rendered the much-hyped American objective of "winning the hearts and minds" of the people of South Vietnam a cruel joke.[65]

As guerrilla resistance persisted (and support from the North intensified), the military-industrial complex in the United States devised an ever-expanding arsenal of anti-personnel weapons. Though ostensibly designed to locate and destroy NLF guerrillas or PAVN units, weapons specifically concocted to shred or scald human flesh were regularly used in assaults on villages providing support for, or deemed to be friendly to, the insurgents. Because most of these devices were dropped from the air or fired from distant bases, peasants who happened to be in the trajectories of bombardments of suspected enemy units frequently bore the brunt of the high-tech killing devices. Such was the fate of the hamlets caught in the middle of the firefight at Ap Bac in January 1963. Both Ap Bac and Tan Thoi were shelled heavily by ARVN artillery, strafed by fighter-bombers, and incinerated by napalm. Whatever their earlier feelings about the NLF, villagers who survived this pummeling but lost their houses, granaries, ancestral shrines, and livestock would almost certainly have emerged deeply hostile to the Saigon government and its American advisors. Napalm, which had been first deployed during the Second World War, was perhaps the most infamous of the anti-personnel weapons used against Ap Bac and Than Thoi. But the hamlets were also bombarded with white phosphorous shells, which scattered volatile acid that burned for days in the flesh of the wounded. As U.S. participation in the conflict escalated, ever more horrific antipersonnel weapons were unleashed over much of South Vietnam, from chemicals and fuel-air explosives to spider mines with nylon trip wires to pineapple bombs, which exploded several feet above the ground and released thousands of steel spikes designed to rip apart any humans and animals within range.[66]

Guerrilla warfare, especially its cadre-based communist permutation of the twentieth century, dissolved the boundaries between civilians and combatants and meshed sociopolitical and military action. After their usually brief and narrowly targeted operations NLF soldiers sought to lose themselves in the peasant population. Villagers—often women or children—supplied and concealed, served as messengers and spies for, and sometimes joined NLF combat units. As in the war against the French, women sometimes commanded NLF units, often ran extensive propaganda campaigns, aided in the transport of supplies, and cared for wounded fighters. As members of the local Women's Liberation Associations organized by the NLF, women were usually responsible for the surveillance and disciplining of village populations. And women as well as children planted bombs in U.S. and ARVN encampments or near bars or brothels frequented by American or government soldiers. Bob Kerrey, who served in Vietnam and later was elected to the U.S. Senate, observed decades after the end of the war: "There are many people [with their names] on the [Vietnam Memorial] wall because they didn't realize a women or a child could be carrying a gun." For men, the opposite assumption applied: in areas controlled by the NLF that were designated Free Fire Zones, Kerrey recalled, it was "standard operating procedure" to regard "any man" as a "target of opportunity."[67]

Not surprisingly, the "promiscuous firepower"[68] that was responsible for the high levels of civilian casualties drove peasants into the communist camp who otherwise would have preferred to remain neutral in the civil war. Although some villagers fled to towns controlled by the Saigon regime, embittered relatives and friends of civilians maimed or killed were more likely to become determined supporters of the NLF. Through posters, vigils for women and children killed in U.S. assaults, and funeral processions along the waterways of the Mekong Delta, local cadres made certain that the "war crimes" of South Vietnamese officials and the American intruders were widely publicized.

As the American assault on rural Vietnam escalated, many peasants found that supporting or joining the NLF greatly enhanced their chances of survival. Collusion with or service in guerrilla units might give them advance warning of attacks on their villages or access to shelter in tunnels once an attack had begun. In the longer term, the NLF proved time and again that it could foil the "assistance" schemes of U.S.-backed regimes that seemed more intent on making war than on improving the living conditions of the people they sought to govern. NLF control meant that landlords, moneylenders, and corrupt officials could not claim the lion's share of the cultivators' produce, demand long hours of unpaid labor, or forcibly recruit their sons into ARVN units. As had been the case in China decades earlier in areas controlled by the Guomindang,[69] peasant youths were more likely to survive as guerrilla fighters than as soldiers forcibly conscripted by the government. Once they were ensconced in the NLF hierarchy, both men and women found opportunities for education and social mobility historically denied to most peasants. ARVN desertion rates were high. And ARVN soldiers, many of them drafted from the cities, were resented for their appropriation of the villagers' limited supplies of food, for their arrogance as overseers of forced labor brigades and resettlement schemes, and for their sexual abuse of women and brutalization of the rural population in general. By contrast, NLF cadres were generally from rural areas, often stationed in their home villages, and severely punished for infractions against peasant supporters.[70]

The lives and homes of peasants were threatened by violent aggression from both sides. NLF methods, which included public executions—at times by beheading—could be brutal in the extreme. And when insurgents were under intense pressure from U.S. and ARVN forces, peasants suspected of siding with the Saigon regime could become the targets of death squads. Not surprisingly, American and Saigon spokesmen publicized these "terrorist" assaults as proof of ruthless, even criminal, nature of the communists. But while the U.S.-ARVN counterinsurgency war depended heavily on massive firepower, communist forces could be more discriminating in targeting adversaries. This meant that the guerrillas were far more in control of the casualties and damage they inflicted on a peasant society that was caught in the middle of a vicious civil war. Whatever the mix of incentives and brute force the NLF employed, it was able to mobilize a substantial portion of the South Vietnamese peasantry to supply and shield guerrillas and to engage actively in violent resistance against the most technologically advanced military force ever assembled to that time.[71]

When high-tech assaults on suspected guerrilla bases and supportive villages proved indecisive at best, U.S. counterinsurgency operations were escalated to a level of warfare that had never before been pursued in such a deliberate, sustained, and routinized manner. Innovative ways of harnessing scientific research and new technology made it possible to wage war on the very environment of South Vietnam, which was so amenable to guerrilla tactics.[72] Research by Dow Chemical and other American corporate giants led to the development of a variety of defoliants and herbicides, of which the most widely employed was agent orange. These chemical concoctions were used to deplete the rainforest cover of areas considered major

NLF sanctuaries. By the late 1960s as much as 25 percent of South Vietnam's highland rainforests, mangrove wetlands in the Mekong Delta, and large swaths of woodland along the Cambodian border had been defoliated. Toxic spraying devastated wildlife, fish and shellfish industries, and Vietnam's hardwood logging industry, which in the French period had become a significant export sector. The slogan adopted by the helicopter crews assigned to spraying operations — "Only we can prevent forests" — was, in its subversion of the virtuous adage of the U.S. forest service's Smokey the Bear, an example of the black humor that American combat troops used to cope with the violation of personal morality that was virtually inescapable for those engaged in counterinsurgency operations. It was also a disturbing reminder of the moral numbing that has invariably been associated with those participating in technowar.

In the more densely populated lowlands, some 5 million acres — or roughly 10 percent of the cultivated area of South Vietnam — were sterilized by repeated chemical spraying. Peasants living in affected areas and soldiers on all sides of the conflict operating there, as well as the U.S. military personnel who carried out the herbicidal campaign, experienced high rates of cancer and contracted respiratory and nervous disorders associated with the chemicals. And such afflictions have persisted through the decades since the end of the war.[73] An additional area the size of Connecticut was rendered incapable of food production by the craterization caused by bombing. The large ponds that formed in the bomb craters in the monsoon season were ideal breeding grounds for malarial mosquitoes. Rats, which thrived in the garbage dumps near American military bases and the fortified hamlets, were the carriers of a rabies epidemic in the late 1960s. A region that had once been one of the great rice-producing and -exporting centers of Asia was importing hundreds of millions of tons of rice by the mid-1960s. Even then the maldistribution of the imported rice, coupled with food shortages in areas where bombing and chemical spraying were concentrated, resulted in widespread brain damage or kwashiorkor among malnourished children.

By 1970 American and U.N. observers estimated that the counterinsurgency war was claiming six civilian casualties for every guerrilla. And of the civilians killed or maimed, four out of ten were children, which meant that a quarter of a million *South* Vietnamese under the age of sixteen had died of combat-related causes during the American phase of the war and over a million had been wounded.[74] All this devastation was being inflicted on the people and the land that U.S. policymakers insisted they were trying to save from the evils of communist domination. The Orwellian doublethink needed to rationalize this obvious contradiction was apparent in a comment by an American air force major after bombing had demolished the town of Ben Tre and caused many civilian casualties: "It became necessary to destroy the town to save it."[75]

In the first years of Lyndon Johnson's presidency after Kennedy's assassination in November 1963, many of the same luminaries who had served in Kennedy's brain trust decided that the success of counterinsurgency efforts in South Vietnam required violent retaliation against the North. They turned to yet another manifestation of presumed American technological supremacy, strategic bombing, to bludgeon the North Vietnamese into relinquishing their aspirations for the revolutionary liberation of all of Vietnam. The aerial assaults on North Vietnam, which began with Operation Flaming Dart in February 1965, only intensified the American reliance on high-tech surveillance and massive firepower. But the bombing campaign marked a decisive shift away from the counterinsurgency experiment that had initially captivated the Kennedy team to a very conventional approach to the war in Vietnam, which suited what was in effect the default combat posture of the U.S. military.

Policymakers' arguments for an air war against the North were driven by the conviction that the NLF insurgency had been instigated and was directed by the communist central committee in Hanoi. These linkages, and longer tendrils of manipulation that allegedly extended from the People's Republic of China and the Soviet Union to Hanoi, were critical components of the rationales that successive U.S. administrations offered for escalating American commitments in Indochina. But they seriously misrepresented the actual political dynamics of NLF resistance to the Saigon regime. North Vietnamese leaders had never given much credence to the promise of the Geneva accords that elections would be held within two years to establish a single government of a united Vietnam. Nonetheless, through the mid-1950s they relied on diplomacy and political pressure rather than insurgency to work toward unification of Vietnam. Diem's brutal repression of dissident groups in the late 1950s led to appeals from leaders in the South, most notably Le Duan, for Hanoi to approve a return to armed struggle. Le Duan's elevation to general secretary of the government in Hanoi in 1960 and the establishment of the National Liberation Front in South Vietnam later that year sealed official backing for the armed revolution against the Diem regime that had actually been in progress for some years.

Although North Vietnamese leaders continued to disagree over timing and strategy, they were confident that the spread of NLF control over the countryside and the weakness of Diem and his successors would pave the way for national unification. But the steady escalation of U.S. military and financial assistance to Saigon from 1960 onward compelled the North Vietnamese to provide the insurgents in the South with food, fuel, arms, and by the mid-1960s, substantial injections of manpower. Through much of this period, the main impetus for resistance to Diem came from dissidents in the South itself, who often had to prod their allies in Hanoi for political and material support. Thus Saigon's brutal suppression of dissent and the American determination to create a permanent anticommunist nation in the South forced the initially reluctant leaders in Hanoi to divert an increasing portion of the North's resources to the insurgency.[76]

Despite the enthusiasm of Kennedy and his advisors for counterinsurgency initiatives, in the early 1960s American responses to guerrilla warfare were little tested. By contrast, strategic bombing, extensively used in the Second World War and in Korea, was regarded as a proven response to enemy aggression. Rostow was particularly persistent in advocating the use of air power against the North Vietnamese. Alarmed by the strength of the NLF-led insurgency at the time of the fall of Diem's regime in October 1963, Rostow and likeminded presidential confidants, including McGeorge Bundy and Robert McNamara, came to view bombing as an essential component of a strategy for containing communism in Indochina. Rostow's faith in air power had been instilled by his service in the air force during the Second World War. The rather equivocal assessment of the effectiveness of the air campaigns against Germany and Japan set forth in the reports of the United States Strategic Bombing Survey apparently did little to unsettle Rostow's conviction that air warfare was the most effective military response available to modern industrial nations.[77]

In a policy paper advancing what became known as the "Rostow Thesis," Rostow argued that a carefully calibrated escalation of precision bombing against North Vietnam would at some (undetermined) time force its leaders to give up the expansionist designs that motivated their support for the insurgency in the South. Rostow predicted that air strikes would greatly reduce North Vietnam's economic and military capacity to support the NLF and would also have a demoralizing effect on the leaders and people, thus forcing the communists to negotiate a settlement. He also contended that they would demonstrate America's resolve in the conflict while building the confidence of the South Vietnamese regimes, whose stability remained a major concern.[78] In the Second World War allied tactical aircraft had been very effective against concentrations of German forces at the front, but there was rarely a front to contest in Vietnam, and PAVN commanders soon learned that concentrating their forces for set-piece battles resulted in appalling casualties. Perhaps most tellingly, the U.S. planners' faith in the psychological impact of massed aerial assault was completely at variance with the responses of the civilian populations of heavily bombed nations, from Britain and Germany to the Soviet Union and Japan. It was to prove misplaced in North Vietnam as well.[79]

From the onset of the air war against the North, CIA analysts argued that adding yet another dimension to America's high-tech conventional war was unlikely to produce a favorable outcome. And as the bombing escalated from the Flaming Dart reprisals to the sustained Rolling Thunder campaign, which began in March 1965 and continued until late October 1968, their contention that strategic bombing would be a costly and largely ineffective way of punishing an underdeveloped country like North Vietnam was resoundingly confirmed.[80] In contrast to the highly urbanized and industrialized societies of Germany and Japan, with their sophisticated transport and communications infrastructures, North Vietnam offered only limited targets for American bombers. U.S. military planners had difficulty identifying substantial numbers of military installations, power plants, or industrial centers whose destruction would seriously impair the communist war effort. Their task was complicated by the fact that most of the available targets were concentrated in relatively populous areas, such as the Hanoi-Haiphong conurbation, where civilian casualties were likely to be heavy. U.S. policymakers feared that extensive destruction in these areas might prod the People's Republic of China to intervene directly in the conflict as it had in Korea.

Even when urban areas and the rather meager infrastructure came under attack as the air war was further escalated in the late 1960s and early 1970s, the North Vietnamese proved adept at devising counter-measures to blunt the destructive power of the advanced weaponry and increasing tonnage deployed against them. Manufacturing and storage facilities for food, fuel, and weapons were dispersed, camou-flaged, and hidden in underground complexes or caves. Much of the North, especially the rural areas, had not yet been electrified, but alternative sources of lighting and power were quickly substi-tuted for areas that had been dependent on bombed-out power sta-tions. Many city dwellers were relocated to the countryside. And hundreds of thousands of men and especially women were orga-nized into labor battalions that repaired roads, bridges, and railway lines or built alternative transport and communications lines when existing facilities were damaged beyond repair. As the main ports in the North came under assault, alternative entrepôts along the border with China, North Vietnam's extensive coastline, and in the Me-kong Delta approaches to Cambodia were developed to maintain the flow of supplies from communist allies.[81]

Although civilian casualties in the North were only a small fraction of those inflicted by aerial assaults in the South, they fed mounting opposition to the war on the part of America's allies, espe-cially in Western Europe, and an increasing proportion of the U.S. citizenry at home. They also provided a major impetus for substan-tial assistance for North Vietnam from China, the Soviet Union, and the countries of the Warsaw Pact. Economic assistance, which in-cluded engineers and in the case of the Chinese tens of thousands of laborers, not only helped offset losses due to American attacks, it made possible significant increases in the GNP of North Vietnam during the first years of the bombing. Military aid, such as the SAM missiles and MiG fighter jets provided by the Russians, enabled the North Vietnamese to exact a heavy toll on the American pilots and planes that carried the war north of the 17th parallel. The number of U.S. aircraft lost over the North increased dramatically. Pilots captured from downed fighters and bombers became a bargaining chip in the intermittent negotiations that were pursued through the final years of the war. And the price of the limited damage the bombing campaign inflicted grew as it escalated. In early 1967 the CIA reported that, *excluding* the cost of aircraft and pilots lost, the amount required to inflict $1.00 worth of damage on North Vietnam had risen from $6.60 to $9.60 in the first year of Rolling Thunder alone.[82]

Despite the 643,000 tons of explosives dropped on North Vietnam by the end of the Rolling Thunder operations, strategic bombing did little to reduce the flow of supplies to NLF and PAVN forces in the South. The communists displayed an unflagging ability to come up with ways to evade American surveillance and military strikes, and to construct alternate routes to those rendered impassable by heavy bombing. Thus, rather than drive the North Vietnamese to the bargaining table, the outcome predicted by Rostow and other presidential advisors, bombing hardened the North Vietnamese leaders' resolve. As in the Second World War and Korea, expanding the bombing to targets that put more civilians at risk had not destroyed North Vietnamese morale but instead had served to rally the population to support what had been to that point an error-prone and demanding communist regime.[83]

Even the most devastating air assaults of Richard Nixon's Linebacker campaigns, which revisionists often stress in their scenarios of how the United States could have won in Vietnam, were of dubious lasting impact. Despite the introduction of laser-guided "smart" weapons and the shift to attacks on heavily populated areas like Hanoi and Haiphong, the Linebacker operations—especially the infamous Christmas bombings in 1972—had precisely the opposite effect of what Nixon and his advisors anticipated. They resulted in sobering losses of American aircraft and pilots, and they further inflamed antiwar sentiment both in the United States and over much of the rest of the world. Contrary to the claims of the "we could have won if only . . ." critics, these last waves of aerial assault were only one of a number of considerations that influenced the North Vietnamese decision to return to the bargaining table. In fact, the main effect of the heightened bombing campaigns was to renew the determination of the leaders in Hanoi to refuse concessions that would pave the way for a rapid, face-saving American retreat from Vietnam. Convinced that Nixon's Vietnamization policy ensured their ultimate victory and intent on sparing their infrastructure and cities further pummeling, the central committee in Hanoi agreed to resume negotiations. But by refusing to negotiate a settlement predicated on the withdrawal of PAVN forces from the South, they made a mockery of the American pretext of turning the war over to the South Vietnamese and of Nixonian rhetoric about salvaging "peace with honor."[84]

Long before the Flaming Dart aerial assaults on North Vietnam were launched in February 1965 and the first U.S. marines sloshed ashore at Danang in March, the violent side of the American crusade to modernize South Vietnam and preserve it from communism had gained the upper hand. In the preceding decade escalating U.S. interventions in Indochina had made for a dramatic resurgence in the centuries-old oscillation between two approaches to civilizing recalcitrant "natives": social engineering projects intended to bring material increase and inculcate western values, and military aggression to punish those who resisted American designs. Not since the wars against the plains Indians on the frontier had the violent side of the civilizing mission so thoroughly eclipsed the impulse to improve nonwestern societies through education, American-style democracy, and the transfer of advanced technologies. In the Philippines the brutality of the initial conquest was gradually superseded by the engineers' imperialism that dominated the early decades of American rule. In Vietnam the sequence was reversed, as a failed project of nation-building gave way to vastly destructive military aggression.

In the early months of 1965, which marked the beginning of both the Americanization of the counterinsurgency war in South Vietnam and the bombing campaign against the North, efforts were under way to redress—at least rhetorically—the imbalance between nation-building initiatives and military responses in Indochina. In a speech in early April, President Johnson sought to mollify growing opposition to the war with a declaration of his renewed commitment to the high-tech, megaproject approach to development. Responding to advisors as diverse as Sir Robert Thompson and Walt Rostow, who insisted that the contest for South Vietnam could not be won if the South Vietnamese peasants did not experience significant improvements in security and living standards, Johnson was anxious to "remind the people that we are doing something besides bombing." The vision he offered to effect that change in perceptions was both grandiose and deeply rooted in his own past. It called for resuscitating the Mekong River development scheme that Johnson had first endorsed as Kennedy's vice president in 1961.[85]

Patterned on the Tennessee Valley Authority, the Mekong River initiative stressed large-scale dams, waterway reconfigurations, and other operations to improve transportation and facilitate electrification across the watershed of the great river. Johnson chose former TVA chairman David Lilienthal as codirector (along with Vu Quoc Thuc, the token representative of South Vietnam) of the Joint Development Group, which was to coordinate implementation of the project. The president believed that New Deal programs enacted to eradicate domestic deprivation could serve as models for American development initiatives overseas. Johnson hoped that his ambitious "Marshall Plan for Asia" would buttress his efforts to preserve his War on Poverty and other Great Society programs at home. As he was acutely aware, his domestic agenda was put at risk by the commitment of American talent and resources to making war in Vietnam. For many Americans engaged in rural improvement programs designed to counter the NLF insurgency in South Vietnam, the appointment of Lilienthal, who had long emphasized the TVA's decentralized administrative structure and grassroots orientation, aroused hope that for the first time U.S. development assistance might take into account local conditions and peasant priorities. Lilienthal was confident that transported to the Mekong, the dam-centric approach he had pioneered in the American south would prove a powerful antidote to communist infiltration throughout Indochina. It is likely that Johnson's pledge to invest billions of dollars in the Mekong scheme was partly a ploy to defuse criticism and partly an expression of his sincere intention to rejuvenate the constructive side of U.S. intervention in Vietnam. But he must have realized that his military escalations left little chance that even a scaled-down version of his Marshall Plan for Asia could actually be funded and implemented.[86]

In the same months when Johnson and his advisors were struggling to cobble together a workable combination of military and developmental responses, John Paul Vann, who had left the military in 1963 after speaking out about the "battle" of Ap Bac, returned to South Vietnam as a civilian working for the Agency for International Development. He found that the situation in the rural Mekong Delta was even worse than before the collapse of the Diem regime. Corrupt, arrogant government officials and ARVN units had further alienated the peasantry. NLF control over much of the area was more blatantly in evidence in villages festooned with communist banners and guerrilla patrols often operating in broad daylight. Even if there had been adequate funding for a TVA-modeled transformation of the Mekong Delta, the government in Saigon had neither the dedicated administrators and engineers nor sufficient control of the countryside to undertake even pilot projects. Vann and other assistance workers were left with the task of building underfunded and poorly supplied clinics, schools, and community centers, which very often became targets of guerrilla raids.[87]

Rural pacification efforts were also undermined by rivalries between civilian and military aid agencies and the absence of a clear command structure. By 1967 the Johnson administration could no longer ignore the need for sweeping reforms. Robert Komer, special deputy assistant to the president, was sent to South Vietnam to impose some semblance of order on rural pacification programs, which he characterized as a "farcical . . . mess." Against considerable bureaucratic opposition, but with strong support from the president and General Westmoreland, in the following months Komer centralized and integrated civilian assistance programs and subordinated them to the military command hierarchy throughout South Vietnam.[88]

The CORDS (Civil Operations and Revolutionary Development Support) agency that Komer created did improve the conditions under which Vann and other aid agents carried out their mission of rural "reconstruction." And there were limited successes in the struggle to win peasant support for the Saigon regime. But the fact that military commanders were in charge of CORDS operations meant that development assistance was oriented even more than before to projects deemed useful to the military campaign. The most blatant instance of the militarization of assistance initiatives was the Phoenix program—designed to improve intelligence gathering and the capacity of U.S. and ARVN forces to infiltrate and eliminate NLF organizations—which was launched in 1968 and sustained through the first Nixon administration. Redirected and insufficiently funded, the modernizing thrust of U.S. intervention was steadily enfeebled and then made irrelevant by the Nixon administration's reliance on the policy of Vietnamization to cover America's retreat from Vietnam.

Postwar critics of the way the United States fought the Vietnam war have argued that measures like CORDS and the Phoenix program offered possibilities for winning the war. But this contention glosses over both the temporary nature of the gains and substantial downside of these stopgap responses to the insurgency. Americans engaged in these operations were encouraged by the expansion of government control in some areas, particularly those where local NLF units had been decimated by the ill-advised shift from guerrilla tactics to conventional assaults during the Tet offensive in early 1968. But PAVN regular forces from the North gradually took over much of the fighting, thereby strengthening Hanoi's control over the struggle in the South. And Tet had seriously undermined public support for the war and revitalized the antiwar movement in the United States.[89]

A number of critics have blamed intensifying antiwar protest in the United States for short-circuiting innovative approaches that they claim would eventually have defeated the insurgency. Drawing their conclusions from contentious readings of problematic data, they argue that potentially fatal damage was inflicted on the NLF by the intelligence inroads made possible by the Phoenix initiatives. But the fact that CORDS and the Phoenix program reached points of diminishing returns long before Saigon was able to reassert effective control over the countryside suggests that the U.S. retreat was much more a product of the ever more costly—but still indecisive—war of attrition being waged in South Vietnam than of the mounting pressure exerted by protest groups at home. Ostentatious corruption on the part of South Vietnamese functionaries involved in the programs produced an epidemic of false accusations, purges of local rivals, and indiscriminate imprisonment, torture, and even execution of peasant suspects. The effects of these pervasive abuses on efforts to garner peasant support only intensified those generated by the continuing high-tech assaults on the land and people of South and North Vietnam. And the Nixon administration's shift to the strategy of Vietnamization exacerbated these trends by making it clear that the communists would be the ultimate victors in the war.[90]

Revisionist critics of the conduct of the war who focus on these programs largely ignore the underlying conditions in the two Vietnams that allowed the insurgency to survive the setbacks of the Tet offensives. This inattention is particularly problematic for pronouncements premised on the notion that U.S. interventions ought to have focused on conventional military operations and left counterinsurgency to the ARVN and the Saigon government.[91] The emphasis placed in these critiques on the military aspects of the Vietnam debacle obscures the crucial role of the nature of the successive Saigon regimes in the demise of America's developmental alternative to communism. The cold war variant of America's civilizing mission that was applied rather haphazardly in Vietnam never meaningfully addressed the political and social issues that were responsible for the civil war in the South. Because support for Diem and his successors came overwhelmingly from landlords in the rural areas and mercantile groups and the military in the cities, land redistribution and other programs for fundamental reform were never seriously considered. Authoritarian and corrupt governments in Saigon rejected or circumvented American suggestions that the maldistribution of land, income, and resources had to be dealt with if the government hoped to win lasting support among the peasantry and laboring classes. Because the NLF insurgents not only claimed to be fighting for revolutionary transformations in the South that would address the injustices inherited from the colonial era, but actually implemented land redistribution programs and other social reforms in areas they controlled, America's client regimes in Saigon stood little chance of gaining support from most of the population.[92]

In the Philippines in the early 1900s, public works projects and educational and democratization programs, rather than military repression, became the focus of American interventions in most regions within two or three years. In Vietnam in the 1960s, by contrast, the war steadily vitiated the modernizing initiatives that were emphasized in the initial stages of U.S. involvement. But even if sufficient development assistance had been available, it is unlikely—given the nature of the South Vietnamese government and the self-serving priorities of leaders from Diem through Nguyen Van Thieu—that large-scale infrastructural improvements, such as those envisioned in the Mekong River scheme, could have been successfully introduced or if introduced would have had much effect on the living standards of the poor. The influence exercised in the Philippines by American officials through their alliances with *compradore* elites, and the patron-client networks that formed the backbone of island society, meant that development projects often brought some material benefits to the mass of the people. In Vietnam internal American control was a good deal weaker and reciprocal links between elite groups and the lower classes were either nonexistent or regularly broken by guerrilla incursions.

The very different perceptions of American engineers held by Filipinos in the early 1900s and South Vietnamese in the 1960s provide a gauge of the sharply contrasting effects of technology transfers and social engineering schemes in the two societies. The visionary, can-do technicians who oversaw the physical and social transformation

of the Philippines evidently bore scant resemblance to their counterparts in Vietnam. By the mid-1960s the South Vietnamese regularly
referred to U.S. army engineers as "The Terrible Ones," according to
the journalist Michael Herr, who notes that even so negative a translation of the Vietnamese phrase did not "approximate [the] odium
carried in the original." Herr recounts a scene that is emblematic of
the debased relationship that had developed between Americans
and the Vietnamese people. He came upon a gang of U.S. engineers
"laughing and shouting" and "gunning" their motorcycles up and
down the steps of "one of the few graceful things left in the country,"
a pagoda that had been built to honor the Vietnamese who had died
in the wars of national liberation.[93]

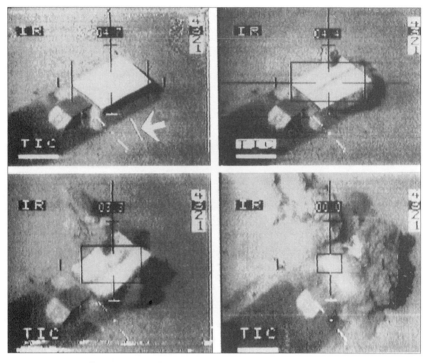

A sequence of targeting and destruction as seen through a sighting device from an air assault vehicle in the 1991 Gulf War exemplifies the video-game images of the war that the American media featured in reporting on the conflict. CameraPress.

Let us have no delusions that advances in smart weapons can make any nation invincible—or that any degree of technology can alter the underlying ghastliness of war.

—GREGG EASTERBROOK, *Newsweek*, February 18, 1991

7

TECHNOWAR IN THE PERSIAN GULF

On the evening of February 28, 1991, the day after a cease-fire was declared in the rout that passed for a war in the Persian Gulf, President George H. W. Bush proclaimed an end to the conflict in a televised address to the nation that was transmitted by satellite across much of the rest of the world. In his assessment, a major obstacle to the realization of his vision of a "new world order" had been removed. Almost from the outset of the Gulf crisis in August 1990, Bush had been determined to use force to reverse Iraq's invasion and annexation of Kuwait. Breaking from a long history of U.S. unilateralist interventionism in regions outside Europe, Bush had assembled a broad international coalition, which included several Arab states, and most of the allied nations had agreed to contribute military forces to the war. With a well-executed diplomatic offensive, Bush and his foreign policy team had won support at the United Nations for military action, brought Iraq's former Russian backers into the coalition, and staved off Israeli threats to take independent action. Proclaiming America the guarantor of Middle Eastern oil sources for the developed world, Bush and his emissaries had convinced Arab, European, and Japanese allies to pay most of the $80 billion cost of the high-tech war. Bush's commanders oversaw superbly planned and executed offensives, beginning with massive air strikes in mid-January, that kept coalition casualties at a minimum. Bush's skillful handling of Saddam Hussein's attempt to establish Iraq as a regional power sent the president's approval ratings soaring above 90 percent.[1] The 1991 Gulf War demonstrated resoundingly that America was the only global superpower left standing after more than four decades of cold war confrontations with communist rivals.[2]

Despite the successes that had brought Bush to his moment of triumph, in the days after the cease-fire announcement the president was incongruously subdued. He remarked that he did not feel the euphoria one might expect after such a resounding victory. Ironically, his most enthusiastic public burst of celebratory pride evoked the memory of the nation's only major defeat in a foreign conflict: at the end of a talk to state legislators on March 2, he abruptly exclaimed, "By God, we've kicked the Vietnam syndrome once and for all!"[3] The president's remark assumed that his preoccupation with the Vietnam debacle was shared by the assembled legislators, and his allusion to a syndrome summoned up more than two decades of controversy about what lessons could be learned from

America's failure in Vietnam and how applicable Vietnam analogies were to subsequent interventions overseas. Before the aerial assault on Iraq began, fears of a recurrence of the high American casualties suffered in Vietnam had been a major source of popular unease at the prospect of the forcible eviction of the Iraqis from Kuwait. Many of the U.S. officers who led the coalition's air and ground offensives had served in Vietnam, and were thus quite aware that the war in the Gulf had little resemblance to that fought in Indochina. Nonetheless, they viewed the rout of the Iraqi forces as a way to restore the American public's esteem for the military, which had fallen precipitously in the aftermath of Vietnam.[4]

In the elaborately orchestrated campaign to win domestic support for war the in the Gulf, Vietnam was not the only historical comparison that the Bush administration deployed. Most of the precedents invoked were misplaced and misleading, but all were charged with memories of earlier threats to international security and the American homeland. Reviving an analogy that had been heavily relied upon during the cold war,[5] Bush's team proclaimed Iraq's aggression in the Gulf a Munich moment, implying that if Iraq were allowed to get away with its aggression as Germany had been allowed to dismantle Czechoslovakia in 1938, it would soon become as great a menace as the Nazis. But comparisons of Iraq to Nazi Germany were wildly misleading. In the size of its population, the education and skill levels of its citizenry, the diversity of its resources, and above all the fact that it was highly industrialized, Germany in the 1930s was a genuine global power. It possessed one of the world's most advanced scientific sectors and a manufacturing base capable of sustaining a war machine that would soon conquer large portions of Europe, the Soviet Union, and North Africa.

In 1990 Iraq had less than one-third the population of 1930s Germany. The Iraqi state that had been artificially demarcated by the great powers after the First World War was deeply divided ethnically and religiously, and the Kurds in the north and the Shi'ites in the south had good reasons to despise the Ba'athist regime, which had seized power in 1968. Although the educational system had improved substantially in the decades before the invasion of Kuwait, only half of Iraq's people were literate, and most of its top scientists had been trained in the West. Its economy was dominated by the production of oil and a rapidly expanding service sector. Industrialization had barely begun, and manufacturing was concentrated in small workshops producing textiles and other consumer goods. Saddam Hussein's military forces were dependent on imports from the Soviet Union, Europe, and the United States for up-to-date weaponry, and even Iraq's oil industry was heavily reliant on technology transfers and scientific expertise from more fully industrialized nations. The steady decline in agricultural production under Ba'ath rule had increased the country's dependence on imports, forcing it to purchase large quantities of food from abroad.[6]

There was little question that Saddam Hussein's armaments programs had made Iraq a growing threat to its Arab neighbors, especially the oil-rich states of the Persian Gulf, as well as to America's close ally, Israel. But as of 1990 the Iraqi military lacked the capacity for sustained offensives. In fact, in the war of 1980–1988 Iraq had barely managed—with a great deal of foreign military assistance (some provided by the United States)—to stave off defeat at the hands of an internally divided, technology-deficient Iran.

If Iraq was not Germany, the equation of Saddam Hussein with Adolf Hitler was equally false. The resemblances between the two had more to do with the generic attributes of absolute dictators than with the level of menace each posed to the world.[7] Nevertheless, the Bush administration's deliberate conflation of Saddam and Hitler were widely persuasive. Bush and his advisors suggested that, like Hitler, Saddam nurtured visions of limitless aggression as well as genocide against ethnic groups like the Kurds. Reports that Saddam's legions were poised to invade Saudi Arabia, which turned out to be false alarms, were used to cajole the Saudis to let the United States turn their kingdom into a staging area for military operations against Iraq.[8] These claims were reported without serious scrutiny by the mainstream American media in spite of evidence, including Russian satellite photos, that the Iraqis were entrenching themselves in Kuwait rather than preparing to attack Saudi Arabia.[9] Not until well after the 1991 Gulf War did military analysts point out that Iraqi forces were clearly already overextended in Kuwait and logistically incapable of the sustained offensives that had made Hitler and the Nazis the masters of Western and Central Europe.

In another comparison with Hitler, Bush administration charges that Saddam Hussein had ordered the gassing of Iraqi Kurds during the war with Iran were offered as proof that a genocidal murderer again threatened humankind. Few in the media attempted to situate the gas attacks in the context of the Iran-Iraq war or to investigate which side had actually gassed the Kurdish village of Halabja in northeast Iraq in March 1988—a question that has been subject to considerable debate, with some compelling evidence suggesting that the Iraqi dictator did not order the attacks. It is noteworthy that the attacks were widely publicized in the United States only after the Iranians had been fought to a stalemate and usually in a manner that left no doubt that Saddam was guilty as accused.[10]

Comparisons with Hitler were part of the effort of Bush and his advisors to personalize the crisis by demonizing Saddam Hussein. By repeatedly insisting that America and its allies had no quarrel with the Iraqi people, Bush made Saddam the sole culprit and the prime target of military action. Focusing on Saddam's iniquity served to obfuscate the very considerable suffering inflicted on the

Iraqi people, first by the massive bombing campaign and later by the international economic sanctions against Iraq, which were extended through the 1990s and into the new century. In his campaign of personal vilification, the president routinely referred to the Iraqi leader as Saddam, which, whether intentionally or not, he pronounced in a way that recalled Sodom, the city of vice destroyed by a wrathful Jehovah in the Old Testament. This not very subtle transmutation of Saddam Hussein's name into a term of ridicule can be seen as one facet of a calculated assault on the dictator's honor, self-esteem, and assiduously cultivated image as a heroic defender of an imagined pan-Arab community.[11] Often in connection with the Hitler analogy, Bush revived the Manichaean metaphor of a cosmic clash between good and evil, which from the time of the Puritans had often provided moral cover for American hostility toward adversaries. The president stressed that Saddam's "brutal aggression" could only be understood in a "black and white" frame of reference, and on a number of occasions he excoriated the dictator's regime as a locus of evil.

The demonization of Saddam Hussein was given a gender twist in the administration's portrayals of the occupation of Kuwait. The tiny sheikdom was in effect feminized as a weak victim, whose defenders had offered little resistance to the invaders (which was in fact not the case) and whose leaders had fled in panic (as in fact they had). The Bush team referred to the Iraqi takeover as the rape of Kuwait—a highly charged metaphor that was also featured in a media blitz, staged by a public relations firm, designed to influence key congressional debates or votes at the United Nations. Both supporters of the use of force and PR consultants added fabricated accounts of Iraqi soldiers dumping babies out of incubators in a Kuwaiti hospital, supposed atrocities that were repeatedly denounced by administration officials. This calculated construction of images of the dictator and his minions summoned up centuries-old orientalist notions of Arab men as devious, cowardly, prone to irrational violence, vengeful, and repressive of women.[12]

As historically flawed as comparisons to Nazi Germany and Hitler were, they have been less of an impediment to our understanding of the causes, nature, and impact of the 1991 Gulf War than the most pervasive analogy associated with that crisis, the debacle in Vietnam.[13] The strong domestic protest against preparations for war in the fall and winter of 1990 made it clear that the Bush administration and its military commanders were not the only ones haunted by Vietnam. Critics of the military buildup evoked images of body bags, demoralized GIs, brutalized civilians, and the humiliating retreat from Saigon to dramatize the perils of intervention in developing countries.[14] Yet, after the crushing victory over Iraqi forces, which was tirelessly contrasted with the failure in Vietnam, many Americans concluded that a formula had at last been found for exerting U.S. hegemony in the developing world. And there was a near consensus that technological innovations were the key to this breakthrough.

Of all the differences between the Vietnam and Persian Gulf conflicts, the most fundamental was that the former occurred at the height of the cold war while the latter came just after the extended superpower stalemate had ended. From the 1950s to the early 1970s policymakers calibrated each escalation of U.S. military operations in Indochina with reference to anticipated responses by the Soviet Union and the People's Republic of China. Particularly during Johnson's presidency, planners sought to avoid actions that might provoke counter-interventions by the Soviets or the Chinese. Successive decisions for escalation were also tempered by the possibility that the conflict might spiral out of control and end in a nuclear confrontation between the superpowers. After the war, military analysts in particular singled out these constraints as a cause of America's defeat. And there is little question that substantial support for North Vietnam from the Soviet Union and its Warsaw Pact allies, as well as the Chinese, was critical to the North's efforts to sustain the insurgency in the South, withstand the U.S. bombing campaigns, and win sympathy from the world community.

Had the Soviet empire not collapsed some months before Iraq invaded Kuwait, it is unlikely that the United States would have dared use overwhelming military force against what had been one of the USSR's most favored client states in a region strategically vital to both superpowers. President Gorbachev and other Russian leaders were stunned by the Iraqi occupation of Kuwait, where the Soviets had expanded their influence considerably in the 1980s. And they were angered by Saddam Hussein's curt, rudely evasive replies to their post-invasion warnings and offers to help him to withdraw without losing face. Symptomatic of the altered relationship between the superpowers, and an ominous portent for Saddam Hussein, was the fact that the American secretary of state, James Baker, was in Russia conferring with his counterpart, Eduard Shevardnadze, when Iraq invaded Kuwait. Though cold war suspicions lingered, Russian leaders joined the United States and its allies in condemning the aggression, and in the following weeks the Russians actively supported American-sponsored resolutions at the United Nations imposing an arms embargo and economic sanctions.[15] After Saddam Hussein rebuffed their offer to broker a negotiated settlement, the Russians joined the U.S.-led coalition. Their commitment to military operations in the Gulf was marginal, but their withdrawal of support for Saddam Hussein's regime deprived Iraq of its main source of weaponry. (In fact the Iraqi weapons that most threatened coalition forces or Israel were outdated Soviet Scud missiles.) The Iraqis also lost their Russian technoscientific specialists and military advisors, who had imparted skills and attitudes that enabled Iraq to maintain air superiority in the Iran-Iraq War and overwhelm Kuwait's tiny armed forces.[16] Saddam Hussein's failure to comprehend the fundamental shifts wrought by the end of the cold war was one of a series of miscalculations that rendered his ill-timed aggression a colossal folly.

The global competition between superpowers had provided the major rationale for expanding U.S. involvement in Indochina in the 1950s and 1960s. By contrast, the Persian Gulf crisis was precipitated by an act of aggression that challenged the multilateralist, liberal-capitalist vision of the new world order that the Bush administration had begun to promote to replace the grim global standoff of the cold war. American determination to contain communism in Southeast Asia had taken little account of the aspirations of the Vietnamese people to determine their own national trajectory. The "line in the sand" metaphor with which President Bush dramatized his determination to resist Iraqi aggression underscored his conviction that if war came it would be a just one, fought to uphold national integrity and global security.[17] Whatever grievances and historic claims Iraq had advanced with regard to Kuwait—and in neither case were they inconsiderable—Bush found international sentiment highly receptive to his call for a stand against Saddam Hussein, a dictator with imperial ambitions.[18] Bush's predecessors had been far less successful in their efforts to turn world opinion against Ho Chi Minh, the avuncular leader of the Vietnamese struggle for national liberation and the socialist transformation of a heavily exploited former colony.

Although technological imperatives—from testing models of development to counterinsurgency weaponry—were embedded in the reasons American cold warriors offered for intervening in Vietnam, they were far more apparent in the Bush team's shifting arguments for ejecting Saddam Hussein's legions from Kuwait. Bush's initial response to Iraq's invasion of Kuwait appeared to have little to do with technology: he insisted that principle was sufficient reason to apply massive force. But he and his advisors soon concluded that they would have to emphasize more reasons than deterring aggression and affirming abstract principles to win domestic support for putting American troops in harm's way. In the months that followed the invasion, the Bush team offered rationales for the march to war that were more fundamentally and explicitly technological than those adduced for any previous U.S. intervention in nonwestern societies.[19]

Not surprisingly, both advocates and critics of military action in the Gulf consistently cited oil as the reason for going to war. Though estimates varied somewhat, it was widely accepted that Saddam Hussein's capture of Kuwait gave him control of as much as one-fifth of the world's known oil reserves. Because most Iraqi-Kuwaiti oil was exported to Western Europe and Japan, Iraq's annexation of Kuwait posed a more direct threat to America's major industrial allies than to the United States itself. But the fear that Iraqi forces would continue to Saudi Arabia, long the primary foreign source of oil for the United States, proved critical in winning domestic and international acceptance for airlifts of American combat forces into the region.

Decades of profligate consumption had steadily drained America's own oil reserves and increased U.S. dependence on imported oil from the Gulf. Critics of military intervention argued that what Bush claimed was a principled stand against Iraqi aggression was actually a campaign to ensure the continued flow of Middle Eastern oil to America and its allies. Prominent economists also disputed the administration's contention that Saddam Hussein's control of Kuwait meant that he could force oil prices to rise. They pointed out that competition would force Iraq to sell its crude at prices roughly in line with those charged by other suppliers. In an apparent attempt to discredit the contention that market mechanisms would frustrate Saddam's ambition to push up prices, the Bush administration refused (until after military operations against Iraq began) to release oil from America's Strategic Petroleum Reserve, which would have had a calming effect on world markets. And Secretary of State James Baker tried, unsuccessfully, to link threats to Middle Eastern oil supplies to rising domestic unemployment.[20]

The perceived threats posed by Iraq's increased control over oil production had not only economic but also political and military dimensions. Saddam Hussein had been amassing an arsenal of advanced weaponry since his seizure of absolute power within the Ba'athist regime in 1979. Iraq's oil had made this buildup possible. The addition of Kuwait's reserves would double Iraq's resources with which to purchase (and increasingly develop itself) sophisticated weapons systems and to expand its air force and armored divisions. During the war with Iran in the 1980s, Saddam Hussein had channeled to the military much of Iraq's oil wealth as well as tens of billions of dollars borrowed from the conservative, oil-rich kingdoms on the Arabian peninsula, which he claimed to be defending from Iran's Shi'ite revolutionaries. Oil revenue enabled Iraq to import fighter planes and missiles from the Soviet Union and France, to improve its communications and surveillance systems, and to upgrade the motorized, armored components of its elite Republican Guard divisions. Iraq's superior weaponry and more highly developed national infrastructure provided the critical edge that allowed it to fight a far larger, more populous Iran to a draw.[21] And despite heavy indebtedness to Kuwait and other oil sheikdoms, Iraq emerged from the war as the most militarily powerful Arab nation in the Middle East and the major adversary of Israel.

Imported technologies and foreign technicians, engineers, and instructors also provided the basis for the potential transformation of Iraq into a strategic threat to an American-dominated new world order. In building the coalition to wage war against Iraq, the Bush administration seized on the aggression against Kuwait as proof of the danger Saddam Hussein's regime posed for the security and stability of the Middle East. In part to broaden domestic support for the use of force, the administration placed increasing emphasis on another high-tech threat: Iraq's programs to produce weapons of mass destruction (WMD). During the 1980s Saddam's regime had invested in specialists and scientific facilities that would enable it to produce chemical, biological, and nuclear weapons. Both the Iranians and the Iraqis had used toxic gas in their long war. Although most experts estimated that Iraq was several years away from producing effective nuclear weapons, by late 1990 the Bush team and hawkish commentators were intimating that their deployment was imminent. When polls showed that allegations regarding Iraq's WMD programs resonated significantly among the electorate, the president went so far as to declare in November that Iraq was only months away from exploding a nuclear device—a claim some of his advisors hoped would be the clinching argument in the administration's ever shifting case for using force to resolve the crisis in the Gulf.[22]

In an address in October 1969, the year after he was relieved (with a promotion to army chief of staff) of his command of U.S. forces in Vietnam, William Westmoreland elaborated on a vision of techno-warfare that he had nurtured during his tour of duty in Indochina. He was certain that the United States would have the capacity to wage such wars in the near future, and lamented that the necessary scientific and technological breakthroughs had not been ready in time to achieve victory in Vietnam. Westmoreland imagined future battlefields "on which we can destroy anything we can locate through instant communications and almost instant application of highly lethal firepower . . . With first round kill probabilities approaching certainty, and with surveillance devices that can continually track the enemy, the need for large forces to fix the enemy will be less important."[23]

In the 1991 Gulf War, Westmoreland's prophecy was fulfilled. But the adversary targeted by the cybertech military machine that America had built since Vietnam had little in common with the insurgents Westmoreland continued to insist could have been crushed with the proper application of U.S. resources and technology. The Vietnamese were masters of the cadre-based guerrilla warfare that had proved so effective against industrialized powers. In stark contrast, the jet fighter squadrons and armored divisions that Saddam Hussein ordered into Kuwait in August 1990 were designed to fight precisely the kind of conventional war at which the United States had historically excelled. The challenge issued by the Iraqi dictator opened the way for the United States to wage the devastating technowar for which its military had begun to prepare even before the retreat from Indochina.[24]

Saddam Hussein's feeble grasp of grand strategy was only one of several factors that accounted for the thoroughly misguided Iraqi approach to the war. In deploying the bulk of their forces in hastily built entrenchments, Saddam and his generals adopted a retrogressive defensive strategy that left their military highly vulnerable to the coalition's surveillance systems, aircraft, and mechanized armored units. It was perhaps predictable that Iraq's military planners would rely on attrition and staying power, which had made it possible for their nation to survive the eight-year onslaught by Iran. This inclination would have been reinforced by the tendency of Saddam's sycophantic advisors, like those of most dictators, to inflate their estimates of the strength of his forces. And Saddam was aware that Bush and his advisors were concerned that Americans were unwilling to risk suffering substantial casualties in a distant conflict, a concern he believed would be the key to Iraq's eventual victory.[25]

Decades earlier the North Vietnamese and the NLF had benefited from the U.S. public's disenchantment with a casualty-ridden war. Unlike the Viet Minh, however, the Ba'athists had come to power through military coups rather than popular insurrection. And

Saddam Hussein's tyrannical repression of political expression, along with the historic hostility of groups like the Kurds and Shi'ites to his regime, meant that, unlike the NLF guerrillas, his forces could not blend in with the general population or rely upon them for sustained support. Also unlike the insurgents in Vietnam, the very conventional Iraqi military was dependent on a fairly developed national infrastructure, complete with modern research facilities and undergirded by communications, power, and highway systems. The regime had to defend these assets or give up its state-centered power base. More mobile troop dispositions might have been adopted, but it is improbable that Iraqi war planners seriously considered anything other than doggedly conventional military responses to the impending offensives.

Several decades of technological innovation had by the late 1980s elevated the United States to military supremacy over any conceivable combination of rivals, including the Soviet Union and its satellites. The "Flexible Response" approach to military interventions overseas that President Kennedy and some of his advisors advanced in the early 1960s required not only a major overhaul of U.S. weapons systems but a thorough rethinking of American military doctrine, particularly as it applied to the developing world. The illusion that the American way of war was being significantly altered to meet the challenge of guerrilla insurgency in Indochina stymied or seriously impeded these transformations through most of the 1960s. But partly in response to setbacks in Vietnam, experimentation with lighter, more maneuverable, multipurpose aircraft, together with the development of laser-guided ordnance and an array of sensing devices, had advanced to the point where these innovative technologies could be deployed in the final stages of that conflict.

As detente and fail-safe measures to prevent an escalation to nuclear war between the superpowers were strengthened in the 1970s, planning for nonnuclear military interventions was increasingly emphasized. Organizational reforms were carried out to reduce the interservice rivalries that had proved so debilitating in Vietnam. The success of these measures was confirmed by the close coordination not only among American forces but also between U.S. and coalition forces in the air war against Iraq. The shift from conscription to an all-volunteer, professional army had by the time of the Gulf crisis produced the most cohesive and disciplined fighting force in American history. And this force was proficient in the use of a new generation of weaponry, from F-16 fighters and Stealth bombers to Tomahawk missiles and M-1A1 tanks, that would overwhelm the Iraqis' mainly obsolescent equipment.[26]

Of all the innovations that transformed the U.S. military in the years after Vietnam, the one most directly linked to the failure in Indochina was the reformulation of strategy for the projection of American power in the developing world. All through the 1970s, retired and active officers had delivered scathing critiques of the conduct of the war in Vietnam, some of which proved influential in the public debate over the reasons for America's defeat and their bearing on the nation's military preparedness. The lessons to be learned from Vietnam were seen as especially critical for future conflicts to check the advance of communism in areas on the "periphery" of the superpower blocks.[27] In the aftermath of the ill-conceived introduction of U.S. military forces into Lebanon in October 1983, which resulted in the death of 240 Marines in a truck-bomb attack on their barracks, there was considerable pressure for the Reagan administration to formulate clear policy guidelines for U.S. interventions overseas. To win congressional and public support these proposals needed to address the criticisms of the conduct of the Vietnam war as well as subsequent indecisiveness in dealing with crises from Iran to Central America.

Planning for interventions in the developing world was overshadowed for a time by big-budget programs for stealth aircraft, the deployment of Pershing missiles in Central Europe, and Star Wars fantasies. But in November 1984 Reagan's defense secretary, Caspar Weinberger, posed a number of questions that he insisted must always be asked when decisions were being made to commit American military forces overseas. The litmus tests that he proposed—national interest, clear objectives, and public support—were later consolidated and clarified in the Bush administration by Bush's chairman of the Joint Chiefs of Staff, Colin Powell. The so-called Powell doctrine dominated American military responses to challenges from the developing world from the late 1980s through the 1990s, with some lapses in the Clinton years. Its stress on a clearly delineated mission, massive application of force, and a well-prepared plan for disengagement shaped all aspects of planning for the campaign to drive the Iraqis from Kuwait, as well as the indecisiveness of the victors as the fighting ended.[28] With attention fixed on stratagems for waging a cybertech conventional war, no one asked whether the Weinberger-Powell approach to regional conflicts on the periphery addressed the question of how to counter revolutionary guerrilla warfare, which had humbled each cold war superpower, first the United States in Vietnam, and later the Soviet Union in Afghanistan.

Once it was evident that the Iraqis had no intention of invading Saudi Arabia, coalition planners had the luxury of preparing for a war that would be fought entirely on their own terms. From the first of the allied air assaults, it was abundantly clear that prewar estimates of Iraq's military strength had been considerably inflated. To a significant degree, the military had taken this public stance to avoid the overconfidence that had proved so problematic in Vietnam.[29] In the run-up to the war, administration spokespersons and supporters never tired of reminding the American public that Iraq had the fourth-largest army in the world, that its ground troops would outnumber those of the coalition, that it could field thousands of armored vehicles, and that the Republican Guards were highly trained, battle-hardened, and fanatically loyal to Saddam Hussein. But once combat began it became apparent that Iraq's technology, from radar systems to weaponry, was mostly outdated and certainly overmatched in the kind of cybertech war that the United States had been preparing to fight against far more formidable adversaries, such as the Soviets and the Chinese.

For decades the Pentagon had been gathering intelligence on Soviet weapons and communications systems. Because Iraq's military was heavily dependent on Soviet technologies, from fighter aircraft to encrypting computers—and because France and the United States had also transferred military technologies to Iraq in the 1980s—U.S. and coalition forces could intercept command-and-control exchanges and plan in advance the optimum ways to deploy their superior weaponry.[30] Even during Iraq's swift conquest of Kuwait, its technological shortcomings were exposed by numerous failures in the equipment of the Republican Guard divisions.[31] Iraqi communications units, anti-aircraft gunners, and fighter pilots were for the most part not adequately trained to operate their own war machines, let alone to cope with the dazzling technologies coalition forces would use against them. In sharp contrast, American and allied service personnel made good use of superb training in deploying their new surveillance technologies and weapons.[32] From the first aerial assaults on desert radar installations and Baghdad, there was little question that Saddam Hussein had blundered into a war he could not win and that his regime would be lucky to survive.

Although few remarked on it at the time and American war planners may not have deliberately intended a symbolic gesture, the fact that the assault on Iraq began at 3:00 A.M. presaged a conflict in which U.S. forces would reclaim the hours of darkness, which had been persistently conceded to the guerrillas in Vietnam.[33] Satellite surveillance devices, drones, infrared sensors, and night-vision equipment prepared the way for sorties by piloted aircraft and laser/computer-guided missile strikes that quickly shut down much of the Iraqi command and control system. Technicians in planes equipped with surveillance technologies tracked enemy aircraft in flight and on the ground, targeted Iraqi radar sites and armored vehicles, and regulated the flight patterns of the bombers, helicopters, F-117 Stealth fighters, and missiles that filled the airspace over the Gulf region during the aerial war. The heavy traffic was due both to the sheer number of aircraft and missiles engaged (1,300 sorties had already been flown by the end of the first 24 hours of the air war) and to their deployment in parallel strikes rather than the serial waves of bombardment that had been standard procedure in earlier conflicts.[34]

Within hours of the first attacks, air defenses over Baghdad, which allied commanders had estimated were among the best in the world, lost the ability to track or fix incoming bombers or missiles.[35] In desperation, the Iraqi high command ordered those operating anti-aircraft weapons to simply mass their fire over the city in the hope of hitting enemy aircraft or missiles. In the weeks that followed, the air offensive either destroyed the Iraqi air force or drove its best planes to neutral Iran; shut down most of Iraq's electrical, telephone, and communications grids; and destroyed critical bridges over the Tigris and Euphrates rivers. It also all but severed the supply lines between Baghdad and the forces in Kuwait and southern Iraq, thus securing an advantage that the United States had never attained in Vietnam.

Even before the coalition land offensives began on February 24, the position of Saddam Hussein's divisions in Kuwait was untenable and much of his high-tech infrastructure had been reduced to smoldering ruins. With the Iraqi air force having conceded the skies to the allies, even surveillance systems still in the testing phase could be deployed without fear of interdiction. Coalition forces seized control of both the night and the day, allowing them to conduct a campaign in which, in the estimate of the U.S. air force chief of staff, "there was no time from day one on, that the Iraqi ground forces were not under heavy air attack."[36] Satellite-guided, computer-directed missiles, capable of driving deep into the earth and armed with delayed explosive devices, also wreaked havoc on weapons and communications systems protected by underground bunkers, denying the Iraqis the subterranean control that had been so critical to the staying power of communist forces in Vietnam.

With the ample time for analysis enjoyed only by combatants not at risk, allied pilots and technicians soon found ways to counter the Iraqis' techniques of camouflaging or embedding tanks and other weaponry in the landscape. Allied pilots also quickly learned to distinguish actual vehicles from the hundreds of decoys the Iraqis had positioned throughout their defensive lines. At twilight or just after nightfall, infrared sensors could detect the heat given off by tanks or personnel carriers, whose armor cooled more slowly than the terrain in which they were concealed. As the date of the ground offensives approached, coalition forces, using mainly laser-guided bombs, destroyed up to 150 armored vehicles in a single night. For the Iraqis, the capacity of the coalition's surveillance technologies and "smart" weapons to locate and destroy their armored vehicles was profoundly demoralizing. One captured Iraqi officer recalled: "During the Iran War, my tank was my friend because I could sleep in it and know I was safe . . . During this war my tank became my enemy . . . none of my troops would get near a tank at night because they just kept blowing up."[37]

With their communications and radar systems all but shut down and their entrenched positions under continuous assault, the Iraqis failed to detect the massive flanking operation that would prove decisive in the ground war. Refusing to be drawn into casualty-ridden, head-on assaults on the Iraqis' trenches, coalition strategists made astute use of their control of the air and their highly mobile mechanized divisions, which were superbly suited for the desert terrain. In the few instances when Iraqi divisions moved into the open to offer significant resistance to coalition ground forces, they were rapidly destroyed by air assaults and the superior range and firing power of American M-1A1 tanks and surface-to-surface missiles. Within hours of the beginning of coalition ground operations, tens of thousands of Iraqi soldiers were dead, retreating toward Iraq, or attempting to surrender en masse. Accounts of the slaughter of fleeing Iraqis that ensued—most terribly in the infamous "turkey shoot" that left miles of burnt-out, mangled vehicles and charred bodies along the highway from Kuwait City to Basra—soon convinced Bush and Powell to declare an end to the one-sided conflict. It had taken only a hundred hours and minimal losses for the coalition to oust the Iraqis from Kuwait and envelop Saddam Hussein's elite Republican Guard divisions.[38] The destruction of what remained of the regime's armed forces and the fall of the dictator were averted only by the American decision to call an early truce. And that decision was prompted largely by fears that regional chaos would result from the sudden collapse of the Ba'athist regime.

The false analogies and demonization tactics that had been mainstays of the Bush administration's efforts to build support for military intervention were also used to justify the prodigious scale and ferocity of the coalition's offensive. But this inflated rhetoric proved problematic for Bush and his commanders when they balked at "finishing the job" by destroying the elite Republican Guard divisions and toppling the "evil" dictator. The decisiveness they had displayed in forging an international coalition and leading it to war contrasted sharply with their poorly coordinated, often inept handling of the aftermath of the conflict. The advanced weapons systems they had so skillfully meshed to overwhelm the enemy would be of little use in dealing with the political chaos that the collapse of the Ba'athist regime threatened to unleash; in fact it was suddenly clear that those weapons' very effectiveness had greatly enhanced the likelihood of that dangerous outcome. And it soon became evident that Bush and his advisors had not thought through the potential fallout from the disintegration of Iraq.

In justifying the decision to end the conflict, administration officials insisted that when Kuwait was liberated the U.N.-mandated mission had been fulfilled, a claim that was technically correct. They also had good reason to believe that if they had pressed to destroy the Republican Guard divisions and overthrow the regime, the coalition would have broken apart.[39] But their contention that the overthrow of Saddam Hussein had never been an objective was disingenuous at best. The air war had been concentrated on military assets and technological systems that undergirded his regime, and only secondarily on his forces in Kuwait. And the rout of his best units in the ground war led the Bush team to conclude that the dictator would either be overthrown by his own military commanders or swept from power by mass revolts from below. On a number of occasions during the conflict, Bush had explicitly urged the Iraqis to end the regime, and administration spokespersons had indicated that it would be fortuitous if Saddam Hussein happened to be killed in one of the air strikes.

There were valid geopolitical reasons for ensuring that Iraq held together, even at the price of allowing Saddam Hussein to remain in power—a fact that is often overlooked by those who insist that the end of the 1991 Gulf War was badly botched. Bush and his advisors certainly knew that the dismemberment of Iraq would be of most benefit to Iran, which American policymakers still regarded as the most formidable locus of resistance to the political order they deemed acceptable in the Middle East. And fears that the fall of the regime would compel the coalition to occupy Iraq, and might provoke guerrilla resistance akin to that in Vietnam, have subsequently been shown to have been well founded.

Ironically, the Bush team's most egregious failure could have been averted by a limited application of the technological supremacy that had to that point been used so effectively. Within Iraq, Kurds and Shi'ites had mounted popular rebellions against Saddam Hussein's regime, which had oppressed both peoples for decades. To a significant degree these uprisings occurred in response to the exhortations of the Bush administration, and the widespread support they received had much to do with the expectation that they would be backed by American forces. But indecision and poor coordination on the part of the Bush administration and the U.S. military opened the way for Saddam Hussein's surviving troops to brutally repress first the Shi'ite rebels in the south and then the Kurdish fighters in the north. General Norman Schwarzkopf, the commander of coalition forces, was sent to negotiate an armistice with Saddam Hussein's defeated generals, but was not given detailed instructions. Uncertain about the intentions of the Bush administration, which had ordered what many in the military considered a premature cease-fire, Schwarzkopf made no effort to prohibit Saddam Hussein's remaining helicopter and tank units from entering the Shi'ite and Kurdish regions that were in revolt. Coalition fighters stood by as disillusioned spectators while the reinvigorated dictator's loyalists suppressed the uprisings with a brutality that appalled even those familiar with his regime's history of human rights abuses.[40]

❖   ❖   ❖

Never before in history had technologies supplanted humans as the heroes of a military victory as thoroughly as they did in media representations and public perceptions of the 1991 Gulf War. To a large extent this technocentrism was determined by the way in which the war was fought. Air assaults all but monopolized nearly six weeks of unequal combat. And even in the last days of the war, the destruction of Iraqi ground forces owed so much to attacks by helicopters, fighters, and laser-guided weapons that it was not unreasonable for military analysts to conclude that the potential of air power to be decisive in combat had at last been realized.[41] What were hyperbolically termed battles consisted mainly of long-distance exchanges of firepower by massed machines—tanks, artillery, and aircraft. With the exception of some skirmishes in the fight to control the town of Khafji, infantry assaults and hand-to-hand combat were notable for their absence from these high-tech clashes. The desert terrain, the highly conventional nature of Iraq's forces, and the coalition's limited objectives were also conducive to the centrality of techno-weaponry. Bush's military planners consciously worked these factors into a strategy designed to take maximum advantage of innovative technologies to minimize coalition and civilian casualties while devastating Saddam Hussein's war machine.

Technowarfare also enabled the military to avoid the intense media scrutiny that many officers were convinced had been pivotal in denying the United States victory in Vietnam.[42] That conviction was unfounded, according to a substantial body of evidence indicating that throughout the 1960s the media lagged behind rather than influenced political shifts in Washington that fed public opposition to the war.[43] Nevertheless, from the onset of the Gulf crisis, the Bush team regarded highly restrictive controls on the media as a critical dimension of their preparations for military action against Iraq. Building on media-management techniques that had proved effective during the 1983 invasion of Grenada and the 1989 police action to remove Manuel Noriega from power in Panama, the military denied reporters access to the battlefields of the Gulf.[44] Correspondents already stationed in Baghdad when the war began were able to describe the air raids against the city from the dramatic opening minutes until they were expelled by Saddam Hussein. But the nightly sound-and-light extravaganzas they broadcast around the world featured the technowar that would overshadow other aspects of the media's coverage of the conflict.

In contrast to Vietnam, where reporters were allowed to accompany American units, contact between reporters and combat forces in the Gulf was highly constricted and controlled. Interviews with GIs in Vietnam were often candid and only rarely censored. As Anthony Swofford relates in *Jarhead*, his profane chronicle of service in the Gulf, the little interaction permitted between the press and American troops was closely supervised, rehearsed, and punctuated by obligatory patriotic platitudes. In anticipation of a visit by reporters, Swofford's platoon sergeant reels off a list of "unacceptable topics" that include "divulging data about the capabilities of sniper rifles or optics and the length and intensity of our training." The sergeant cautions his men to keep their responses vague, banal, and gung-ho—rather like those of a professional athlete being interviewed before a big game back home: "Basically, don't get specific. Say you can shoot from far away. Say you are highly trained, that there are no better shooters in the world than marine snipers. Say you're excited to be here and you believe in the mission and that we'll annihilate the Iraqis. Take off your shirts and show your muscles."[45]

Many reporters bridled at being denied access to the battlefront, and some risked their lives in unauthorized forays into the desert. But public protest about the restrictions from mainstream news providers was surprisingly rare, at least until the war was over. On the contrary, high-profile media figures vied for places in the press pools organized by the military, which were generally located in Saudi Arabia far from even the rear lines of coalition operations. The chief sources of information for correspondents assigned to the pools were official news releases and elaborately staged press briefings, both of which emphasized the coalition's use of high-tech weaponry. It was clear that the war planners had devised effective ways of co-opting the satellite systems and other advances in communications technology that some pundits had predicted would permit the media to cover future wars in real time with unprecedented independence.[46]

A mix of motives accounted for media acquiescence. The long-distance nature, speed, and firepower of the coalition's weaponry was reflected in high rates of casualties due to friendly fire;[47] thus the military's assertion that restrictions were necessary to ensure reporters' safety was not entirely without foundation. Confronted with tight budgets and intense competition, the corporate conglomerates that dominated newsmaking were content to let the Pentagon cover a substantial share of production costs. And broadcasters and reporters were often as captivated as their audiences by the visuals provided by the military, with cruise missiles streaking into the night sky or blowing apart Iraqi command centers. In the absence of live coverage from the front, a good deal of air time was devoted to disquisitions by retired generals and former security advisors on the weaponry deployed, the intended targets, and preliminary assessments of the damage inflicted.[48]

Media compliance with the restrictions allowed the military to construct the narrative of the conflict that was projected to the American public and a global audience. And that narrative, as General Charles Horner summarized it in one of the more memorable briefing sessions for the press pool at coalition headquarters, was the story of "a technology war, although it is fought by men and women."[49] The charts, video images, and analysis of weapons performance that dominated the press briefings featured sophisticated tracking and killing machines obliterating enemy weapons systems that were shown to be decidedly inferior. Most sessions showcased videos shot by cameras installed on fighter planes or in the nose cones of laser-guided missiles or bombs. The briefings were shown in news broadcasts around the world, and the videos made possible relentless replaying of the destruction of inanimate objects—radar stations, Ba'ath Party buildings, Saddam's palaces, bridges, tanks—by cybertech weaponry.[50]

The human presence in this imagery of the war often consisted of American mechanics and technicians servicing weapons systems or reporters surveying the devastation in areas "neutralized" or "cleared" by coalition assaults. As intended, these images projected an offensive that was superbly planned, precise—even surgical—in targeting enemy "assets," and proceeding according to a predetermined schedule toward well-defined objectives. The Bush administration never tired of reminding the global audience that the military was conducting a police action that was restrained, rational, and fought according to the civilized standards befitting an international coalition pursuing a just cause.

Particularly in the early stages of the conflict, the images of Operation Desert Storm were deliberately fashioned to underscore the differences between the 1991 Gulf War and the failed intervention in Vietnam. The technocentric depictions of Gulf operations provided stark contrasts with the intensely human strife that had been the focus of broadcast and press reporting in Vietnam. Images of patriotic Green Berets and increasingly disgruntled GIs, indefatigable guerrillas, protesters and veterans, wounded soldiers, and Vietnamese victims, especially women and children, continue to resonate in the collective memory Americans have so contentiously constructed of that failed crusade as well as "the Sixties" as a distinct historical era. Those who managed the dissemination of information during Operation Desert Storm sought to purge images of body bags and mangled corpses, human brutality, and mortar-riddled helicopters whirling out of control. Filming or photographing at Dover air base in Delaware, where the American dead were brought home in both wars, was prohibited during the 1991 conflict and in subsequent overseas interventions.[51]

Survivors of the Vietnam debacle, both American and Vietnamese, have left searing accounts of that ordeal in memoirs, novels, poetry, and films that taken together are equaled only by those left by soldiers and nurses caught up in the First World War. Few works of comparable power by those who served in or reported on the 1991 Gulf War have yet appeared. The war machines, mostly helicopters and gunboats, that move in and out of reconstructions of the Vietnam war seem out of place in the rainforests and rice paddies, and as vulnerable in combat as the human protagonists. In sharp contrast, the cyberweaponry deployed against the Iraqis is at home in the expansive desert landscape and well able to exploit the adversary's conventional defenses and obsolescent technologies.[52] Representations depicting the differences between the two wars were foregrounded even after the Iraqis had been driven from Kuwait. In a rather obvious attempt to affirm closure, the coalition's triumph was signified by a helicopter landing on the roof of the U.S. embassy in Kuwait City in an image intended to counter perhaps the most infamous symbol of defeat in Vietnam—the last helicopter to escape, crammed with Vietnamese refugees, taking off from the roof of what was reported at the time to be the embassy in Saigon in 1975.[53]

As Operation Desert Storm demonstrated, a war fought mainly by machines yields few human heroes. But unlike the demoralized and divided conscript army that fought in Vietnam, American troops who served in the Gulf were celebrated in news broadcasts and editorials as well-trained professionals with careers on the line. Rather than macho warriors, they were routinely depicted as competent technicians doing their jobs. The military force that won the war was "glorified not for its strength, aggression, or courage but rather for its education, training, and mastery of 'high' technology."[54] In all the services, technospeak was pervasive, and the capacity to operate complex communications or weapons systems was valued far more than sheer physical strength or fighting skills.

The technocentrism in media representations is illustrated by *Operation Desert Shield*, a "special issue" magazine in which photos and text are devoted—as the cover promises—to "America's Awesome Firepower," "Newest High-Tech Lethal Weapons," and "The Most Devastating War Machine EVER!" Virtually all the humans depicted are operating or servicing high-tech weapons. Even the infantrymen who pose in a handful of the photos in front of palm trees or desert expanses are less the focal point than the automatic weapons they brandish (though one caption reminds us that "you still have to have infantrymen . . . if you are going to win"). A rather different sort of Gulf War memorabilia, Desert Storm trading cards, also stress high-tech weaponry and intelligence gathering rather than the service personnel who wielded them in the conflict. In the Topps Company's Pro Set version of Desert Storm trading cards, those depicting "Military Assets," from advanced fighters to Tomahawk missiles, are both more vivid and far more numerous than the rather pedestrian photos of "Leaders" and the uninformative "Geography" selections. "Control Centers" and "Intelligence File" cards generally focus on computer technologies and high-tech spying devices. The numbers of cards available in each category leave little doubt that Lockheed F-117A Nighthawks and Apache attack helicopters were in far greater demand than snapshots of any of the human luminaries, with the possible exception of Schwarzkopf, who were responsible for winning or losing the war.[55]

Because the 1991 Gulf War was the first in which women were assigned to U.S. combat units, they shared in the celebration of technical proficiency in the media's coverage of the war. In interviews and special reports, typified by a *Newsweek* cover story on "Women Warriors," they were represented as experts in the technologies their units were deploying. And like the men they served with, they were depicted as confident, can-do professionals who, as army specialist Sandra Chisholm affirmed, "have a job to do here and . . . will do it." The *Newsweek* article portrays them piloting "giant C-141 transport" planes, repairing everything from Huey helicopters to Humvees, serving as naval navigators, and processing intelligence data. Statistical tables document their disproportionate concentration in communications and intelligence, medical support, mechanical repair, and administration, as well as the engineering corps. Thus the women who served with combat units in Operation Desert Storm challenged, at least implicitly, the longstanding assumptions that both military and technological pursuits were properly gendered male. Some made that challenge quite explicit: Lieutenant Stephanie Shaw insisted that she "could fly that F-15 just as well as a man," and an anonymous interviewee dismissed the argument that women lacked the physical strength to participate fully in combat because "on a ship, war is high tech . . . Men aren't any better at video games than women."[56]

Similar depictions of African-American men and women operating or maintaining advanced weaponry strengthened public perceptions of their capacity to master complex technologies and command large units in combat. The media's effusive praise for Colin Powell, who as chairman of the Joint Chiefs of Staff was responsible for coordination of the sophisticated operations mounted against Iraq, had much to do with Powell's emergence as one of the few human heroes of the war. Powell's insistence that professional career opportunities for minorities in the military far surpassed those attainable in other American social sectors[57] was a subtext in the 1996 film *Courage under Fire*. In this compelling dramatization of the Gulf conflict, Denzel Washington portrays a much-decorated army colonel who first directs a tank battalion in a disorienting night clash with Iraqi armored divisions and later unravels the often highly technical details in conflicting accounts of a female officer's death in combat. The film powerfully conveys the centrality of advanced technologies in the war and the massive firepower advantage they gave to American forces in clashes with outgunned Iraqi soldiers. At the same time it underscores the heightened dangers of human error that in the Gulf conflict resulted in high levels of friendly fire casualties owing in large part to the frenetic pace at which technowars are fought.

The technocentric blitzkrieg projected by the media led to inflated estimates of the significance of the coalition's victory and peripheralized the staggering price the war exacted from the Iraqi people. The prewar claim that Iraq had the world's fourth-largest army was undercut by the fact that much of it consisted of poorly trained, forcibly conscripted soldiers who had little loyalty to the Ba'ath regime.[58] This was especially true of the units entrenched in Kuwait, which were expected to bear the brunt of the coalition offensive. Pummeled for weeks by air strikes, and deprived of food and water by air assaults on their supply lines, Iraqi troops had the good sense to surrender or desert in the tens of thousands. Media images of bedraggled Iraqi soldiers supplicating or prostrating themselves before stalwart American fighters abounded in the final days of the war. But the media never showed video recordings of Iraqi soldiers caught in the lens of night-vision cameras "like ghostly sheep flushed from a pen—bewildered and terrified" and then "blown to bits," or of the "turkey shoot" massacre of panicked conscripts fleeing along the "Highway of Death" toward Basra, and clearly no longer a threat to coalition forces.[59]

In contrast to the obsessive focus on body counts in the Vietnam conflict, neither U.S. commanders nor the media showed much interest in the number of Iraqi soldiers killed, though according to even conservative estimates it was in the tens of thousands.[60] As military briefings and television specials made apparent, what mattered to the media about Operation Desert Storm was not human losses—

in part because those inflicted on coalition forces were fewer than 200—but the performance of the new technologies for war. When Iraq's soldiers were represented as anything other than minions of the evil Saddam Hussein, they were often dismissed as incompetent in deploying advanced weaponry, servile, and unmanly (if not cowardly). Mike Ettore, a captain in a marine tank-killing company, summed up this assessment: "They were an inferior enemy. Pathetic . . . They didn't want to fight. And anyway, their reaction time is really bad. The first tank you hit stops everything for them. And they just sit there. And our guys are just banging tanks faster than hell . . . two, three a minute. So, these guys, they see this, they couldn't surrender fast enough. They just wanted out of the whole thing."[61]

The military sought to portray the war as a "clean," surgical action that was directed against military targets and minimized civilian casualties. Bush repeatedly declared that the U.N.-backed coalition was fighting a police action against Saddam Hussein (and implicitly those misguided enough to fight to defend his regime), not against the Iraqi people. American war planners may well have been sincere in their intent to wage a war that limited what they euphemistically termed "collateral damage." In contrast to many of their European counterparts, earlier U.S. advocates of strategic bombing had consistently emphasized the importance of precision instrumentation and—during the Second World War—of daytime sorties to ensure that high percentages of ordnance found military targets.[62] Technological limitations, and in Vietnam the nature and disposition of enemy forces, had hitherto made it difficult to even approximate these ideals. But those in command in Operation Desert Storm were confident that the new communications technologies and "smart" weaponry devised since Vietnam would reduce civilian casualties to levels that would not undermine the Bush administration's contention that it was waging a just war. The calculated presentation in press briefings of the air offensive as the ultimate in science-driven, cybertech warfare was designed to foreground the civilian-friendly dimensions of this innovative approach to armed conflict. And with rare dissent, this was the predominant view of the war that American and other Western (though not Arab) media projected to viewers and readers.

When "smart" projectiles strayed from their programmed trajectory or zeroed in on civilian targets because of faulty intelligence, the military's version of the Gulf saga was called into question, though briefly and in muted misgivings rather than more fundamental objections. Footage of the grieving relatives of hundreds of women, children, and elderly men reduced by a laser-guided bomb to steaming heaps of flesh in a Baghdad air raid shelter seriously compromised the claim that few Iraqi civilians had been killed or injured in the aerial assault that had rained nearly 90,000 tons of explosives on their country in just six weeks. But the military insisted that the enemy had been using the shelter as a communications center, a claim that was generally accepted by both broadcast and print media, which reported the incident as an unfortunate accident of war and blamed Saddam Hussein for his duplicitous double use of the structure. The news story and its disturbing images were quickly relegated to sound bites or back pages, and an incipient debate over the impact of the war on civilians dissipated.[63]

As in the case of Iraqi military casualties, there was never a serious attempt to count civilians killed, injured, or made homeless by coalition bombing. In any case, most civilian losses were caused not by direct hits but by the indirect effects of the aerial assaults. Reflecting the Bush administration's uncertainty regarding the ultimate objective of the war, the bombing raids had from the outset extended far beyond Saddam Hussein's palaces, command-and-control centers, and military installations to target the nation's basic infrastructure. Unlike North Vietnam, Iraq offered plenty of targets. Allied attacks destroyed power stations, water and sewage treatment plants, food storage facilities, and bridges. In combination with the economic embargo that remained in place until after Iraq's defeat in the 2003 Gulf War, this destruction caused or contributed to the death of hundreds of thousands of civilians, many of them children. These ravages were rarely mentioned in the mainstream media in the United States, even though they were widely reported in Europe and elsewhere.[64]

One fundamental contrast between the Vietnam and 1991 Gulf wars was the level of development each nation had achieved at the time of American military intervention. Vietnam was a war-torn, divided country, impoverished by a century of French colonization. Its citizenry was overwhelmingly made up of poor peasants; its urban growth was stunted and concentrated near Hanoi-Haiphong and Saigon; what infrastructure it had was largely obsolescent; and its industrial sector was confined mostly to processing raw materials for export. Iraq in 1991 also had a large peasantry, and its people had only begun to recover from the war with Iran. But unlike Vietnam, Iraq had made substantial advances toward sustained development in many sectors. Successive Ba'athist regimes—including that presided over by Saddam Hussein—had channeled the revenues generated by the nation's oil reserves not only into weapons programs and the military but also into ambitious infrastructural projects, socioeconomic reforms, and programs to improve living standards. By the time of the outbreak of the war with Iran in 1980, large segments of the Iraqi population in both urban and rural areas benefited from state-funded health care and education systems, large-scale housing projects, low levels of infant mortality, steadily increasing life expectancy, and extensive electric, highway, and communications systems. The socialist cast of many of these programs ought to have been offset, from the American perspective, by the Iraqi government's stress in the 1980s on promoting growth through market incentives and the privatization of some industries.[65] In addition, the Ba'ath regime, unlike its more conservative, deeply religious Muslim neighbors such as Saudi Arabia, was firmly secular and committed to improving the status of women, whose access to education, technical training, and professional employment it had steadily expanded.[66]

Acknowledging Iraq's successes in development initiatives would have compelled the Bush administration and the American media to confront the fact that the combination of the air war and economic sanctions had thrust Iraq backward into the ranks of the underdeveloped nations. The refusal to even consider the Iraqi path to development as a viable alternative owed a great deal to the tyrannical nature of Saddam Hussein's regime, under which significant social and economic improvements had been made while state terror and political oppression had increased by several orders of magnitude. But there is obvious hypocrisy in the claim that this is the main reason for the failure to move beyond the demonization of Saddam Hussein to a serious appraisal of the considerable development that Iraq had achieved. The United States had frequently backed dictatorial regimes that had done far less for the welfare of their people, and had turned a blind eye to the repressive excesses of allies such as Chiang Kai-shek, Ngo Dinh Diem, Sese Seko Mobutu, and Augusto Pinochet. In fact, the importance of "strongmen"—usually a euphemism for brutish dictatorial leaders—had come to be widely accepted by those who still adhered to the modernization paradigm. Affirming Walt Rostow's insistence that political stability was essential for emerging states to make the transition from underdevelopment to take-off, American policymakers demonstrated a remarkable tolerance for even the cruelest and most rapacious tyrants, as long as they adhered to the policy positions of the United States in its global contest with communism.[67] Saddam Hussein himself had received substantial support from the CIA and other U.S. agencies both in the 1960s and on a far grander scale in the 1980s during his war with Iran. His well-known brutality had little effect on the willingness of successive U.S. administrations to accommodate him when doing so suited American objectives in the region.[68]

Ignoring Ba'athist initiatives for national development was consistent with the Bush administration's propensity to finesse challenges to American approaches to aid for emerging nations. The collapse of command communism in Eastern Europe and China's turn to "market-Leninism"[69] eased the pressure on Bush, and his successor Bill Clinton, to fashion a coherent strategy for overseas development programs from the array of alternatives that had vied for favor since the early 1970s. Domestically, the waning of communism also gave impetus to an assault from the far right on the expanded social safety net that had been the defining legacy of the New Deal. Lyndon Johnson's failed crusade to eradicate poverty at home while simultaneously vanquishing communist insurgency in Vietnam had bolstered the case made by conservatives opposed to government-funded welfare programs, which they linked to the development schemes of the modernizers that had done so little to shore up the American-sponsored regime in Saigon. Reagan and Bush dismissed state-sponsored social engineering schemes at home and abroad as wasteful, futile endeavors that should be consigned—along with communism—to history's dustbin. In their place both administrations favored the promotion of laissez-faire, free market options.

In the years after the retreat from Vietnam, American aid programs and foreign assistance policies were in disarray. By the early 1970s public support for agencies, such as USAID, that provided financial and technical support to emerging nations had fallen to an all-time low. This trend was reflected in the decline in the following decade of the proportion of U.S. government revenues allotted to foreign aid from an already meager 0.31 percent to 0.25 percent. At the same time, longstanding hostility on the part of conservatives in Congress to U.S. financial backing for international development agencies, particularly those sponsored by the United Nations, intensified markedly. On occasion allied with more moderate legislators, they sought to channel foreign aid mainly to security-oriented support for regimes deemed compliant with America's broader policy objectives. They also voted to constrict economic assistance to emergency relief efforts, particularly food exports that often served the interests of their constituents in the farm belt.[70]

By the mid-1970s even the support of many liberals for overseas aid programs had waned. As the *New York Times* columnist Paul Krugman later recalled, he had intended to specialize in development economics in graduate school, but had been dissuaded by the sense that it "was just too depressing to pursue."[71] Legislators who had once been staunch proponents of modernization for impoverished nations had been disillusioned by the failure of development projects to win hearts and minds in Vietnam. They were also critical of what they considered the high cost and scant returns from programs launched in the Kennedy era, including the Peace Corps and the Alliance for Progress. Politicians across the ideological spectrum indicted international agencies, such as UNESCO and UNICEF, as little more than boondoggles for bureaucrats recruited disproportionately from the "Third World." Critics charged that these institutions engaged in exorbitantly expensive projects that mostly benefited dominant elite groups in developing societies. Echoing Lyndon Johnson's insistence prior to direct U.S. military intervention in Vietnam that it was a "mistake" to attempt to "reform every Asian country in our image," some critics even dared to suggest that America's path to take-off was perhaps not the most appropriate model for emerging nations.[72]

From the early 1970s domestic dissatisfaction with America's approach to development assistance was also fed by environmental concerns. Many of these were set forth in reports by an international team of academic specialists and entrepreneurs called the Club of Rome. Making extensive use of computer-based projections, these reports warned that population growth and the spread of industrialization were rapidly outstripping the resources of the planet as well as generating devastating levels of pollution. Furthermore, the Club of Rome research revealed that despite the depletion of irreplaceable natural resources and severe degradation of the environment, living standards for most people in emerging nations had improved little or even declined. The authors of the reports concluded that unless the rates of human consumption slowed and eventually stabilized, environmental stress and social strife might well lead to the collapse of "modern civilization."[73]

Not surprisingly, the champions of the global marketplace and modernization theory disputed the findings of the Club of Rome. The computers of the conservative Hudson Institute churned out counter-projections, and its director, Herman Kahn, assured the world that technological innovations would allow humankind to surmount the resource and pollution barriers to never-ending progress, American-style.[74] Innovative technologies, Kahn and other techno-enthusiasts contended, were making it possible to exploit hitherto untapped deposits of key resources or making substitutes economically viable—and thus the predictions of global shortages were unfounded.

The shift in the outlook of many developmentalists was reflected in the perspectives on the global predicament that Robert Heilbroner articulated in *An Inquiry into the Human Prospect*, published in 1974. In the 1960s Heilbroner had favored a statist, social engineering approach to the problems of underdevelopment that shared a good deal with the course pursued by communist regimes. In *Inquiry* he called into question both the communist and the capitalist formulas for achieving economic increase and social equity in emerging nations. Neither approach, he pointed out, had significantly improved the condition of the poor or promoted sustainable growth while seriously taking into account environmental repercussions. Heilbroner predicted that if current rates of industrial production, economic growth, and population increase continued, the planet would become uninhabitable within two and a half centuries.[75]

In the late 1970s and the 1980s there was a deluge of reports and essays on "the greenhouse effect," global warming, the disappearing rainforests, accelerated extinction of species, threats to the ozone layer, and other resource and environmental issues. Invariably linked to growth trends, approaches to development, and at least implicitly the underlying imperative of sustainability, these issues became the focal points of public debate and government policy in many countries and international forums. The Earth Summit in Rio de Janeiro in 1992 and the follow-up conference in Tokyo in 1995 marked the culmination of over two decades when the convergence of dissatisfaction with development strategies and demographic-environmental concerns gave rise to major efforts to reorient both U.S. and international assistance programs.

Largely because of the failure of development projects in Vietnam, the rethinking of U.S. aid to emerging nations was under way by the early 1970s. In 1973 Congress passed two landmark acts mandating major changes in strategies and procedures for foreign assistance. The sponsors of both measures sought to ensure that funding and technical advising were channeled away from elites and mega projects to "the poorest strata of recipient nations." Rather than emphasizing the promotion of long-term growth, aid was to be concentrated on immediate concerns, such as famine relief and health initiatives. And in implementing each development project, local conditions as well as grassroots feedback were to be taken fully into account.[76]

Over the next two decades, though substantial resources continued to be committed to airport and hydroelectric dam construction (and siphoned off by bureaucrats in the developing nations), landless peasants, women, and small-scale, local projects were much more meaningfully factored into U.S. development assistance programs and those of other industrialized nations. Specialists in both government and private international lending agencies—including Robert McNamara at the World Bank—advocated diverting substantial development resources away from large-scale, high-tech projects to investments in appropriate technologies and human resources.[77] Increasing emphasis was placed on assessing local needs and concerns, identifying environmental constraints, and determining which technologies would be best suited for specific sites. E. F. Schumacher's insistence on the importance of small and relatively simple machines—from improved irrigation pumps to fuel-efficient stoves—was heeded by many development planners, some of whom explicitly acknowledged the influence of the controversial advocate of alternative technologies. By 1983 USAID reported that as much as $164 million was being spent on making low-capital, labor-intensive technologies available to developing countries. The shift toward more manageable technologies went hand in hand with the expansion of programs for training indigenous technicians and with a heightened appreciation by development specialists of local knowledge and techniques, which had long been dismissed as "traditional" obstacles. This enhanced sensitivity to local responses and capabilities as well as to the ecological effects of development schemes was symptomatic of a fundamental shift in the way the development process itself was conceived. Many development planners and field consultants moved beyond a preoccupation with economic growth to finding ways to raise living standards that would be socially and environmentally sustainable.[78]

In part because of planners' recognition of the connections between the empowerment of women in developing nations and the reduction of population growth, women's education and job training moved to the foreground of many assistance programs. The 1973 U.S. legislation that reoriented development assistance mandated that USAID give priority to fostering the integration of women into national economies, thereby promoting their social welfare and status. These objectives were also stressed by development agencies in other industrial nations, such as Canada and Japan. Feminist activists, who in the post-Vietnam era increasingly sought to bridge the divide between women in industrial and developing societies, also played vital roles in this shift. Women's advocates began to assume positions in the bureaucratic hierarchies of international lending agencies, and women's support for highly vocal grassroots organizations in developing nations became more pronounced. In the following decades women's empowerment in many developing countries was reflected in their rising literacy rates, commitment to family planning, and participation in entrepreneurial ventures from bus companies in Kenya to cell phone services in India.[79]

Rhetorically at least, the high tide of the rethinking of development strategies came in a 1989 World Bank study on sub-Saharan Africa, where neither the command communist nor the modernization approach had made much headway in promoting sustained economic growth or raising living standards. The study's report dismissed four decades of high-tech, large-scale projects imposed from above as ill-considered and often ineffectual:

> The postindependence development efforts failed because the strategy was misconceived. Governments made a dash for "modernization," copying, but not adapting, Western models. The result was poorly designed public investments in industry; too little attention to peasant agriculture; too much intervention in areas in which the state lacked mana-

gerial, technical, and entrepreneurial skills; and too little effort to foster grass-roots development. This top-down approach demotivated ordinary people, whose energies most
needed to be mobilized in the development effort.[80]

But as the report made disturbingly clear, alternative development
approaches were not yet widely implemented in Africa, the most
impoverished continent in the world. In most postcolonial societies sustainable approaches remained ideals rather than realized
projects.

From the early 1980s, even the little momentum such approaches
had achieved began to be dissipated by the Reagan administration's
insistence—through policy pronouncements if not always actual
program implementation—on shifting to private rather than government-funded assistance, and its determination to channel U.S. foreign aid funding into security (usually meaning military) assistance
for client states.[81] In combination with the economic downturn of
the late 1980s, the laissez-faire inclinations of the Reagan and Bush
administrations erased many of the limited gains made by women,
peasants and landless laborers, and the urban poor under alternative approaches to development. Programs to help subsistence cultivators afford the seeds, chemical fertilizers, and water allotments
needed to grow the Green Revolution's more productive strains of
rice and maize were cut back or never implemented.[82] Women in
much of the developing world saw their recently improved access to
education, health care, and career opportunities deteriorate, partly
because of structural adjustments imposed by international lending
agencies, particularly the World Bank and the International Monetary Fund. In the 1980s the World Bank increasingly made its loans
to developing countries contingent on the willingness of recipient
governments to cut back public spending, which often meant reducing subsidies for staple foods and necessities such as cooking oil and

kerosene and scaling back health care, family planning, and career training programs or access to small business loans. As the global economy slumped, women often lacked the political clout to hold on to the economic and social advances they had achieved. The United Nations and other agencies documented a decline in the living conditions of most of the women and the poor in non-industrial societies in the late 1980s.[83] The growing reliance on the trickle-down effects of expanding markets to improve living standards would do little to reverse—and would often give impetus to—these trends in the boom years of globalization in the 1990s.

In the Persian Gulf in 1991, as earlier in Korea, Vietnam, and elsewhere, the constructive, nation-building side of America's mission in the developing world was overwhelmed by the devastation and human suffering inflicted by war. And years of postwar punitive sanctions—enforced by the power of the United States—extended and exacerbated the damage caused by the military action.

For the American public, meanwhile, the scant media coverage of the "collateral damage," from civilian deaths to environmental degradation, combined with the absence of conscription, the brevity of the war, the remarkably few coalition casualties, the negligible effects on life on the home front, and the fact that America's allies paid the lion's share of the bills, made the 1991 Gulf War a high-comfort conflict. America's long-cherished faith in its technological superiority and its ability to manage the turbulent developing world seemed affirmed by the crushing victory. And only the most implacable environmentalist critics pointed out that waging war to deny Saddam Hussein control over Kuwait's oil reserves had done nothing to reduce U.S. dependence on fossil fuels imported from volatile regions of the globe. The postwar intensification of the craze for gas-guzzling SUVs was fed by a reinforced can-do confidence in America's ability to retain, by force if necessary, unfettered access to the imported resources and overseas markets on which the nation's affluence depended.

There can be little doubt that Bush's masterly coalition building and oversight of the successful war to roll back the Iraqi occupation of Kuwait will stand as the most memorable and historically significant undertakings of his presidency. And the president's insistence on multilateralism and working through the United Nations reflected his commitment to pursuing foreign relations in ways that privileged collective security and mutual interests wherever possible. The 1991 Gulf War was also a watershed conflict in terms of both military and technological development and the ways in which they were combined. Westmoreland's battlefield of the future became present reality during the six-week rout of Iraq's armed forces. The technically proficient allied combatants made a compelling case for reliance upon professional, all-volunteer military services. The war also suggested that armies of the future could be a good deal smaller, at least in terms of actual combatants, than the bloated *levées en masse* that had fought the wars between great powers since the time of the French Revolution. But the allied offensives also demonstrated that fighters in wars of the future would require massive, expensive logistical systems to supply and service their high-tech weapons, surveillance and communications systems, and mechanized forces.

The asymmetry in killing power between the coalition's technowarriors and the outgunned, poorly trained Iraqis cautions against drawing broader conclusions from this war about the adaptability of American soldiers or lasting U.S. military supremacy. Those could be tested only in conflicts with technically proficient adversaries armed with comparable communications and weapons systems. In addition, Operation Desert Storm, however overwhelmingly successful, could not exorcize the specter of America's failure in Vietnam. Because the Iraqis fought an archetypically conventional war, their quick defeat has little bearing on the potential for insurgents to win guerrilla wars against the United States or other industrial powers. And in part because the turmoil in the Middle East was considerably intensified by the ambiguous ending and aftermath of the 1991 Gulf War, guerrilla warfare would soon mutate in ways that threatened the United States itself as well as Europe and America's allies in the developing world.

September 11, 2001. The South Tower of the World Trade Center in New York City collapses within an hour of being struck by a hijacked jetliner. Associated Press, Valeo Clearance License 3.5721.4626108-95882.

We as Americans like to put our template on things. And our template's good, but it's not necessarily good for everyone else.

— JAY GARNER, First Administrator of Occupied Iraq, 2003

EPILOGUE: THE PARADOX OF TECHNOLOGICAL SUPREMACY

On September 11, 2001, the North American sanctuary was caught up for the first time in one of the most disturbing currents of twentieth-century history. A Boeing 767 jetliner, which had departed Boston bound for Los Angeles, smashed into the north tower of the World Trade Center in New York City. Minutes later a second passenger plane hurtled into the south tower, which was instantly engulfed in red-hot clouds fringed with thick black smoke. Speculation that the initial collision was accidental—a private plane off course or a similar mishap—was rudely dispelled by the blow to the south tower. Within two hours of the first strike, both towers had collapsed. Horrified onlookers and stunned news analysts struggled to comprehend the meaning of a slaughter without precedent in American history. The civilian population of the United States had been deliberately targeted by foreign aggressors determined to inflict mass casualties and traumatic devastation.

Their nation's previous exemption from military attacks had set Americans off from much of the rest of humanity throughout the twentieth century. The unparalleled advances in scientific discovery and technological innovation of that era, which had made possible exponential increases in the killing power of weaponry, had also contributed to the erosion, and ultimately the near dissolution, of the boundaries between military forces and civilian populations. This blurring had transformed the conduct of war. From the 1930s through the end of the Second World War, military planners' steadily increasing (though largely unwarranted) faith in the decisiveness of massed aerial assaults on enemies' cities and infrastructure had resulted in appalling devastation beyond the battlefield for all major industrialized powers except the United States. And the cold war interventions of the American and Soviet superpowers in troubled postcolonial regions extended the range of the new weaponry to much of the globe. The dramatic shift in the ratio of civilian to military casualties in major conflicts from the First World War through Vietnam underscores this deeply unsettling trend. From less than 10 percent of the fatalities due to military operations in the First World War, civilian deaths soared to over 60 percent of the total in the Second World War, to over 70 percent in the war in Korea, and by some estimates to more than 80 percent in the Vietnam war.[1] Even though the United States was a major combatant in each phase of the twentieth-century march to total war, and by far the world's major source of the innovative weaponry that propelled this transformation, its home front remained inviolable. Even Japan's attack on the U.S.-ruled Territory of Hawaii in December 1941 involved targets and losses that were overwhelmingly military.

As a new millennium opened, America's apparent immunity from the "wholesale violence convulsing the rest of the globe"[2] fed a resurgence of the triumphalism that had peaked at the end of the Second World War, then dissipated during the cold war, particularly after the failure in Vietnam. The collapse of the Soviet Union in 1991, which seemed to diminish the threat of nuclear war, greatly enhanced Americans' sense of national security and engendered a mood of complacency in the boom decade of market-based globalization that followed. The terrorist assaults on New York and Washington in September 2001 shattered the exceptionalist conceit that the United States was impervious to the carnage that had been such a prominent feature of world history in the preceding century. And as the initial shock gave way to somber appraisal of the attacks' larger ramifications, Americans were forced to recognize that the vulnerability of their society had increased even as their nation had emerged as the hegemon of the global order. This paradox between the unprecedented power of the United States and the fact that the civilian population has since been at unprecedented risk of assault by foreign adversaries encapsulates fundamental challenges that are likely to shape public discourse and temper U.S. foreign policy decisions well into the twenty-first century.

Long before the rise of the United States to global hegemony, the nation's dependence on advanced technological systems made it increasingly vulnerable to weapons designed to inflict mass devastation on civilian society. By 1930 the acceleration of suburban sprawl,

encouraged by relatively inexpensive automobiles turned out on assembly lines, had provided the main impetus for the concentration
of nearly half the population in urban areas. In the decades after
1945, regional conurbations, or nearly continuous urbanized areas,
came to define much of the American landscape from Boston to
Washington, D.C., on the western shore of Lake Michigan, and on
the southern coasts of California and Florida. Largely because the
technological systems devised to supply food, water, electricity,
and consumer amenities to these congested zones were normally so
effective, few analysts gave much thought to the ever more pronounced dependence of urban America on external sources of sustenance. By 1970, just over 4 percent of the U.S. population was engaged in producing the food that, along with imports, sustained the
other 96 percent. By the 1990s, both the number of Americans living
on farms and the percentage employed in agriculture had dropped
below 2 percent. Most Americans, who by the post-1945 era were
overwhelmingly concentrated in urban-suburban conurbations, had
little idea where their cellophane-wrapped food or the natural gas
that heated their homes came from, but they assumed that such necessities would be available at the supermarket or summoned by the
turn of their thermostats.[3]

In part because they had not experienced wartime assaults on
their national infrastructure, Americans had not invested a great
deal in providing security for the complex technological systems on
which they depended. Investigations after the September 11 attacks
revealed that even highly sensitive components of these systems,
such as power stations and chemical plants, often had only minimal surveillance and protection. With the important exception of
the threat of nuclear devastation during the cold war, most Americans evinced little concern over the safety of such facilities from assault by external aggressors or internal subversives. Monitoring and
damage control were concentrated on containing breakdowns and

quickly restoring services lost to malfunctioning technology or acci-
dents—such as the near meltdown of the nuclear reactor at Three
Mile Island in Pennsylvania—or natural calamities such as hurri-
canes and floods.

The World Trade Center epitomized the vulnerability of the
highly urbanized and technology-dependent U.S. society.[4] The outsized towers, which had become defining features of the New York
megatropolis, were culminating nodes of a national technological
grid that sustained the material abundance most Americans had
come to regard as an entitlement. Conceived in the 1950s amid "the
technological optimism of the early space age," the proposed skyscraper complex had a "catalytic bigness" that some of its promoters
saw as a dramatic way to reassert New York's commercial preeminence, which had been imperiled by the decline of the city's port facilities after the Second World War. By the 1980s the World Trade
Center was the hub of the New York financial district. And with the
collapse of command communism at the end of that decade, the
twin towers emerged as the investment and trading epicenter of an
ascendant capitalist global order. Like the sprawling urban complex
they anchored, they were dependent on elaborate technological systems, extending far into the surrounding region, that supplied them
with power, water, waste disposal, and the very air breathed by those
who worked in them. The first terrorist assault on the World Trade
Center, in February 1993, revealed that the towers were neither well
protected nor impervious to potentially catastrophic damage.

A number of critics insisted even in the early 1970s, as the twin
skyscrapers began to rise above lower Manhattan, that their size
and radically innovative architecture made them high-risk structures. Ada Louise Huxtable warned in the *New York Times* that the
buildings' untested "gigantism" made them a huge "gamble of triumph or tragedy," liable to a range of problems that were impossible
to anticipate. Lawrence Wien, the owner of the Empire State Build-

ing, speculated that fires or explosions in the towers could cause thousands of casualties, and he predicted that a plane would inevitably crash into one or the other. Defying such cautions, the project's backers made sheer scale a priority. Minoru Yamasaki, the Japanese architect who won the commission to design the structure, was explicitly instructed to come up with what would later be dubbed a "swagger building"—one that would be a landmark for New York City and inspire imitation (presumably on a much smaller scale) in city centers around the globe.[5]

The escalating costs of designing and constructing the towers compelled Yamasaki and his team to find ingenious—and largely untested—ways to create more rentable floor space. By enveloping each tower in massive curtains of steel columns, which carried most of the vertical load and were locked together laterally by steel spandrels, Yamasaki was able to use far fewer vertical girders in the interior than in traditionally designed skyscrapers. And the use of bridge-like steel trusses both to support the floors and to tie in the exterior columns made it possible to dispense with horizontal steel beams in the interior as well. Combined with an innovative elevator system that limited the number of shafts running through all 110 floors, Yamasaki's external support structure conserved an unheard-of 75 percent of the space at each level for office rentals.[6]

Many of the innovations devised to maximize scale and enhance profitability rendered the twin towers highly vulnerable to the attacks in September 2001.[7] Soaring above the forest of skyscrapers on Manhattan, the towers presented even the inexperienced al Qaeda pilots with ready targets. The impact of the jetliners shattered the external support columns over large sections of multiple stories of each tower. Despite this damage, the structures might have survived had it not been for the intense heat of the fires fed initially by high-octane gasoline from the airliners' fuel tanks, which were still nearly full in the early stages of their transcontinental flights.

Because the fires were too high in the buildings to be reached by firefighters wielding heavy equipment, they raged unchecked—replenished by inflammable office materials—through several floors of each tower. The sprinkler systems—inadequate in any case—shut down, as did the grids that provided power for the elevators, lighting, and fresh air. The elevator shafts and stairwells in the central cores (particularly in the north tower), enclosed with wallboard rather than the customary concrete, were ripped open and quickly transformed into giant chimneys, drawing in air to fan the flames and denying escape routes to survivors on the floors above the fires. But the trusses may have been the fatal source of vulnerability. Inadequately fireproofed and fabricated of light steel, they simply melted in the massive ovens that the crash areas had become. When they buckled, both support for the floors they undergirded and the structures that locked in the external steel columns were lost. As the upper stories crashed downward, they precipitated the vertical implosions that obliterated each tower within a matter of seconds.

A good deal of the trauma inflicted on the American public by the September 11 attacks was due to their combination of concentrated devastation with what had seemed to be a high level of implausibility. Despite U.S. involvement in localized conflicts in Africa and the Middle East in the 1990s, the threat of armed aggression against its civilian population, which had loomed for most of the cold war, had appeared to dissipate with the breakup of the Soviet Union. During the cold war decades both superpowers had built huge arsenals of nuclear weapons and intercontinental missile systems to deliver them. According to the doctrine of Mutual Assured Destruction (MAD), which had come to dominate the strategic planning of both adversaries, these weapons systems were specifically designed to inflict mass devastation on civilian populations. By the mid-1950s the nuclear standoff had transformed the great ocean barriers, which for centuries had protected the United States from external assault, into hiding places for Soviet submarines carrying missiles capable of obliterating the densely populated conurbations clustered along America's Atlantic and Pacific coasts.

Ironically, the technologies that put the U.S. civilian population at imminent risk for the first time in history were largely of America's own devising. By the middle decades of the cold war, American scientists and technicians had proved to be the supreme architects of the sophisticated weapons systems whose overkill potential and global reach defied any logic imaginable. Consistent with the illogic of the continuing proliferation of ever more lethal nuclear weaponry, U.S. military planners gave little serious thought to shielding the civilian population that had become the primary target of late twentieth-century warfare. Defensive systems were in fact banned by agreements aimed at reducing risks of nuclear war between the two superpowers by treating their civilian populations as hostages. Defense shelters and evacuation plans focused on protecting weapons for reprisal and ensuring the survival of political and military leaders. Despite a great deal of hype about civil defense early in the cold war,[8] the overwhelming majority of civilians, as well as the complex technological systems on which they depended, were simply written off because there was no effective way to protect them. And contingency planning for national recovery after a full-scale nuclear exchange ranged from feeble to absurd.

Decades before the end of the cold war, the possibility of a nuclear exchange between the superpowers had begun to recede. The Cuban missile crisis in October 1962 and the push for détente in the Nixon-Brezhnev years encouraged the creation of structures of consultation and accommodation—backed up by increasingly sophisticated fail-safe procedures—designed to reduce the possibility of miscalculations by either superpower resulting in a nuclear exchange or localized confrontations escalating into global clashes. The dramatic thaw in U.S.-Soviet relations following the Reagan-Gorbachev summit in October 1986 was widely seen as the beginning of a process that would put an end to the threat of nuclear apocalypse. With the collapse of the command communist systems of the Soviet Union and Eastern Europe, the doomsday prognostications of the preceding decades seemed abruptly to lose relevance—to become merely sensationalist, rather curious artifacts of bygone era. Despite gleeful celebrations in some intellectual and political circles of the triumph of U.S.-style democracy and capitalism, post–cold war triumphalism was generally muted. But few Americans wanted to dwell on the perils of what George H. W. Bush styled "the new world order" or to be reminded that the nuclear weapons systems were still in place, and that those in the former Soviet Union were in fact far less secure from espionage and accident than they had been during the cold war.

The dangers so cruelly brought home to Americans in 2001 by the attacks on the World Trade Center and the Pentagon contrasted in a number of unsettling ways with those of the cold war era. The terrorist attacks seemed to come out of the blue, without warning or the buildup of tensions that had been associated with cold war crises—although Americans who followed world news closely must have been aware of the rising anger against the United States in many developing societies. Investigations since September 2001 have revealed that FBI agents and other intelligence operatives had explicitly reported the possibility of impending terrorist activities within the United States,[9] but U.S. policymakers greatly underestimated the perils posed by alienated groups overseas. The clandestine nature of the organizations that colluded in the September 11 conspiracy meant that there was no possibility for the intense negotiations and offers of accommodation that had become standard procedure in crises between the cold war superpowers. There were no emergency sessions at the United Nations, somber addresses from the Oval Office, or air raid sirens. All the rules and agencies that in the past had mediated confrontations between the United States and external adversaries were contemptuously dismissed by the assailants of September 11.

Consistent with the terrorists' disciplined maintenance of the element of surprise, little was known about the source and motivations of the attacks at the time they occurred. The surprise was even greater because the assaults were planned and executed by operatives from Arab societies that few U.S. policy analysts had imagined were capable of producing aggressors so adept at exploiting the vulnerabilities of a highly industrialized superpower.[10] Subsequent inquiries have indicated that the larger menace that September 11 foreshadowed will prove to be more amorphous and insidious than the communist threat of the cold war decades. But as the small bands of determined attackers demonstrated, they and the cause they killed and died for represent very lethal, direct, and enduring threats to American society.

Americans had long been accustomed to intervening with impunity—and often with military force—in the affairs of non-European peoples. From the earliest clashes between English settlers and the Amerindian peoples of North America through the 2003 Gulf War, technological advantages had given Anglo-American or U.S. forces an overwhelming preponderance of killing power, even in situations, such as the Vietnam war, where determined resistance blunted the American military edge. Excepting localized and temporary victories by Amerindians fighting for *their* homelands and the early Japanese offensives against American colonies in the Pacific in the Second World War, no non-European people had ever responded to confrontations with the United States with attacks on American territory. From the time of America's emergence as a global superpower with both conventional and nuclear arsenals, retaliation of that sort had been all but inconceivable.

The devastation of September 11 was wrought not by a great power rival, which the military and security systems of the United States have historically been oriented to defend against, but by a handful of operatives from several nonwestern, postcolonial nations. The analysis that has appeared thus far on the causes of the attacks has stressed their origins in terrorist organizations professing fundamentalist versions of Islamic belief. But it is critical to assess both the assaults and the larger struggle against the United States in the broader context of American policies toward postcolonial societies, particularly the U.S. campaign to defeat peasant-based revolutionary movements, which was under way in China, Greece, and the Philippines by the late 1940s. Confronted by military interventions launched in the name of anticommunism, dissident peasants and their leaders found that guerrilla warfare was the mode of resistance that stood the best chance of success against the daunting firepower of American military forces. Maoist approaches to guerrilla war, modified by local communist leaders from Vo Nguyen Giap to Fidel Castro, often proved a potent counterforce against military interventions by industrialized nations in Asia, Africa, and Latin America.

Guerrilla tactics have been deployed since ancient times — usually by military forces decisively weaker than their adversaries. But Maoist innovations from the 1930s, which included highly centralized command structures and cadre-centered social organizations, greatly enhanced the capacity of rebel forces to mobilize support and to engage forces armed with industrial weaponry. The three-phase Maoist strategy of revolutionary operations also provided a formula for moving from reactive counterthrusts to offensive operations to the seizure of state power. Recruitment, cadre indoctrination, and terrorist attacks were to be the focus of stage one, which was to build into full-scale guerrilla warfare in stage two. In stage three guerrilla conflict would give way to conventional battles to destroy the incumbent regime's armed forces, thereby clearing the way for its displacement by the forces of revolution.

In the cold war decades, communist leaders in postcolonial states found that if they were adequately supplied with weapons by their allies and had recourse to ample sanctuaries in neutral, neighboring countries, they could topple American-backed regimes and—most emphatically in Vietnam—defeat even direct U.S. military interventions. The international mystique generated by guerrilla victories in China, Vietnam, and Cuba tended to overshadow both the high cost of these successes and the numerous defeats suffered by agrarian rebels deploying Maoist tactics. From at least the early 1960s, the threat to U.S. security posed by the spread of communism through guerrilla warfare was taken as seriously by American policymakers as the possibility of a Soviet nuclear attack. And in the 1980s the Soviets, who had done so much to nurture peasant revolutions against American-backed regimes in the "Third World," were themselves humiliated by a rather different sort of guerrilla resistance in Afghanistan. Their unexpected and thorough defeat played a not inconsiderable role in the breakup of the Soviet empire and the fall of communist dictatorships in Eastern Europe.

Ironically, the crisis of command communism that ended the cold war rendered the crusade to spread the American version of democracy and market capitalism to developing nations more problematic and more treacherous. Because the crisis made the United States the sole global hyperpower, its more sobering ramifications for America's self-appointed global mission have been underestimated or largely ignored. For one thing, the demise of the Soviet Union and its satellite regimes deprived postcolonial countries of major sources of financial assistance, technological transfers, and technical expertise. The sharp drop in aid from the communist bloc was compounded by the severe economic crisis and chronic instability that gripped Russia and many other formerly Soviet states, which led to the diversion to Eastern Europe of a good deal of the Western and Japanese development assistance once channeled to nations in Africa, the Middle East, and Latin America.[11]

The dissolution of the Soviet Union, America's major rival for the political and ideological loyalty of the "Third World," resolved few of the controversies regarding which path to modernity was best suited to developing nations, but it did seem to create new opportunities for U.S. involvement in developing areas. Many of these, however, have proved illusory when actually pursued, and in fact are likely to increase threats to U.S. security in the foreseeable future. For example, the breakup of communist regimes unleashed an epidemic of ethnic clashes in areas as diverse as the Balkans, East Africa, and Central Asia. These conflicts have provoked U.S. intervention in often peripheral locales, where deeply rooted historical enmities defy resolution by even highly informed and well-intentioned mediators.

Despite these decidedly cautionary post–cold war trends, ambitious schemes — often involving calls for the resuscitation of modernization theory[12] — for advancing development priorities in accord with American precedents resurfaced in a number of regions from the Balkans and the Middle East to Somalia and Haiti. But President George W. Bush, in his inaugural address in early 2001, set forth clear limits to American intervention, adhering (until September 11, 2001) to the basic approach to postcolonial societies that the Clinton foreign policy team had pursued in a much lower key.[13] Eschewing full-blown nation-building commitments, both administrations preferred to allow market mechanisms and a robust capitalist-driven globalization to advance a trickle-down approach to economic growth and social improvement in the developing world.[14] Serious research on the impact of this laissez-faire version of America's civilizing mission is as yet fragmentary and selective. Nonetheless, critics of the unrestrained competition fostered by the accelerated extension of the global market economy in the 1990s have marshaled a substantial body of evidence suggesting that it has often led to further deterioration in the quality of life endured by the poor, by women, and by the unskilled working classes, who make up the great majority of the citizenry of postcolonial nations.

Even enthusiastic advocates of globalization have warned that unless opportunities for entry into international market networks are expanded and social safety nets buttressed, unbridled capitalism could fan widespread social unrest in the developing world. In a global system in which American hegemony has become so pervasive, U.S. investment and even U.S. economic assistance, which is widely viewed in developing nations as synonymous with that provided by international lending agencies, provide compelling targets for dissidents. These include religious fundamentalists of varying persuasions, who excoriate U.S.-led globalization as the main cause of the growing gap between wealthy minorities and the impoverished masses. And though the often-conflicting information amassed thus far indicates that the sources of resistance to capitalist globalization are complex and varied, investigations into the backgrounds of the perpetrators of the September 11 attacks underscore their determination to engage in aggression against the United States, which they perceived as the prime mover behind the commercialization, secularization, and homogenization of the global order.[15]

With the decline of Soviet Union, the United States also lost the countervailing force that had provided significant constraints on

American military interventions in the postcolonial arena. Not surprisingly, many U.S. policymakers have seen this diminution of checks on the exercise of American power overseas as a positive development. But military forays into troubled areas like Somalia, Afghanistan, and Iraq have generated new sources of hostility and revealed unsettling levels of vulnerability on the part of both America's high-tech forces deployed overseas and—as September 11 made clear—civilians within the United States. The fact that the United States was the only remaining military superpower heightened the expectation of both America's leaders (if not the broader public) and substantial segments of the world community that it could be counted on to intervene in crisis situations where force was deemed essential. Iraq's invasion of Kuwait in 1990 and the U.S. leadership of the coalition that crushed Saddam Hussein's expansionist ambitions in 1991 appeared to validate America's self-appointed mission as global policeman. But as calls for intervention from the Balkans to the Caribbean to Africa proliferated in the 1990s, the policeman's propensity to become a target for dissidents and renegades became increasingly evident to policymakers in both the George H. W. Bush and Clinton administrations. At the same time, despite the rout of Iraqi military forces, memories of the political fallout from the heavy U.S. casualties in Vietnam instilled a deep reluctance, particularly in the Clinton administration, to put American forces in harm's way unless vital U.S. interests were at stake. These rationales for caution were accentuated by the ascendancy of the Powell doctrine after its successful application in the 1991 Gulf War.[16]

These often conflicting imperatives ensured that the United States would be a highly selective and reluctant enforcer. On the one hand, it refused to intervene to put an end to genocidal slaughter in Rwanda and responded belatedly to attempted "ethnic cleansing" in Bosnia. In sharp contrast, it was willing to use its high-tech arsenal to head off the threat of these sorts of outrages in Kosovo and to punish—through an indefinite extension of a crippling embargo and intermittent cruise missile strikes—a defeated Iraq that had failed to overthrow the regime of the unrepentant Saddam Hussein. Fomenters of multiplying resistance movements to U.S. hegemony, including Osama bin Laden and his lieutenants, were prone to emphasize in their recruiting rhetoric that these inconsistencies revealed the tentative and self-serving nature of American foreign policy. They also charged that the United States applied the same double standard when deciding to abet or do little about the development of weapons of mass destruction and military occupations by its regional surrogates, such as Pakistan and Israel, while denying Iraq or Iran similar latitude to pursue perceived national interests.

In the aftermath of the 1991 Gulf War it was clear that America enjoyed unprecedented military superiority over any potential adversary.[17] This capacity to deploy overwhelming force has proved critical to U.S. aspirations to enforce global law and order, but it also has the potential to engender arrogant policies and ill-considered punitive strikes. It certainly provided impetus for the move away from the multilateralism that George H. W. Bush and his advisors had pursued so skillfully in expelling Saddam Hussein's legions from Kuwait. Despite globalist rhetoric to the contrary, Bill Clinton increasingly distanced the United States from its erstwhile allies in the coalition Bush had forged. The blatantly unilateralist posture adopted by George W. Bush's administration after the September 11 attacks only enhanced the visibility of widely unpopular U.S. approaches to a wide range of global issues, particularly those of great importance to developing nations. America's deepening isolation on questions ranging from the Kyoto protocols on global warming to the establishment of an International Criminal Court made it more than ever the focus of "Third World" resentment.

Even under the comparatively internationalist administration of the first President Bush, American intransigence had surfaced in the refusal to honor assurances to the leaders of Saudi Arabia that U.S. troops stationed in their kingdom would be withdrawn as soon as the danger of Iraqi aggression had passed and Kuwait had been liberated. For many devout Muslims the continuing presence of these forces in the land of Islam's most sacred shrines was deeply defiling. And the pledge that three administrations left unfulfilled came back to haunt the global hegemon. The "infidel occupation" of Saudi Arabia was repeatedly cited by Osama bin Laden and his followers, including some of the operatives involved in the terrorist outrages on September 11, as one of their most important motivations for waging war against the United States.[18]

Laser-guided weapons, coordinated by a panoply of surveillance devices, made it possible for the American military to track, fix, and attack enemy forces from great distances and with stunning rapidity. The low levels of American casualties that, so far at least, have resulted from cybertech military responses—including full-scale invasions of varying magnitudes—have made them a preferred option for American politicians and planners coping with perceived threats from the developing world. But reliance on high-tech quick fixes in crisis situations, where deep-seated and complex interests converge, has led on a number of occasions to military strikes launched on an ad hoc basis with insufficient consideration of their longer-term repercussions. The failed attempt to capture two key lieutenants of the Somali strongman Muhammad Farrah Aidid, for example, which touched off the deadly "Black Hawk down" firefight in Mogadishu in October 1993, was the product of policy drift. American famine relief efforts, initiated by the George H. W. Bush administration, had gradually devolved in the early Clinton presidency into an ill-informed, desultory, and inadequately prepared exercise in political manipulation in a disintegrating postcolonial state. Even more disturbing in terms of hasty implementation and apparent lack of grounding in broader policy objectives were the cruise missile attacks in August 1998 against alleged al Qaeda training camps, launched in reprisal for terrorist bombings of American embassy compounds in Kenya and Tanzania months earlier. Not only did several of the seventy-five Tomahawk missiles go astray and kill civilians in Pakistan, at the time a wavering ally of the United States, those on target destroyed what turned out to be a pharmaceutical factory in Sudan—a neutral nation—and an abandoned al Qaeda camp in Afghanistan.[19]

Despite the fact that both of the political organizations targeted in these strikes had been officially declared potential long-term threats to the United States itself, its overseas forces, or its global interests, each strike was an isolated reprisal, without sustained follow-up operations. In Somalia, the forced retreat of American forces from Mogadishu in 1993 precipitated the withdrawal of U.S. combat troops from that strife-ridden country in March 1994. More ominously, the ineffectual cruise missile diplomacy directed at al Qaeda facilities is reported to have prompted Osama bin Laden's second in command, Ayman al-Zawahiri, to declare: "The war has started. Americans should wait for an answer." It also strengthened Bin Laden's assumption that the United States lacked the will to engage his fighters on a sustained basis.[20] Both of these ill-considered interventions confirmed the admonitions of Owen Harries, one of the more insightful critics of the aggressive stance toward the developing world advocated by Secretary of State Madeleine Albright during Clinton's second term, which were summed up in Harries's prescient maxim "Tactical energy in a strategic and conceptual void is a recipe for disaster."[21]

Although waging cybertech wars against vastly outgunned pre-industrial societies has been a surefire way to raise flagging presidential approval ratings (at least in the short term), the civilian casualties such wars have inflicted have greatly intensified anti-American sentiment overseas. Massive American firepower, deployed in the Middle East and elsewhere in the postcolonial world, has generated widespread support in targeted societies for even corrupt and despotic leaders calling for sustained, violent aggression against the United States and its allies.[22] Americans' claims that their advanced surveillance systems and the pinpoint accuracy of their futuristic weaponry minimize civilian losses ring false in much of the world. While in reporting on these wars the mainstream American media focus almost exclusively on military operations, in other countries hostility to the United States has been fanned by vivid and extensive press and television coverage of the suffering imposed on civilians. America's high-tech responses—which in the early twenty-first century are even more reliant on the overwhelming firepower and distancing of combat forces from the victims of their aggression that proved so counterproductive in Vietnam—may well not only feed anti-American hostility but also lead to "the perception of the United States as a cowardly nation unwilling to back up its principles with genuine sacrifice."[23]

The collapse of the Soviet Union, and thus the demise of a credible balance of global power, appeared to presage an era in which the United States could intervene with impunity anywhere on the globe where its interests were threatened or its designs frustrated. But America's drubbing of the conventional forces arrayed against it in Iraq in 1991 and 2003, as well as in Yugoslavia and Afghanistan, forced major changes in approaches to resistance on the part of militant dissidents in the developing world. As the policy analyst Matthew Brzezinski observed in 2003: "Taking away any chance of the enemy's inflicting losses on the battlefield might also spur a more ominous development. Adept enemies will search for weaknesses, and if those weaknesses can't be found on the traditional field of battle, that might mean exporting the fight to other places . . . It could mean co-ordinated attacks on the civilian population of the United States, state-sponsored terrorism on a scale we have not yet seen."[24] Though Brzezinski sees these adaptations as future possibilities, September 11 made it evident that they are already well under way.

Beginning with Islamic revivalist resistance to the Soviet occupation of Afghanistan in the 1980s, major alterations were made in the organization and strategy of guerrilla warfare. These changes were increasingly globalized following the rout of largely Soviet-armed Iraqi conventional forces in the 1991 Gulf War by an alliance empowered by cybertech American weaponry. In response to the rather different outcomes of these challenges to the global superpowers, militant opponents of U.S. hegemony deemphasized the second and third phases of Maoist guerrilla strategy—the shift to *offensive* guerrilla operations and conventional war for capture of state power—and gave precedence to tasks delineated for the first, defensive phase: recruiting, indoctrination, and terrorist operations. In part this shift was a logical outgrowth of the successes achieved in the decade-long war of resistance in Afghanistan. The Soviet withdrawal came while rebel groups were still mounting defensive guerrilla operations; the move to full-scale offensives and conventional combat occurred only when the Taliban seized state power in a civil war against internal rivals.

Equally critical in shaping the decision to concentrate on guerrilla terrorism was the unlikelihood that militant organizations could precipitate the collapse of American democracy, much less seize control of the United States. Insofar as al Qaeda has set the agenda for violent anti-Americanism since the end of the cold war, the main objectives have been to frustrate the hyperpower's efforts to project its influence overseas, particularly in the Middle East and the Islamic belt more generally, and to strike at American targets both in the United States and abroad in reprisal for U.S. military interventions in the developing world. In contrast to revivalist Islamic movements, such as the Muslim Brotherhood in Egypt and the revolution led by the Ayatollah Khomeini in Iran, guerrilla terrorist organizations like al Qaeda have not been dedicated to seizing state power and building Islamic utopian societies. Their operations fall between the first and second stages of Maoist-style guerrilla warfare. To this point, al Qaeda and its militant Islamic allies have demonstrated a considerable capacity to threaten and destroy, but have offered little in terms of enduring sociopolitical organization or a more balanced and just international order.

In addition to reconfiguring their revolutionary strategies, militant Islamic organizations, including al Qaeda, have internationalized their networks through alliances among regionally focused cells. From finances and recruiting to communications and logistical backup, guerrilla terrorism in the 1990s was designed to be global in scope and dedicated primarily to training and indoctrinating militant cadres. Broadening the range of their transmuted version of guerrilla warfare allowed terrorist organizations to strike far from their home bases and to do so without offering clear targets for high-tech reprisals.[25]

As the invasion of Afghanistan provoked by the September 11 attacks amply demonstrated, even the supreme exemplar of guerrilla terrorism, al Qaeda, has neither the numbers nor the weapons to stand a chance in open combat against the sophisticated surveillance systems and killing machines that the United States and its allies, particularly Great Britain, can deploy anywhere in the world. Consequently, terrorists, like guerrilla fighters through the ages, operate clandestinely much of the time, emerging periodically for sudden assaults on carefully chosen targets. The shock caused by such operations publicizes the dissidents' grievances and alternative vision; demonstrates the weakness of their enemies; undermines civilian support for adversary regimes; and attracts recruits who join their cause out of fear or admiration.

All these objectives were apparently involved in the planning for September 11, 2001. The terrorist operations were carefully prepared over several years at an estimated cost of half a million dollars.[26] They were carried out by teams of four or five well-trained and intensively indoctrinated conspirators, who were the only al Qaeda members lost in the day's operations. Evading U.S. intelligence agencies and security precautions with disconcerting ease, the terrorists inflicted what until that time were unimaginable civilian casualties. They also very likely surpassed their own expectations in destroying the twin towers, which symbolized America's global economic influence, and severely damaging the Pentagon, which stood for U.S. military power. The attacks were carried out by organizations that could not have begun to conceive, much less manufacture, the high-tech weaponry that secured America's global hegemony. But the conspirators managed to capture some of the world's most sophisticated civilian technology. Wielding the crudest of weapons—cutters used for cardboard boxes—they hijacked four airliners and converted them into jet-propelled missiles capable of wreaking terrible devastation.

The understandable anger and determination to exact revenge for the losses of September 11 have fed a predictable revival of the militant rhetoric that has been used to justify America's civilizing mission for centuries. Although President George W. Bush's initial calls for a global "crusade" against terrorists were quickly reworked into more tactful exhortations to wage an extended war on terrorism, the lethal assaults on the twin towers and the Pentagon continued to be condemned as barbaric or savage acts, perpetrated by cruel and cowardly zealots striking at the heart of the civilized world from anarchic, backward lands. In the run-up to the war against al Qaeda and its Taliban hosts in Afghanistan, the president was prone to mix such imagery from the imperialist era with gunslinger taunts reminiscent of the western frontier, including promises to run the varmints into the ground, smoke them out, and bring them back dead or alive.

But the most persistent, rousing, and problematic of the civilizing mission themes resurrected by the Bush administration has been the depiction of the war against terrorism as a struggle of good versus evil. Accompanied by an often explicitly articulated assumption that Islamic fundamentalism—and by inference all varieties of Muslim revivalist or resistance movements—are synonymous with terrorism, this vilification of selected states as components of an "axis of evil" has undermined forces for moderation in developing societies that seek ways of accommodating their national interests to American global hegemony. It may also intensify the threat of nuclear warfare, which had appeared to recede with the passing of the cold war. The current inclination of American politicians and media to demonize nations resistant to incorporation into the U.S.-dominated global system may well push Korea, Iran, and other developing countries to accelerate research programs leading to the production of nuclear weaponry. As an alternative to guerrilla terrorism as a means of coping with America's cybertech arsenal, the temptation for postcolonial societies to acquire weapons of mass destruction could well prove dangerous to international security.[27]

The tendency to see current Islamic resistance to America's path to "civilized modernity" as part of a more general, centuries-old and implacable hostility to the Christian West—what is known as the "clash of civilizations"[28]—may prove even more detrimental to America's relations with the developing world than the rhetoric of the civilizing mission. Not only is this explanation for the surge of anti-Americanism in recent decades based on a highly selective and questionable reading of the historical record, it is premised on a number of problematic presuppositions. It assumes that discrete and well-integrated civilizations can be isolated, and that the concept of civilization itself is the proper way to categorize highly variegated culture systems. This perspective privileges intercultural barriers—which were and are more vague and porous than it assumes—and cross-cultural conflict over the structures of accommodation and networks of peaceful exchange that have predominated through most of human history.

These distortions have been of particular relevance since the eclipse of America's communist rivals, as Muslims have become a main focus of America's insecurities and an object of its censure for alleged bigotry and militant aggression against non-Muslims. Though intolerance and calls for violent confrontation with religious and secular infidels have been central features of militant Islamic movements, and although the numbers of their adherents have been on the increase, within the global Muslim community these predilections have historically been the exception rather than the rule. In the history of Islam, which spread largely through peaceful conversion in the millennium after the appearance of the prophet Muhammad in the seventh century, internal divisions have often been more intense than clashes with non-Muslims. As a substantial corpus of scholarship has demonstrated, Jews and Christians, considered by Muslims to be fellow "people of the book," have worshiped freely and flourished economically and culturally under Muslim regimes from al-Andalous to Anatolia.[29]

Especially in the United States, the current obsession with Islamic militancy is not a product of a deeply entrenched opposition between monolithic civilizations. Much of the cross-cultural animosity and violence in the contemporary world has been generated by militant fundamentalism, which has been on the rise in recent decades not only in Islam but in a number of religious traditions, including Hinduism, Christianity, and Judaism. And it is important to keep in mind that for much of the Muslim world—as well as a sizeable portion of the more secularized citizenry of Europe, East Asia, and other regions—George W. Bush, with his base of support on the Christian right and his widely publicized claims of personal divine council, epitomizes a militant fundamentalist leader.[30]

The resurgence of America's civilizing rhetoric since September 11, 2001, has been accompanied by the restoration of the mix of nation-building projects and violent interventions that had been widely deemed discredited since the humiliating retreat from Vietnam. The Vietnam debacle was not the only instance in which American attempts to shape the course of development of non-western societies by deploying this blend of material incentives and punitive reprisals yielded only limited gains and resulted in severe setbacks. The long history of these interventions, embarked upon with can-do enthusiasm and unbounded faith in the power of U.S. technology to transform even the most ancient and sophisticated cultures, should provide cause for caution at the very least. And, paradoxically, the overwhelming technological superiority of the United States over countries that may become objects of the "nation-building" variant of its civilizing mission ought to heighten rather than mollify concern about such ventures.

Even during the cold war, its innovative high-tech arsenal often enabled the United States to project its military power with considerable success into contested areas of the postcolonial world. The more recent quantum leap in superiority made possible by cyber-technologies has rendered resistance by conventional forces largely self-destructive. Yugoslav air defenses could do little to counter strikes by Stealth bombers and cruise missiles; and in 2003 Saddam Hussein's air force dared not risk a single fighter plane in defense of his overmatched regime. But after both those conflicts, American triumphalism obscured shortcomings of the victories and encouraged miscalculations about the future projection of U.S. military power in the world. Following the forced withdrawal of Yugoslav troops from Kosovo in the summer of 1999, Clinton aides and military analysts declared that air power had finally been vindicated as the key to victory in modern warfare. This assessment overlooked the inconvenient fact that Clinton's public pledge that he would not put American or NATO ground troops at risk had allowed Yugoslav forces to proceed with "ethnic cleansing" in Kosovo. And the fierce aerial bombardment left Slobodan Milosevic and his henchmen in power in Belgrade.[31] Equally ironically, the rapid U.S. victory in the 2003 invasion of Iraq called into question the G. W. Bush administration's claims that war had been imperative because Saddam Hussein's Iraq posed a major threat to its neighbors and to the world. The poor performance of Iraqi forces should also caution against using the 2003 conflict as a gauge of the U.S. military's preparedness for a major war.

Consistent with the pattern of earlier U.S. interventions in Asia and Latin America, forced entry continues to be much less difficult than short-term occupation, much less the building of viable nations. President Bush and his administration have cited the reconstruction of Japan after the Second World War as a demonstration of America's capacity for successful nation-building, but thus far the occupations of both Afghanistan and Iraq suggest that Vietnam may be a closer analogy than Japan.[32] Long before their descent into military dictatorship in the 1930s, the Japanese had striven to establish democratic institutions. They had also achieved a high level of industrialization decades before the military seized power. In addition, the Japanese population was remarkably ethnically homogeneous and deeply attached to symbols of national unity, particularly those associated with the imperial household. When the American occupation began in 1945, the Japanese were exhausted by nearly a decade and a half of conflict with China and world war, and their industrial infrastructure and their cities were in ruins. When American strikes on Afghanistan began in October 2001, the country's much less developed infrastructure and urban areas had already been reduced to rubble by guerrilla resistance to Soviet occupation and by civil war. Even more than in North Vietnam in the late 1960s, U.S. air and naval commanders were hard-pressed to find significant targets for their high-tech weapons.[33]

Osama bin Laden's alliance with the Taliban proved a major miscalculation for al Qaeda and a disaster for the Afghan regime because it provided the American military with a state center to target. Al Qaeda's support for the Taliban also drew its guerrilla-terrorist fighters into a conventional war in which the United States could effectively deploy its advanced weaponry. In the U.S.-led invasion of Iraq in 2003, Saddam Hussein and his advisors again demonstrated the lack of strategic vision that had ensured Iraq's defeat in 1991. Their pig-headed determination to wage conventional war resulted in the rapid and decisive destruction of Iraq's regular armed forces. Even though North Vietnam had offered a centralized state target in the 1960s, and its armies had almost invariably suffered costly defeats whenever they engaged U.S. units in conventional battles, guerrilla tactics had allowed PAVN and NLF forces to conduct a parallel war of attrition that proved too lengthy, costly, and indecisive for America to persevere.

The occupations of Afghanistan and Iraq after the Taliban and the Ba'ath party were driven from power necessitated a shift away from high-tech warfare—designed to reduce direct combat between adversary forces—to demanding and dangerous operations essential to establishing control on the ground, which have strong similarities to those mounted against the NLF in Vietnam. Occupation troops are far less able to deploy their cybertech weaponry because it is likely to cause high levels of casualties among civilians and friendly forces.

While patrolling crowded streets or traveling through contested terrain, occupying forces, as well as civilian advisors, are highly vulnerable to guerrilla assaults and terrorist bombings. As the ambush of American special forces units in Mogadishu in 1993 had revealed, when combat is at close quarters, the most advanced helicopters and the best-trained soldiers are vulnerable to handheld grenade launchers and submachine guns, even when those weapons are wielded by poorly trained fighters.

In Afghanistan, and even more egregiously in Iraq, post-conquest occupations have devolved into quagmires that in some ways recall Vietnam. All these debacles have owed a great deal to the overweening can-do confidence—grounded in exaggerated assessments of the potential of American technology to surmount all obstacles—that has been an enduring feature of U.S. efforts to dominate nonwestern societies. Techno-hubris goes far to explain the miscalculations of the civilian planners in the Pentagon who were the main architects of the 2003 invasion of Iraq. It accounts for the inadequacy of preparations for the postwar phase, which were premised on inflated expectations regarding Iraqi support for the invaders and a corresponding tendency to downplay the strength and staying power of resistance forces. It has contributed to serious underestimations of the spiraling costs of the occupation, of the size of the military and police forces needed for effective control, and of the difficulties of rebuilding a partially modernized society that has been reduced to a shambles. It has left the often ill-informed occupying powers at a loss as to how to democratize societies with deep ethnic and religious divisions, or how to counter the local power brokers, such as the warlords in Afghanistan and the Sh'ia clergy in Iraq, who oppose the occupiers' intent to centralize power in the hands of a regime of their own choosing.[34]

The guerrilla-terrorist outrages of September 11, 2001, which exposed the vulnerability of the U.S. citizenry, provoked the invasion and occupation of Afghanistan. The fears and jingoism aroused by the September 11 attacks, in conjunction with allegations (as yet unsubstantiated) of links between al Qaeda and Saddam Hussein's regime, opened the way for the war on Iraq. In the current Afghan and Iraqi incarnations of America's civilizing mission, as in Vietnam, especially after 1965, military aggression and forcible occupation have thus far taken precedence over development projects and nation-building initiatives. If the present course of these two occupations is maintained, they may well prove even more debilitating for the United States than the defeat in Vietnam. They will certainly contribute substantially to upward-spiraling national indebtedness and, in the longer term, inflation. Both trends will intensify U.S. dependence on foreign investors, whose capital transfers to cover the ballooning government deficits and whose commitment to the dollar as the standard of international exchange have for some decades been essential to national solvency and the maintenance of American living standards.[35]

The dangers of overcommitment arising from the revival of America's civilizing mission are apparent. As in the Vietnam war, they are magnified by the pugnacious unilateralism of recent U.S. foreign policy, which has alienated many traditional allies and drastically reduced the military and economic contributions that they, and the United Nations, are willing to provide for American interventions in the developing world.[36] The contrast between the levels of international backing for the first and second American-led wars in Iraq also suggests that unilateralism has greatly increased the U.S. citizenry's share of the casualties and financial burdens of wars, occupations, and reconstruction efforts. And going it alone in ways that deliberately flout world opinion makes the United States the preferred target for future guerrilla-terrorist operations. All these ill-considered postures have contributed to American overextension in the world, which was already apparent during the cold war. Unilateralism and indiscriminate interventionism will almost certainly demand increases in the proportion of national resources allocated to the military rather than the civilian sectors of the economy. With the reduction of both capital and ingenuity for research and development for consumer production, the United States may again fall behind its competitors in such key markets as automobiles, audiovisual equipment, and personal computers, as it did after the end of the Indochina wars. These trends are likely to renew fears of economic decline that preoccupied financial pundits and members of Congress in the 1980s, and they could in the long term undermine the nation's global dominance. If these concerns again become central to debates about America's position in the world system, it will be critical to include the costs of popular resistance and guerrilla-terrorist reprisals emanating from the developing world—issues that were largely neglected in the debates over the causes of the decline and fall of empires in the last decade of the cold war.[37]

The drain of resources from the civilian to the military sectors of the U.S. economy could also spark a resurgence of the civil unrest and ethnic strife of the Vietnam years, which owed much to the disparities in living standards, opportunities, and burdens of military conscription made apparent by that prolonged conflict. In the twenty-first century, such dissent could well be intensified by the threat that unilateralism and overextension overseas pose for American social entitlements, which have already been significantly reduced in the past two decades. And "blowback" from America's resurgent mission to civilize in the developing world will almost certainly require an expansion of measures for "homeland security," a trend that may prove the greatest danger of all to the citizenry of the global hegemon. Heightened fears of guerrilla-terrorist strikes after the September 11 attacks prompted a significant erosion of rights and freedoms that have been foundational for American democracy. By 2003, two years into the war on terror, public opinion polls and con-

gressional soundings indicated that increasing numbers of the populace were beginning to have second thoughts about this erosion of
rights, even though to that point it had mainly affected noncitizens
residing in the United States.[38] But future terrorist incidents could
rekindle support for intrusive surveillance technologies and long
incarcerations on terms that violate constitutional protections and
deny prisoners even their basic right to know who their accusers are
and what crimes they are alleged to have committed. If these dangers become realities, the guerrilla-terrorists will have succeeded in
delivering mortal blows to American democracy. In the broader perspective of global history, the paradoxical vulnerability of the first
hyperpower—fixed in human memory by the collapse of the twin
towers—may both undermine the civil rights and unsettle the multiethnic accommodation that are among the supreme achievements
of the American experiment in representative government.

ABBREVIATIONS

AWST   *Aviation Week and Space Technology*

BIA   Records of the Bureau of Insular Affairs, U.S. National Archives

CRMJ   *The Chinese Recorder and Missionary Journal*

DSB   *Department of State Bulletin*

GPO   U.S. Government Printing Office

LCMD   Library of Congress, Manuscript Division

NYT   *New York Times*

PCR   *Philippine Commission Reports*

PP   *Pentagon Papers,* Senator Gravel edition (Boston: Beacon Press, 1971)

RCIA   U.S. Department of the Interior, *Report of the Commissioner of Indian Affairs*

NOTES

Introduction: A Train for the Shogun

1. Samuel Eliot Morison, *"Old Bruin" Commodore Matthew C. Perry, 1794–1858* (Boston: Little, Brown, 1967), 284–285, 332; John H. Schroeder, *Matthew Calbraith Perry: Antebellum Sailor and Diplomat* (Annapolis: Naval Institute Press, 2001). For biographical information on Perry and accounts of the origins of and preparations for the Japan expedition, I have relied mainly on Morison and Schroeder, as well as Peter Booth Wiley with Korogi Ichiro, *Yankees in the Land of the Gods: Commodore Perry and the Opening of Japan* (New York: Viking, 1990); and Arthur Walworth, *Black Ships off Japan: The Story of Commodore Perry's Expedition* (New York: Knopf, 1946).

2. See H. Paul Varley, *Japanese Culture* (Honolulu: University of Hawaii Press, 1984), 109–113, 143–145; Ronald P. Toby, *State and Diplomacy in Early Modern Japan: Asia in the Development of the Tokugawa Bakufu* (Princeton: Princeton University Press, 1984), 184–209. Williams quoted in Walworth, *Black Ships*, 96, 104–105.

3. Allan B. Cole, ed., *With Perry to Japan: The Diary of Edward Yorke McCauley* (Princeton: Princeton University Press, 1942), 29, 85, 90.

4. James Morrow, *A Scientist with Perry in Japan* (Chapel Hill: University of North Carolina Press, 1947), 136–138; Cole, *Diary of McCauley*, 27–28.

5. Walworth's *Black Ships*, ch. 16 title. An exception is Walter LaFeber's *The Clash: A History of U.S.-Japan Relations* (New York: Norton, 1997).

6. Thomas Parke Hughes, *American Genesis: A Century of Invention and Technological Enthusiasm* (New York: Penguin, 1989), 1–2 (quotations). See also see Walter LaFeber, "Technology and U.S. Foreign Relations," *Diplomatic History* 24/1 (2000), 1–19; Nick Cullather, "Research Note: Development? It's History," ibid. 24/4 (2000), 640–653.

7. Walworth, *Black Ships*, 23. My overview of the U.S. domestic transformations draws on the superb syntheses in Allan Cole's introduction to McCauley's diary, esp. 15–25; and Carroll Pursell, *The Machine in America: A Social History of Technology* (Baltimore: Johns Hopkins University Press, 1993), chs. 3–5.

8. Cole, *Diary of McCauley*, 15–16.

9. See, e.g., Emily S. Rosenberg, *Spreading the American Dream: American Economic and Cultural Expansion, 1890–1945* (New York: Hill and Wang, 1982); Frank Ninkovich, *The Diplomacy of Ideas: U.S. Foreign Policy and Cultural Relations, 1938–1950* (Cambridge: Cambridge University Press, 1981). Revisionist diplomatic historians have stressed internal social and economic instability as the driving force behind U.S. overseas expansion, and more recent scholars have focused on domestic cultural anxieties to explain cross-cultural perceptions in international relations. For the revisionists' position see William Appleman Williams, *The Roots of the Modern American Empire* (New York: Random House, 1969). For expositions of the recent emphasis on cultural factors, which often converge with social and economic interpre-

tations, see Amy Kaplan, "Domesticating Foreign Policy," *Diplomatic History* 18/1 (1994): 97–106; Laura Hein, Free-Floating Anxieties on the Pacific: Japan and the West Revisited," ibid. 20/3 (1996): 411–437.

10. Morison, *Old Bruin*, 127.

11. Robert Tomes, *The Americans in Japan: An Abridgement of the Government Narrative of the U.S. Expedition to Japan* (New York: D. Appleton, 1857), 239–240; Cole, *Diary of McCauley*, 24; S. Wells Williams, "A Journal of the Perry Expedition to Japan (1853–1854)," *Transactions of the Asiatic Society of Japan* 37/2 (1910), 148; Peter Duus, *The Japanese Discovery of America: A Brief History with Documents* (Boston: Bedford, 1997), 9–10, 12–13, 39–40; Schroeder, *Matthew Perry*, 228.

12. H. F. Moorehouse, *Driving Ambition: An Analysis of the American Hot Rod Enthusiasm* (Manchester: Manchester University Press, 1991). The German experience suggests that Americans were not alone in seeing the automobile as the repository of numerous possibilities for personal fulfillment; see Wolfgang Sachs, *For Love of the Automobile: Looking Back into the History of Our Desires* (Berkeley: University of California Press, 1992).

13. Cole, *Diary of McCauley*, 86, 105; Tomes, *Americans in Japan*, 265.

14. See, e.g., Cole, *Diary of McCauley*, 22, 27, 29, 83, 87, 158; Walworth, *Black Ships*, 24–25; Morison, *Old Bruin*, 327–329.

15. On these connections see, e.g., essays by A. R. Hall and Peter Mathias in Mathias, ed., *Science and Society, 1600–1900* (Cambridge: Cambridge University Press, 1972), and those by Peter Drucker and James Feibleman in *Technology and Culture* 2/4 (1961). On the convergence of science and technology in the decades on either side of the Perry embassy, see Edwin Layton, "Mirror-Image Twins: The Communities of Science and Technology in 19th-Century America," *Technology and Culture* 12/4 (1971): 562–580.

16. Cole, *Diary of McCauley*, 20.

17. Ibid., 85, 90, 100; Cole, *Yankee Surveyors in the Shogun's Seas* (Princeton: Princeton University Press, 1947); Roger Pineau, *The Japan Expedition: The Personal Journal of Commodore Matthew C. Perry* (Washington: Smithsonian Institution Press, 1968), 176–180, 194; William Heine, *With Perry to Japan* (Honolulu: University of Hawaii Press, 1990), 74–75; Morrow, *A Scientist with Perry*, 128–143.

18. Cole, *Diary of McCauley*, 30. Japanese artisans had performed a similar feat centuries earlier, when, within a generation after the Portuguese introduced firearms to the islands, they were manufacturing superior guns based on the imported prototypes. See Noel Perrin, *Giving Up the Gun: Japan's Reversion to the Sword, 1543–1879* (Boston: Godine, 1979), 16–17.

19. See, e.g., Duus, *Japanese Discovery*, 16–18, 30–33; Morison, *Old Bruin*, 332–333 (quoting translations of records left by Japanese officials); William Elliot Griffis, *Matthew Calbraith Perry* (Boston: Cupples and Herd, 1887), 333, 344–348, 367–370. On the internal turmoil in Japan see Victor Koschmann, *The Minto Ideology* (Berkeley: University of California Press, 1987); Conrad Totman, *The Collapse of the Tokugawa Bakufu, 1862–1868* (Honolulu: University of Hawaii Press, 1980).

20. See Duus, *Japanese Discovery*, 39, 68, for earlier and later observers in agreement with Perry and his entourage.

21. Charles C. Gillispie, *Edge of Objectivity: An Essay in the History of Scientific Ideas* (Princeton: Princeton University Press, 1960), 9. For more on Rostow see Chapter 5.

22. Morison, *Old Bruin*, 332–333.

23. John S. Haller, *Outcastes from Evolution: Scientific Attitudes of Racial Inferiority, 1859–1900* (New York: McGraw-Hill, 1970); Stanford M. Lyman, "Slavery and Sloth: A Study in Race and Morality," *International Journal of Politics, Culture, and Society* 5/1 (1991): 49–79; Barbara Jeanne Fields, "Slavery, Race and Ideology in the United States of America," *New Left Review* 181 (1990): 95–118; George Fredrickson, *Racial Supremacy: A Comparative*

Study in American and South African History (Oxford: Oxford University Press, 1980), ch. 5.

24. Stephen Jay Gould, *The Mismeasure of Man* (New York: Norton, 1981), ch. 2; Glenn Antony May, *Social Engineering in the Philippines, 1900–1913* (Westport, CT: Greenwood, 1980), xviii, 89–92; Christopher Fisher, "'The Hopes of Man': The Cold War, Modernization Theory, and the Issue of Race in the 1960s" (Ph.D. diss., Rutgers University, 2002).

25. Priscilla Wald, "Legislating Subjectivity in the Emerging Nation," in Amy Kaplan and Donald E. Pease, eds., *Cultures of United States Imperialism* (Durham: Duke University Press, 1993), 75–80. American officials and literati in the nineteenth century had difficulty even conceptualizing the Indians in ways that did not emphasize their "otherness"; see Lucy Maddox, *Removals: Nineteenth-Century American Literature and the Politics of Indian Affairs* (New York: Oxford University Press, 1991).

26. See C. D. Sheldon, *The Rise of the Merchant Class in Tokugawa Japan* (Locust Valley, NY: Augustin, 1958); George Sansom, *A History of Japan, 1615–1867* (Stanford: Stanford University Press, 1963), chs. 9–10; John Whitney Hall, *Japan from Prehistory to Modern Times* (New York: Dell, 1970), 199–213; Peter Duus, *The Rise of Modern Japan* (Boston: Houghton Mifflin, 1976), ch. 3.

27. Sansom, *History of Japan*, ch. 15; Chushichi Tsuzuki, *The Pursuit of Power in Modern Japan, 1825–1995* (Oxford: Oxford University Press, 2000), 26–30; Donald Keene, *The Japanese Discovery of Europe* (Stanford: Stanford University Press, 1969); Hall, *Japan to Modern Times*, 223–225.

28. Pineau, *Personal Journal of Perry*, 198; Cole, *Diary of McCauley*, 43, 88–89, 113; Walworth, *Black Ships*, 36.

29. The surmise is Captain David Porter's, quoted in Walworth, *Black Ships*, 7.

30. Schroeder, *Matthew Perry*, 234–235; Cole, *Diary of McCauley*, 103–106.

31. Morrow, *A Scientist with Perry*, 144.

32. See, e.g., Annette Kolodny, *The Land before Her: Fantasy and the Experience of the American Frontiers, 1630–1860* (Chapel Hill: University of North Carolina Press, 1984); Jane Hunter, *The Gospel of Gentility* (New Haven: Yale University Press, 1984); Andrew Rotter, "Gender Relations, Foreign Relations: The United States and South Asia, 1947–1964," in Peter L. Hahn and Mary Ann Heiss, eds., *Empire and Revolution: The United States and the Third World since 1945* (Columbus: Ohio State University Press, 2001).

33. Here I employ a variant—conceptualized somewhat more broadly—of Thomas Holt's useful notion of "marking." See Holt, "Marking: Race, Race-making, and the Writing of History," *American Historical Review* 100/1 (1995): 1–20.

34. My conception of ideology has been influenced in important ways by Terry Eagleton's *Ideology: An Introduction* (London: Verso, 1991), esp. 5–19, 28–29, 45–48. Undergirding Eagleton's arguments and my own approach are Antonio Gramsci's writings on the dynamics of hegemonic idea systems working across classes, which are summarized with exquisite clarity by Raymond Williams in *Marxism and Literature* (Oxford: Oxford University Press, 1977). Quoted phrase from Williams, ibid., 110.

1. *"Engins" in the Wilderness*

1. Thomas Morton, *New English Canaan* (1637) (New York: B. Franklin, 1967 ed.), 179–181; William Bradford, *Of Plymouth Plantation, 1620–1647* (New York: Knopf, 1952), 61–62; Ralph Lane, "An Extract of [his] Letter to . . . Hakluyt," in David Freeman Hawke, ed., *Hakluyt's Voyages to the New World: A Selection* (Indianapolis: Bobbs-Merrill, 1972), 90–91; John Smith, "A Map of Virginia," in *The Jamestown Voyages under the First Charter, 1606–1609*, ed. Philip L. Barbour (Cambridge: Cambridge University Press, 1969), 2:345. On the sweet smells, see Philip Amadas and Arthur Barlowe, "The first voyage made to the coasts of America . . . ," in Richard Hakluyt, ed., *The Principal Naviga-*

tions, *Voyages, Traffiques and Discoveries of the English Nation* (Glasgow: J. MacLehose, 1903–1905), 8:298; on the New World as Paradise, Bernard W. Sheehan, *Savagism and Civility: Indians and Englishmen in Colonial Virginia* (Cambridge: Cambridge University Press, 1980), ch. 1.

2. William Wood, *New England's Prospect* (1634), ed. Alden T. Vaughan (Amherst: University of Massachusetts Press, 1977), chs. 1–9; Thomas Hariot, *A Briefe and True Report of the New Found Land of Virginia* (New York: Dover, 1972 facsimile of 1590 De Bry ed.), 11–24; Morton, *New English Canaan*, 2d book.

3. Martin H. Quitt, "Trade and Acculturation at Jamestown, 1607–1609: The Limits of Understanding," *William and Mary Quarterly*, 3d ser., 52/2 (1995): 227–258. The settlers' dependence on supplies from overseas was starkly demonstrated by the loss of the entire colony at Roanoke, which was out of contact with England for several seasons in the late 1580s because of the threat of the Spanish Armada. See *The Roanoke Voyages, 1584–1590*, ed. David Beers Quinn (New York: Hakluyt Society, 1955).

4. Winthrop, "Reasons for the Intended Plantation," in Alan Heimert and Andrew Delbanco, eds., *The Puritans in America: A Narrative Anthology* (Cambridge, MA: Harvard University Press, 1985), 71–74. On the sense of being God's agents, see Michael Zuckerman, *Almost Chosen People: Oblique Biographies in the American Grain* (Berkeley: University of California Press, 1993), 91; Peter Carroll, *Puritans and the Wilderness: The Intellectual Significance of the New England Frontier, 1629–1700* (New York: Columbia University Press, 1969), 13–14, 37–38, 89–94, 119–126. For the chronology of their settlement as divine history, see Bradford, *Plymouth*, 58, 70, 122; and *Tracts and other Papers, relating principally to the Origin, Settlement, and Progress of the Colonies in North America*, collected by Peter Force (Gloucester, MA: P. Smith, 1963 ed.), 4:4–5.

5. Winthrop, "A Model of Christian Charity," in Heimert and Delbanco, *Puritans in America*, 90–91.

6. Gray, *A Good Speed to Virginia* (New York: Da Capo, 1970 facsimile ed.). See also George Peckham Knight, "A true Report of the late discoveries" (1583), in Hakluyt, *Principal Navigations*, 8:119–120; Robert Johnson, *The New Life of Virginea* (1612) (New York: Da Capo, 1971 facsimile ed.), f. E3. And see John Parker, "Religion and the Virginia Colony," in K. R. Andrews, N. P. Canny, and E. H. Hair, eds., *The Westward Enterprise* (Detroit: Wayne State University Press, 1979), 245–270; Sheehan, *Savagism*, ch. 5.

7. Gray, *Good Speed*, ff. C2, C3, D2, D3; James Axtell, *The Invasion Within: The Contest of Cultures in Colonial North America* (New York: Oxford University Press, 1985), 133–135; William Kellaway, *The New England Company, 1649–1776* (London: Longman, 1961), 91. Hariot, *Brief and True Report*, 27. But some writers, most notably Cotton Mather, observed that the Indians' enthusiasm for European material goods did not usually prompt them to convert to Christianity. See Axtell, *Invasion Within*, 132–133.

8. Wood, *New England's Prospect*, 95–96. See also James Rosier, *True Relation of the Most Prosperous Voyage Made in this Present Yeere 1605*, in *The New England Voyages, 1602–1608*, ed. David B. Quinn and Alison M. Quinn (London: Hakluyt Society, 1983), 270, 272, 274, 277; Roger Williams, *A Key into the Language of America* (1643; Providence: Rhode Island Tercentenary Committee, 1936), 158, 164; and (for more mixed responses) Joyce Chaplin, *Subject Matter: Technology, the Body, and Science on the Anglo-American Frontier, 1500–1676* (Cambridge, MA: Harvard University Press, 2001), 15, 22, 36–38, 42, 56, 69–74, 142–143.

9. Quitt, "Trade and Acculturation," 231; Bradford, *Plymouth*, 207; William Strachey, *The Historie of Travell into Virginia Britania* (1612; London: Hakluyt Society, 1953), 75, 115; Rosier, "True Relation," 277; Sheehan, *Savagism*, 140, 165.

10. Gray, *Good Speed*, f. C2; Hariot, *Briefe and True Report*, 27; Rosier, "True Relation," 274; Gesa Mackenthun, *Metaphors of Dispossession: American Beginnings and the Translation of Empire, 1492–1637* (Norman: University of Oklahoma Press, 1997), 146–148; Axtell, *Invasion Within*, 10–12, 19.

11. E.g., Wood, *New England's Prospect*, ch. 8; Strachey, *Travell*, 74; Rosier, "True Relation," 268, 276.

12. Alden T. Vaughan, "From White Man to Redskin: Changing Anglo-American Perceptions of the American Indian," in Vaughan, *Roots of American Racism* (New York: Oxford University Press, 1995), 3–33; Karen Kupperman, "Presentment of Civility: English Reading of American Self-Preservation in the Early Years of Colonization," *William and Mary Quarterly* 54/1 (1997), 193–195, 226–228. For early descriptions of physiological differences that suggested recognition of innate divergence, see Chaplin, *Subject Matter*, 177–191.

13. James Axtell, *Beyond 1492: Encounters in Colonial North America* (New York: Oxford University Press, 1992), 53–56.

14. Winthrop, "Reasons for the Intended Plantation," 72–73. Winthrop's view was widely shared. See, e.g., Carroll, *Puritans and Wilderness*, 14, 181–182; Roderick Nash, *Wilderness and the American Mind* (New Haven: Yale University Press, 1974), 31; Karen Kupperman, *Settling with the Indians: The Meeting of English and Indian Cultures in America, 1580–1640* (Totowa, NJ: Rowman and Littlefield, 1980), 89.

15. Wood, *New England's Prospect*, 95; Carroll, *Puritans and Wilderness*, 11–12, 72–79, 111–114; Nash, *Wilderness and American Mind*, 28–29, 33–37; Zuckerman, *Almost Chosen People*, 82–85; Alan Heimert, "Puritanism, the Wilderness, and the Frontier," *New England Quarterly* 26/3 (1953), 370–372, 377–378. And see Smith, "Map of Virginia," 448.

16. Richard Bernheimer, *Wild Men in the Middle Ages: A Study in Art, Sentiment, and Demonology* (Cambridge, MA: Harvard University Press, 1952), 19–20 (quotation), 16–20; Benjamin Rowland Jr., *Art in East and West* (Boston: Beacon, 1954), 71–84; Nash, *Wilderness and American Mind*, 17–20; Keith Thomas, *Man and the Natural World: Changing Attitudes in England, 1500–1800* (Harmondsworth: Penguin, 1984), 192–195.

17. London Council of the Virginia Company, "Instructions given by way of advice . . . for the intended Voyage to Virginia . . . ," in

Barbour, *Jamestown Voyages*, 1:52 (quotation); Edmund S. Morgan, *The Puritan Dilemma: The Story of John Winthrop* (Boston: Little, Brown, 1958), 65; Carroll, *Puritans and Wilderness*, 136–137; Bradford, *Plymouth*, 62; Nash, *Wilderness and American Mind*, 28–29; Heimert, "Puritanism," 362, 370, 373.

18. *Leviathan*, ed. Richard Tuck (Cambridge: Cambridge University Press, 1996), 88–90. And see Richard Ashcraft, "Leviathan Triumphant: Thomas Hobbes and the Politics of Wild Men," in Edward Dudley and Maximillian E. Novak, eds., *The Wild Man Within: An Image in Western Thought from the Renaissance to Romanticism* (Pittsburgh: University of Pittsburgh Press, 1972), 141–181.

19. See, e.g., Lane, "Letter to Hakluyt," 91; Rosier, "True Relation," 297; John Smith, "Description of Virginia," in Lyon G. Tyler, ed., *Narratives of Early Virginia, 1606–1625* (New York: Scribner, 1907), 81–82; Robert Cushman, "Reasons and Considerations Touching the Lawfulness of Removing out of England into the Parts of America" (1622), in Heimert and Delbanco, *Puritans in America*, 43–44; Sheehan, *Savagism*, 34–35, 68, 107; Kupperman, *Settling with Indians*, 80.

20. Gray, *Good Speed*, f. D2 (quotation), D3. See also Strachey, *Travell*, 24; Rosier, *True Relation*, 297–298; Kupperman, *Settling with Indians*, 80–81, 89–90; Gary Nash, "The Image of the Indian in the Southern Colonial Mind," *William and Mary Quarterly* 29/2 (1972), 203, 208, 222–223; Sheehan, *Savagism*, 161.

21. Knight, "True Report," 112 (quotation); Anonymous (William Bradford and Edward Winslow), *Mourt's Relation: A Journal of the Pilgrims at Plymouth* (1622), ed. Dwight B. Heath (New York: Cornith, 1963), 84; Parker, "Religion and Virginia," 268; and esp. Carol Shammas, "English Commercial Development and American Colonization," in Andrews et al., *Westward Enterprise*, 167–168.

22. Knight, "True Report," 110–119; Rosier, "True Relation," 288–297; Gray, *Good Speed*, ff. B, B2, B3, D3; Smith, "Map of Virginia," 345–350, 353–354; Hariot, *Briefe and True Report*, pt. 1; Gabriel

Archer, "Description of the River and Country" (1607), in *James-town Voyages*, 1:98–102; Richard Hakluyt, "The Discourse of Western Planting," in *The Original Writings of the Two Richard Hakluyts*, ed. E. G. R. Taylor, 2:313–319; "Notes Framed by a Gentleman," in Richard Hakluyt, comp., *Divers Voyages Touching the Discovery of America* (1582), ed. John W. Jones (London: Hakluyt Society, 1850), 132–138.

23. Strachey, *Travell*, 24; Henry Spelman, "Relation of Virginia," in *Travels and Works of Captain John Smith*, ed. Edward Arber and A. G. Bradley (Edinburgh: J. Grant, 1910), cvi–cvii.

24. Alan I. Marcus and Howard P. Segal, *Technology in America: A Brief History* (San Diego: Harcourt Brace Jovanovich, 1989), 14, 22–28; Brooke Hindle, *Technology in Early America* (Chapel Hill: University of North Carolina Press, 1966), 26–27; Timothy Silver, *A New Face on the Countryside: Indians, Colonists, and Slaves in the South Atlantic Forests* (Cambridge: Cambridge University Press, 1990), 117–138.

25. *Mourt's Relation*, 91–92; Strachey, *Travell*, 25–29; Francis Jennings, *The Invasion of America* (New York: Norton, 1976), 80–84; Roy H. Pearce, *Savagism and Civilization: A Study of the Indian in the American Mind* (Baltimore: Johns Hopkins University Press, 1965), 7–8, 20–21; Kupperman, *Settling with Indians*, 89.

26. Anthony Padgen, *Lords of All the World: Ideologies of Empire in Spain, Britain and France, 1500–c. 1800* (New Haven: Yale University Press, 1995), 76–77, 82–84; Jennings, *Invasion of America*, ch. 8.

27. Patricia Seed, "Taking Possession and Reading Texts: Establishing the Authority of Overseas Empires," *William and Mary Quarterly* 49/2 (1992), 186–187.

28. Knight, "True Report," 119–121, 123, 130–131; Gray, *Good Speed*, ff. B, B2, B3, C2; Heimert, "Puritanism," 378–379; Parker, "Religion and Virginia," 255–256, 267.

29. Though perhaps less decisive than Anthony Padgen argues, in *The Fall of Natural Man* (Cambridge: Cambridge University

Press, 1982), 13, 19, was the case in Spanish interaction with Indian peoples farther south in the Americas.

30. This fact had much to do with the English disinclination to push conversion efforts in Africa or Asia in this era in sharp contrast to their determination to proselytize among the Amerindian peoples. See Loren E. Pennington, "The Amerindian in English Promotional Literature, 1575–1625," in Andrews et al., *Westward Enterprise*, 177–178; Louis B. Wright, *Religion and Empire: The Alliance between Piety and Commerce in English Expansion, 1558–1625* (Chapel Hill: University of North Carolina Press, 1943), 57, 62–64.

31. See Amadas and Barlowe, "First Voyage," (1584), 298–300; Lawrence C. Wroth, ed., *The Voyages of Giovanni da Verrazzano, 1524–1528* (New Haven: Yale University Press, 1970), 134–135. See also Robert Johnson, *Nova Britannia* (1609) (New York: Da Capo, 1969 facsimile ed.), f. B3.

32. Quotations from Samuel Eliot Morison, *Admiral of the Ocean Sea: A Life of Christopher Columbus* (Boston: Northeastern University Press, 1983), 230. Translations of these passages vary considerably, but all juxtapose the comments on nudity with those on deficient technology.

33. Edmund Morgan, "The Labor Problem at Jamestown," *American Historical Review* 76/2 (1971), 597–600; Pennington, "Amerindian in Promotional Literature," 179–184, 192–193; Nicholas P. Canny, "The Ideology of English Colonization: From Ireland to America," *William and Mary Quarterly* 30/4 (1973), 593–595; Nash, "Indian Image," 200–203.

34. On associations with nudity, see Strachey, *Travell*, 24; Johnson, *Nova Britannia*, f. B3. As Kupperman observes in *Settling with Indians*, 39–41, and "Presentment of Civility," 199–204, naked meant varying states of undress depending upon the observer. On Indians as gullible traders, see Morton, *New English Canaan*, 177; Jennings, *Invasion of America*, 98–104; James H. Merrell, "'The Customes of Our Countrey': Indians and Colonists in Early

America," in B. Bailyn and P. D. Morgan, eds., *Strangers within the Realm: Cultural Margins of the First British Empire* (Chapel Hill: University of North Carolina Press, 1991), 131–134; on their helplessness in the face of natural forces, see William Cronon, *Changes in the Land: Indians, Colonists, and the Ecology of New England* (New York: Hill and Wang, 1983), 40–41.

35. Smith, "Map of Virginia," 354; Strachey, *Travell*, 74; Bradford, *Plymouth*, 25; Morton, *New English Canaan*, 121; Sheehan, *Savagism*, 101–102; Cronon, *Changes*, 33, 41; Pearce, *Savagism and Civilization*, 7–8, 12.

36. Cronon, *Changes*, ch. 2; Kathleen M. Brown, *Good Wives, Nasty Wenches and Anxious Patriarchs: Gender, Race and Power in Colonial Virginia* (Chapel Hill: University of North Carolina Press, 1996), 34–35, 55; Canny, "Ideology of Colonization," 587; Sheehan, *Savagism*, 52, 56. On vagrancy and vagabondage, see Paul Slack, *Poverty and Policy in Tudor and Stuart England* (London: Longman, 1988), esp. 91–104, 114–127; A. L. Beier, *Masterless Men: The Vagrancy Problem in England, 1560–1640* (London: Methuen, 1985).

37. See Michel Foucault, *Discipline and Punish* (Harmondsworth: Penguin, 1982), 135–163; Robert Muchembled, *Popular Culture and Elite Culture in France, 1400–1750* (Baton Rouge: University of Louisiana Press, 1985), pt. 2; Peter Perdue, "Boundaries, Maps, and Mobile People: Chinese, Russian, and Mongolian Empires in Early Modern Central Eurasia," *International History Review* 20/2 (1998).

38. Gray, *Good Speed*, ff. C, C3; Brown, *Good Wives*, 57; Cronon, *Changes*, 55–56; Sheehan, *Savagism*, 21, 52, 99–100, 107–109; Peter A. Thomas, "Contrastive Subsistence Strategies and Land Use as Factors for Understanding Indian-White Relations in New England," *Ethnohistory* 23/1 (1976), 5; Heimert, "Puritanism," 376.

39. Strachey, *Travell*, 24; Gray, *Good Speed*, ff. C2, C3; Axtell, *Invasion Within*, 157–158, 178.

40. Daniel C. Beaver, "The Great Deer Massacre: Animals, Honor

and Communication in Early Modern England," *Journal of British Studies* 38/2 (1991), 187–216. Roger B. Manning, *Hunters and Poachers: A Social and Cultural History of Unlawful Hunting in England, 1485–1640* (Oxford: Oxford University Press, 1993), ch. 1.

41. David D. Smits, "The 'Squaw Drudge': A Prime Index of Savagism," *Ethnohistory* 29/4 (1982), 284.

42. Hariot, *Briefe and True Report*, 14; Spelman, "Relation of Virginia," cxi–cxii; Kupperman, *Settling with Indians*, 82–86, 172–174, 182; Smits, "Squaw Drudge," 283–284; Carolyn Merchant, *Ecological Revolutions: Nature, Gender, and Science in New England* (Chapel Hill: University of North Carolina Press, 1989), 84–86; Cronon, *Changes*, 36–37; Sheehan, *Savagism*, 102, 109–110, 115. On the importance of maize and on the first thanksgiving, see Bradford, *Plymouth*, 65–66, 85, 90, 131–132. And on tobacco and the Virginia colony, see Shammas, "English Commercial Development," 151; Morgan, "Labor Problem at Jamestown," 595, 610–611.

43. Morton, *New English Canaan*, 172–173; Cronon, *Changes*, 48–51. As Timothy Silver notes in *New Face*, 104–106, settlers in the southern colonies followed a variant of this pattern for much of the seventeenth century. They gradually killed the trees by cutting off the bark, burned off the undergrowth, and then planted the areas between the rotting trunks.

44. Cronon, *Changes*, 53, 77–78, and ch. 7; Merchant, *Ecological Revolutions*, 86–89, 98–109; Axtell, *Invasion Within*, 153–156; Jennings, *Invasion of America*, 63–66; Marcus and Segal, *Technology in America*, 14.

45. Brown, *Good Wives*, 20, 24–29, 57–58, 83–88; Quitt, "Trade and Acculturation," 235.

46. Smits, "Squaw Drudge," 294–295; Axtell, *Invasion Within*, 153–155; Jennings, *Invasion of America*, 63; Merchant, *Ecological Revolutions*, 81, 84; Herndon, "Indian Agriculture," 288–289.

47. Smits, "Squaw Drudge," 285–286; Kathleen M. Brown, "The An-

glo-Algonquian Gender Frontier," in Nancy Shoemaker, ed., *Negotiators of Change: Theoretical Perspectives on Native American Women* (New York: Routledge, 1995), 32–33; Brown, *Good Wives*, 57–58; Cronon, *Changes*, 43–45; Quitt, "Trade and Acculturation," 236.

48. On the contributions of hunting, gathering, and farming to the diet of the peoples of southern New England, see Merchant, *Ecological Revolutions*, 74–83. On such patterns in Virginia, see Sheehan, *Savagism*, 99–102.

49. Jennings, *Invasion of America*, 90–92; Smits, "Squaw Drudge," 295.

50. George Percy, "Observations by Master George Percy, 1607," in Tyler, *Narratives of Early Virginia*, 18; Strachey, *Travell*, 114; Smith, "Map of Virginia," 356–357; Smits, "Squaw Drudge," 280–297; Brown, *Good Wives*, 46–49, 58; Herndon, "Indian Agriculture," 289.

51. Wood, *New England's Prospect*, 96, 112; Smith, "Description of New England," in Karen Kupperman, ed., *Captain John Smith: A Select Edition of His Writings* (Chapel Hill: University of North Carolina Press, 1988), 251; Nash, "Indian Image," 220; Smits, "Squaw Drudge," 282; Cronon, *Changes*, 45, 52–53, 55. As Martin Quitt notes ("Trade and Acculturation," 232–233), in the case of Virginia the charge of indolence was at the very least ironic, since in the first years of settlement most of the gentlemen colonists refused to do physical labor.

52. Sheehan, *Savagism*, 37, 56, 107, 109; Axtell, *Invasion Within*, 133; Strachey, *Travell*, 74; Smith, "Description of Virginia," 251–252; Smits, "Squaw Drudge," 289–290; Kupperman, *Settling with Indians*, 123–124; Nash, "Indian Image," 220; Jennings, *Invasion of America*, 101–102. Some contemporary writers, notably Thomas Morton, viewed the Indians' storehouses as an indication of their capacity to save and plan for the future. See *New English Canaan*, 160–161.

53. Strachey, *Travell*, 108–109; Williams, *Language of America*, 131, 164; Wood, *New England's Prospect*, 94. On Indian and settler

military adaptations see Patrick M. Malone, *The Skulking Way of War: Technology and Tactics among the New England Indians* (Lanham, MD: Madison, 1991); and Malone, "Changing Military Technology among the Indians of Southern New England, 1600–1677," *American Quarterly* 25/1 (1973): 48–63.

54. This discussion of English representations of Indian warfare is based upon Malone, *Skulking Way of War*; Jennings, *Invasion of America*, ch. 9; Strachey, *Travell*, 108–111; Smith, "Map of Virginia," 360–361; Spelman, *Relation of Virginia*, cxiii-cxiv; and Wood, *New England's Prospect*, 94–97 and ch. 13.

55. Wood, *New England's Prospect*, 103; and see Morton, *New English Canaan*, 248. Brown, "Gender Frontier," 34–36; Heimert, "Puritanism." On European warfare see Geoffrey Parker, *The Military Revolution: Military Innovation and the Rise of the West, 1500–1800* (Cambridge: Cambridge University Press, 1988); William H. McNeill, *The Pursuit of Power: Technology, Armed Force, and Society since A.D. 1000* (Chicago: University of Chicago Press, 1982).

56. See Michael Adas, *Machines as the Measure of Men: Science, Technology and Ideologies of Western Dominance* (Ithaca: Cornell University Press, 1989), ch. 1.

57. For African parallels, see ibid., 32–41, 59–68; and W. G. L. Randles, *L'image du sud-est Africain dans la littérature européenne au XVIe siècle* (Lisbon: Centro de Estudos Históricos Ultramarinos, 1959), 151–154. Some travelers did selectively admire aspects of African or Amerindian cultures. But as Kupperman has concluded (*Settling with Indians*, 147), "No writer seriously considered the possibility that England might copy Indian society in any fundamental way."

58. Knight, "True Report," 119–120.

59. Strachey, *Travell*, 24. On Bacon's highly gendered vision of man's mastery of nature, see Carolyn Merchant, *The Death of Nature: Women, Ecology and the Scientific Revolution* (New York: Harper and Row, 1982).

60. Strachey, *Travell*, 75, 108–109; Rosier, "True Relation," 269; Smith, "Map of Virginia," 354; Hariot, *Briefe and True Report*, 25; Alexander Whitaker, *Good Newes from Virginia* (London: Kyngson and Welby, 1613), 25–26; Morton, *New English Canaan*, 161–165; Wood, *New England's Prospect*, 81, 91; Malone, "Changing Military Technology," 48, 50–53, 56–57; Bradford, *Plymouth*, 206–208. Gray, *Good Speed*, f. C2.

61. Axtell, *Invasion Within*, Ch. 7; W. Stitt Robinson, "Indian Education and Missions in Colonial Virginia," *Journal of Southern History* 17 (1952), 152–168; Norman E. Tanis, "Education in John Eliot's Indian Utopias, 1646–1675," *History of Education Quarterly* 10/2 (1970), 318–320; Kellaway, *New England Company*, 91.

62. Tanis, "Education," 308, 312–313, 316–317; Neal Salisbury, "Red Puritans: The 'Praying Indians' of Massachusetts Bay and John Eliot," *William and Mary Quarterly*, 3d ser., 31/1 (1974), 33–34, 42; Kellaway, *New England Company*, 85–86, 107–108, 116–117; Henry W. Bowden, *American Indians and Christian Missions: Studies in Cultural Conflict* (Chicago: University of Chicago Press, 1981), 127–128; Axtell, *Invasion Within*, esp. 159–163; Malone, "Changing Military Technology," 55–56.

63. An outcome that recent historians stress was the conscious object of Eliot and his disciples. See Axtell, *Invasion Within*, 177–179; Salisbury, "Red Puritans," 28–29, 42.

64. Richard White, *The Middle Ground: Indians, Empires and Republics in the Great Lakes Region, 1650–1815* (Cambridge: Cambridge University Press, 1991).

65. "Instructions," in Barbour, *Jamestown Voyages*, 1:51–53.

66. Jennings, *Invasion of America*, 39–42, 85–87, 102–104; Merrell, "Indians and Colonists," 132–137; Kupperman, *Settling with Indians*, 94; Sheehan, *Savagism*, 140–141.

67. Axtell, *Invasion Within*, 159–161, 165–167; Kellaway, *New England Company*, 86, 108–109.

68. See Salisbury, "Red Puritans," 31; Kellaway, *New England Company*, 91.

2. Machines and Manifest Destiny

1. Martyn J. Bowden, "The Great American Desert and the American Frontier, 1800–1882: Popular Images of the Plains," in T. K. Hareven, ed., *Anonymous Americans: Explorations in Nineteenth-Century Social History* (Englewood Cliffs, NJ: Prentice-Hall, 1971), 48–79; C. W. Dana, *The Great West or the Garden of the World* (Boston: Wentworth, 1858), 2 (quotation).

2. John C. Ewers, *Artists of the Old West* (Garden City, NY: Doubleday, 1965), 183.

3. For a similar treatment of an Indian encampment in a mountain setting, see Alfred Miller's *Indian Village* (1850).

4. "Destiny of the United States, A speech delivered at St. Paul, Minnesota, September 18, 1960, and reprinted in the *Albany Evening Standard*," in Seward, *Tracts*, no. 16, 1. Nancy K. Anderson and Linda S. Ferber, *Albert Bierstadt: Art and Enterprise* (New York: Hudson Hills, 1990), 24–25, 31–34, 74.

5. See Albert Boime, *The Magisterial Gaze: Manifest Destiny and American Landscape Painting* (Washington: Smithsonian Institution Press, 1991), 10–12, 82–83; Kenneth W. Maddox, "Asher B. Durand's *Progress*: The Advance of Civilization and the Vanishing Indian," in Susan Danly and Leo Marx, eds., *The Railroad in American Art: Representations of Technological Change* (Cambridge, MA: MIT Press, 1988), 62–63, 66.

6. Emerson, "The Young American," in *Essays and Lectures* (New York: Library of America, 1983), 213, 228.

7. Ruth Miller Elson, *Guardians of Tradition: American Schoolbooks of the Nineteenth Century* (Lincoln: University of Nebraska Press, 1964), 25–27, 37–39, 71–80. Peter Novick, *That Noble Dream: The "Objectivity Question" and the American Historical Profession* (Cambridge: Cambridge University Press, 1988), ch. 4.

8. Emerson, "Young American," 217, 230.

9. Richard White, *The Middle Ground: Indians, Empires, and Republics in the Great Lakes Region, 1650–1815* (Cambridge: Cambridge University Press, 1991); Henry Howe, *The Great West* (New York: Tuttle, 1858), 215–216, 227; Marvin C. Ross, *The West of Alfred Jacob Miller* (Norman: University of Oklahoma Press, 1968).

10. David E. Nye, *America as Second Creation: Technology and Narratives of New Beginnings* (Cambridge, MA: MIT Press, 2003), 98–111; Michael William, *Americans and Their Forests: A Historical Geography* (New York: Cambridge University Press, 1989), pt. 3, esp. 201–216, 300–304, 315–320; Carroll Pursell, *The Machine in America: A Social History of Technology* (Baltimore: Johns Hopkins University Press, 1995), 110–112.

11. Pursell, *Machine in America*, 72–78 and chs. 3–5; Williams, *Americans and Forests*, 300–304; Frieda Knobloch, *The Culture of Wilderness: Agriculture as Colonization in the American West* (Chapel Hill: University of North Carolina Press, 1996), 18–23, 51–56; Merritt Roe Smith, *Harpers Ferry Armory and the New Technology: The Challenge of Change* (Ithaca: Cornell University Press, 1977).

12. Parkman quoted in Barbara Novak, *Nature and Culture: American Landscape and Painting, 1825–1875* (New York: Oxford University Press, 1995), 145.

13. Ernest L. Schusky, *Culture and Agriculture: An Ecological Introduction to Traditional and Modern Farming Systems* (New York: Bergin and Garvey, 1989), 100, 105–107. On the waves of new farming technologies see John Schlebecker, *Whereby We Thrive: A History of American Farming, 1607–1972* (Ames: Iowa State University Press, 1975), 80–82, 97–123, 174–182, 190–205.

14. Boime, *Magisterial Gaze*, 43–47 and the full-color reproduction opposite 85.

15. Leo Marx, *The Machine in the Garden: Technology and the Pastoral Ideal in America* (London: Oxford University Press, 1964). For American pride see, e.g., the discussion of the railroad in the 1873

compendium of *The Great Industries of the United States* (Hartford, CT: J. B. Burr and Hyde) assembled by Horace Greeley and "other eminent writers."

16. Matthew Simon, "The Pattern of New British Portfolio Foreign Investment, 1865–1914," and Brinely Thomas, "Migration and International Investment," in A. R. Hall, ed., *The Export of Capital from Britain, 1870–1914* (London: Methuen, 1968), 26–27, 53–54. Brooke Hindle, *Technology in Early America* (Chapel Hill: University of North Carolina Press, 1966), 26–27; Samuel Bowles, *Our New West* (New York: J. D. Dennison, 1869), 72–73. As Merritt Roe Smith *(Harpers Ferry Armory)* and others have argued, the correspondence between labor scarcity and technological innovation was neither invariable nor always direct, but without question new tools and machines were critical determinants of the pace of frontier expansion, and in many areas made sustained settlement and resource exploitation practicable.

17. Peter Hassrick, *The Way West: Art of Frontier America* (New York: Harry N. Abrams, 1983), 103, 110–11, 114–115; John C. Hudson, "Settlement of the American Grasslands," in Michael Conzen, ed., *The Making of the American Landscape* (Boston: Unwin, Hyman, 1990), ch. 9; *RCIA* (1856), 65; Theodore Roosevelt, *The Winning of the West* (New York: Putnam, 1900), 1:15.

18. The apt phrase is Leo Marx's; see "The Railroad in the Landscape," in Danly and Marx, *Railroad*, 198–199. On early railway paintings, see Susan Danly, "Introduction," in Danly and Marx, *Railroad*, 1–16, 26–30. For contemporary responses, Marx, *Machine in the Garden*, 11–19, 227–229, 242–265.

19. Danly, "Introduction," 7–9, 26–30; Nicholai Cikovsky Jr., "George Inness's *The Lackawanna Valley*: Type of the Modern," in Danly and Marx, *Railroad*, 57–91.

20. Marx, *Machine in the Garden*, 202–209; John F. Kasson, *Civilizing the Machine: Technology and Republican Values in America, 1776–1900* (New York: Penguin, 1977), 36–51; Thomas L. Karnes, *William Gilpin: Western Nationalist* (Austin: University of

Texas Press, 1970), 216–220; Novak, *Nature and Culture*, 164, 167–168; Nye, *Second Creation*, 152–173; Nye, *American Technological Sublime* (Cambridge, MA: MIT Press, 1994), 56–64; Bowles, *New West*, 74.

21. *Great Industries of the U.S.*, 26, 71–76; Marx, *Machine in the Garden*, 208–209; Pursell, *Machine in America*, 87–97, 124–126; Smith, *Harper's Ferry Armory*, esp. ch. 8; Kasson, *Civilizing the Machine*, ch. 2.

22. Nye, *Technological Sublime*, 119–123; Cecelia Tichi, *Shifting Gears: Technology, Literature, Culture in Modernist America* (Chapel Hill: University of North Carolina Press, 1987), 137–164.

23. Josiah Strong, *Our Country: Its Possible Future and Its Present Crisis* (New York: Baker and Taylor, 1885), 22; Henry Nash Smith, *Virgin Land: The American West as Symbol and Myth* (Cambridge, MA: Harvard University Press, 1950), 182–183.

24. For Marsh, see Steven Stoll, *Larding the Lean Earth: Soils and Society in Nineteenth-Century America* (New York: Hill and Wang, 2002), 178–184; for Powell, see Roderick Frazier Nash, *American Environmentalism: Readings in Conservation History* (New York: McGraw-Hill, 1990), 63–68.

25. Knobloch, *Culture of Wilderness*, 18–23, 54–57, 61–64, 73–78; Richard A. Bartlett, *The New Country: A Social History of the American Frontier, 1776–1890* (New York: Oxford University Press, 1974), 250–260, 263–272; Aldo Leopold, *A Sand Country Almanac* (New York: Ballantine, 1966), 12–16; Donald Worster, *Nature's Economy: A History of Ecological Ideas* (New York: Cambridge University Press, 1977), 221–230; Worster, *An Unsettled Country: Changing Landscapes in the American West* (Albuquerque: University of New Mexico Press, 1994), 67–72; R. Douglas Hurt, *The Dust Bowl: An Agricultural and Social History* (Chicago: Nelson-Hall, 1981), 17–23; Roderick Frazier Nash, *Wilderness and the American Mind* (New Haven: Yale University Press, 1974), 23–27, 40–41, 114, 135–136.

26. Roosevelt, *Winning of the West*, 1:80–81, 90; 3:44–45.

27. Ibid., 1:41–42, 4:242–243; Daniel Usner, "Between Creoles and Yankees: The Discursive Representation of Colonial Louisiana in American History," paper presented at the Faculty-Graduate History Seminar, Cornell University, Feb. 1999.

28. *Winning of the West*, 1:vii-xiii; Thomas R. Hietala, *Manifest Design: Anxious Aggrandizement in Late Jacksonian America* (Ithaca: Cornell University Press, 1985), 132, 153–161, 195–196; Reginald Horsman, *Race and Manifest Destiny: The Origins of Racial Anglo-Saxonism* (Cambridge, MA: Harvard University Press, 1981), 229, 232–235.

29. Turner, "The Significance of the Frontier in American History," in George Rogers Taylor, ed., *The Turner Thesis* (Boston: Heath, 1956), 1, 3–4, 11–12 (quoted portions), 14–18; Smith, *Virgin Land*, 3–4 and chs. 11–12; Marx, *Machine in the Garden*, 135–141, 203–207.

30. Elwell S. Otis, *The Indian Question* (New York: Sheldon, 1878), 229–233, 240, 268; Roy Harvey Pearce, *The Savages of America: A Study of the Indian and the Idea of Civilization* (Baltimore: Johns Hopkins University Press, 1953), 85, 99–103, 113–114, 120–121, 126, 127, 149, 158–166, 181; Brian Hosmer, "Rescued from Extinction? The Civilizing Program in Indian Territory," *Chronicles of Oklahoma* 68/2 (1990), 147–150; Dana, *Great West*, 14 (quoted phrases); Horsman, *Race and Manifest Destiny*, 206–207.

31. R. Douglas Hurt, *Indian Agriculture in America: Prehistory to the Present* (Lawrence: University of Kansas Press, 1987), ch. 4; Preston Holder, *The Hoe and the Horse on the Plains: A Study of Cultural Development among the North American Indians* (Lincoln: University of Nebraska Press, 1970), 58, 66–72, 90–99, and ch. 3; Richard White, *The Roots of Dependency: Subsistence, Environment, and Social Change among the Choctaws, Pawnees, and Navajos* (Lincoln: University of Nebraska Press, 1983), 199–207; Andrew C. Isenberg, *The Destruction of the Bison: An*

Environmental History, 1750–1920 (Cambridge: Cambridge University Press, 2000), chs. 1–2.

32. Emerson, "Young American," 213. For a similar faith in technology to bind the Union together, see William Seward's oration "The Destiny of America" (Sept. 14, 1853), in George E. Baker, *The Life of William H. Seward* (New York: Redfield, 1855), 327–329. For sod house construction, see Joanna L. Stratton, *Pioneer Women: Voices from the Kansas Frontier* (New York: Simon and Schuster, 1981), 52–56; for denigration of Indian camps, see Otis, *Indian Question*, 237, 240; Margaret Carrington, *Ab-sa-ra-ka, Home of the Crows* (Philadelphia: Lippincott, 1868), 192–193; Richard Irving, *Our Wild Indians* (Hartford, CT: Worthington, 1882), 240–245.

33. Bowles, *New West*, 31, 316 (quotation); William W. Fowler, *Pioneer Women of America* (Hartford, CT: S. S. Scranton, 1891), 31, 33, 175–194, 524–525; E. F. Ellet, *Pioneer Women of the West* (New York: Scribner, 1852), 41–42, 59–60; Stratton, *Pioneer Women*, ch. 2 and 112–113, 131–132, 144–148, 152–153; Karen Sachs, *The Invisible Farmers: Women in Agricultural Production* (Totowa, NJ: Reisman and Allanheld, 1983), 17–20; Mary W. M Hargreaves, "Women in the Agricultural Settlement of the Northern Plains," *Agricultural History* 50/1 (1976), 184–187; Pearce, *Savages*, 93, 156; David D. Smits, "The 'Squaw Drudge': A Prime Index of Savagism," *Ethnohistory* 29/4 (1982), 286–288; 291–292, 299–301.

34. Pursell, *Machine in America*, 110–111, 115, 120–121; Marx, *Machine in the Garden*, 205–206; Smith, *Virgin Land*, ch. 14.

35. Patricia Nelson Limerick, *The Legacy of Conquest: The Unbroken Past of the American West* (New York: Norton, 1987), esp. 78–87, 124–133; William C. Robbins, *Colony and Empire: The Capitalist Transformation of the American West* (Lawrence: University of Kansas Press, 1994), 62–65, 73–82; Pearce, *Savages*, 73, 120, 160, 209–210. For comparisons with frontier expansion around the globe, see John Weaver, in *The Great Land Rush: And the*

Making of the Modern World, 1650–1900 (Montreal: McGill-Queen's University Press, 2003).

36. Thomas Jefferson, *Notes on the State of Virginia* (Chapel Hill: University of North Carolina Press, 1982), 140; Bernard W. Sheehan, *Seeds of Extinction: Jeffersonian Philanthropy and the American Indian* (Chapel Hill: University of North Carolina Press, 1973), 7–11, 25–26, 44, ch. 5.

37. Quotations from Horsman, *Race and Manifest Destiny*, 202. See Francis Paul Prucha, *American Indian Policy in the Formative Years* (Lincoln: University of Nebraska Press, 1962), ch. 9; Ronald N. Satz, *American Indian Policy in the Jacksonian Era* (Lincoln: University of Nebraska Press, 1975); Herman J. Viola, *Thomas L. McKenney: Architect of America's Early Indian Policy* (Chicago: Sage, 1974), chs. 10–11.

38. William G. McLoughlin, *Cherokees and Missionaries, 1789–1839* (New Haven: Yale University Press, 1984), 124–146; Satz, *Indian Policy*, 18–20; John Ehle, *Trail of Tears: The Rise and Fall of the Cherokee Nation* (New York: Doubleday, 1988); White, *Roots of Dependency*, 116–146.

39. Stephen Jay Gould, *The Mismeasure of Man* (New York: Norton, 1981), 56–60.

40. *Crania Americana* (Philadelphia: J. Dobson, 1839), esp. 62–82; *RCIA* (1869), 700–701; Robert E. Bieder, *Science Encounters the Indian, 1820–1880: The Early Years of American Ethnology* (Norman: University of Oklahoma Press, 1986), 68–76, 85–88.

41. Morgan, *Ancient Societies* (Cambridge, MA: Harvard University Press, 1964), 11–45, quotation 40; Bieder, *Science*, 239, 241–242; and, for a bowdlerized, popular expression of these scientific pronouncements, Otis, *Indian Question*, 6, 98, 221, 228–229, 232–233, 238–242, 248–249.

42. Hassrick, *Way West*, 112–113; Ewers, *Old West*, 204–209.

43. Danly, "Introduction," 17, 20–21; Robert M. Utley, *The Indian Frontier: 1846–1890* (Albuquerque: University of New Mexico Press, 2003), 19, 43, 94, 110–112, 166.

44. Hassrick, *Way West*, 116–117, Danly, "Introduction," 27–28; Donald Worster, *An Unsettled Country: Changing Landscapes of the American West* (Albuquerque: University of New Mexico Press, 1994), 67–69; Isenberg, *Destruction of the Bison*, 97–113, 128–143, 151–156, 162–163; Bartlett, *New Country*, 260–263; Frank B. Linderman, ed., *Pretty Shield: Medicine Woman of the Crows* (Lincoln: University of Nebraska Press, 1972), 250.

45. See Evan S. Connell, *Son of the Morning Star: Custer and the Little Bighorn* (New York: Harper and Row, 1984), 257–259.

46. Smith, *Harpers Ferry Armory*, chs. 8–9; Walter Millis, *Arms and Men: A Study of American Military History* (New York: New American Library, 1956), 108–116; James M. McPherson, *Battle Hymn of the Republic: The Civil War Era* (New York: Ballantine, 1988), 471–477; *Great Industries of the U.S.*, 944 (quoted phrase).

47. Unless otherwise noted this overview of the wars on the plains is based on Utley, *Indian Frontier*, chs. 4–6; Russell F. Weigley, *The American Way of War: A History of United States Military Strategy and Policy* (Bloomington: Indiana University Press, 1973), ch. 8; and Ralph K. Andrist, *The Long Death: The Last Days of the Plains Indian* (New York: Macmillan, 1964). On the social antecedents and enduring cultural impact of frontier violence, see Richard Slotkin, *The Fatal Environment: The Myth of the Frontier in the Age of Industrialization, 1800–1890* (New York: Harper, 1985).

48. Roosevelt, *Winning of the West*, 1, 17 (quotation), 79–81. Connell, *Morning Star*, 258–259.

49. Weigley, *American Way of War*, 158–160; Utley, *Indian Frontier*, 124–125, 171–181.

50. Frances Paul Prucha, *American Indian Policy in Crisis: Christian Reformers and the Indian, 1865–1900* (Norman: University of Oklahoma Press, 1976), chs. 6–7; Utley, *Indian Frontier*, ch. 8.

51. RCIA (1856), 63–67; (1868), 461–466, 470–471, 477–479; (1869), 445, 451–458, 504, 582, 585, 610, 656, 730; Otis, *Indian Question*, 241–243, 246, 270; Prucha, *Indian Policy in Crisis*, ch. 8; Robert

Berkhofer, *Salvation and the Savage: An Analysis of Protestant Missions and American Indian Response* (Westport, CT: Greenwood, 1965), 6–11, 70–83; Justina Parsons Bernstein, "'Hope We Be a Prosperous People': Ethnic Reorganization and the Indian Way of Living Through" (Ph.D. diss., Rutgers University, 2000), 46–119, 127–150, 195–211.

52. *RCIA* (1869), 559–560, 563–564, 574, 581, 627, 679, 686, 689, 696, 705; Otis, *Indian Question*, 221, 227–228, 238, 240–246; Berkhofer, *Salvation*, 74–79; Prucha, *Policy in Crisis*, chs. 7, 9, 10; David W. Adams, *Education for Extinction* (Lawrence: University Press of Kansas, 1995), 28–36, 51–59, 100–118, 141, 149–156; Clyde Ellis, "'A Remedy for Barbarism': Indian Schools, the Civilizing Program and the Kiowa-Comanche-Apache Reservation, 1817–1915," *American Indian Culture and Research Journal* 18/3 (1994): 85–120; Carol Devins, "'If We Get the Girls, We Get the Race': Missionary Education of Native American Girls," *Journal of World History* 3/2 (1992): 219–237; Parsons-Bernstein, "Prosperous People," 188–195, 212–220. For moving accounts of the lives of Indian students after leaving the boarding schools, see *In the White Man's Image*, a videorecording produced and directed by Christine Lesiak (Alexandria, VA: PBS, 1991).

53. Michael Hunt, *The Making of a Special Relationship: The United States and China to 1914* (New York: Columbia University Press, 1983), 5–12, 20–23.

54. A. Owen Aldridge, *The Dragon and the Eagle: The Presence of China in the American Enlightenment* (Detroit: Wayne State University Press, 1993), 36–37, 81–82, 87–92, 144, 150–156; Michael Adas, *Machines as the Measure: Science, Technology, and Ideologies of Western Dominance* (Ithaca: Cornell University Press, 1989), 79–95.

55. Smith, *Virgin Land*, 11; Norris, *The Octopus: A Story of California* (Garden City, NY: Doubleday, 1901), 651.

56. Bowles, *New West*, 71–74, 314; Henry Heusken, *Japan Journal, 1855–1861* (New Brunswick, NJ: Rutgers University Press, 1964),

156; Smith, *Virgin Land*, 13, 20–30, 37, 46; Karnes, *Gilpin*, 132–
136, 216; Ernest N. Paolino, *The Foundations of the American Empire: William Henry Seward and U.S. Foreign Policy* (Ithaca: Cornell University Press, 1973), 28–30, 35–36, 146–147; Anders
Stephanson, *Manifest Destiny: American Expansion and the Empire of Right* (New York: Hill and Wang, 1995), 58–63, 105.

57. Dana, *Great West*, 14.

58. Robbins, *Colony and Empire*, chs. 5–7; William Appleman Wil-
liams, *The Roots of the Modern American Empire: A Study of the
Growth and Shaping of Social Consciousness in a Marketplace
Society* (New York: Random House, 1969).

59. Robbins, *Colony and Empire*, chs. 4–5; Douglas C. North, "International Capital Flows and the Development of the American
West," *Journal of Economic History* 16/4 (1956): 493–505; Clark C.
Spence, *British Investments and the American Mining Frontier,
1860–1901* (Ithaca: Cornell University Press, 1958); Hunt, *Special
Relationship*, 144–151, 169–183; Jay Luvaas, *The Military Legacy of
the Civil War* (Chicago: University of Chicago Press, 1959), 229–
233; and the essays in E. B. Potter et al., *The United States and
World Sea Power* (Englewood Cliffs, NJ: Prentice-Hall, 1958).

60. Hunt, *Special Relationship*, 14–24, 169–170.

61. Ibid., 19–20; Arthur Smith, *The Uplift of China* (New York: Presbyterian Board of Missions, 1907), 130–131.

62. The overview that follows is based primarily on the works of
Thomas Parke Hughes, including *Networks of Power: Electrification in Western Society, 1880–1930* (Baltimore: Johns Hopkins University Press, 1983) and *American Genesis: A Century of Invention
and Technological Enthusiasm, 1870–1970* (New York: Penguin,
1990); as well as on David A. Hounshell, *From the American System to Mass Production, 1800–1932: The Development of Manufacturing Technology in the United States* (Baltimore: Johns Hopkins
University Press, 1984); and David Nye, *Electrifying America: Social Meanings of a New Technology* (Cambridge, MA: MIT Press,
1990).

63. John M. Levy, *Urban America: Processes and Problems* (Saddle River, NJ: Prentice-Hall, 2000), 26–27.

64. Pursell, *Machine in America*, ch. 6; Siegfried Giedion, *Mechanization Takes Command* (Oxford: Oxford University Press, 1948); Thomas Schlereth, *Victorian America, 1876–1915* (New York: Harper, 1991); Ruth Schwartz Cowan, *More Work for Mother: The Ironies of Household Technology from the Open Hearth to the Microwave* (New York: Basic Books, 1983).

65. These revealing statistics provide a fitting preface for A. J. P. Taylor's *The Struggle for the Mastery of Europe* (Oxford: Oxford University Press, 1954), xxviii–xxix.

66. Carnegie, *Triumphant Democracy* (New York: Scribner, 1886), 1–22. See also Kasson, *Civilizing the Machine*, chs. 4–5; Henry Nash Smith, ed., *Popular Culture and Industrialism, 1865–1890* (New York: New York University Press, 1967).

67. For reactions to the Chinese in the U.S., see Robert Lee, *Orientals: Asian-Americans in Popular Culture* (Philadelphia: Temple University Press, 1999), 32–50, 61–62, 89–91. For negative reports on Chinese society, Henry Norman, *The Peoples and Politics of the Far East* (London: Fisher, Unwin, 1895), 201–206, 209–210, 287; Thomas F. Millard, *The New Far East* (New York: Scribner, 1906); Eliza Scidmore, *China: The Long-Lived Empire* (New York: Century, 1900), 5–8, 16, 91, 146, 297, 374; R. H. Graves, *Forty Years in China* (Baltimore: R. H. Woodward, 1895), 226–227; William Barclay Parsons, *An American Engineer in China* (New York: McClure, Phillips, 1900), 167–168.

68. Samuel Wells Williams, *The Middle Kingdom* (New York: John Wiley, 1847), 2:143–145, 153–154, 178; *CRMJ* (July–Aug. 1875), 206–207; (Nov.–Dec. 1877), 466–467; Parsons, *Engineer in China*, 22–24, 131, 136–139, 166–168, and ch. 8; Michael C. Coleman, "Presbyterian Missionary Attitudes toward China and the Chinese," *Journal of Presbyterian History* 56/3 (1978), 189–190, 194–196 (quoted phrase).

69. *The Complete Journal of Townsend Harris*, ed. Mario Emilio

Cosenza (Rutland, VT: Charles Tuttle, 1959), 362–363; Sidney Gulick, *Evolution of the Japanese: Social and Psychic* (New York: Revell, 1903), 117–118, 195–196, 429–430, 433; Millard, *New Far East*, 38–39; M. L. Wakeman Curtis, "Japanese Rivalry as Studied on the Ground," *Overland Monthly* 30 (Spring 1897), 228–230, 234–235.

70. William Griffis, *Townsend Harris: First American Envoy in Japan* (New York: Houghton Mifflin, 1895), 335–342; I. W. Wiley, *China and Japan: A Record of Observations* (Cincinnati: Hitchcock and Walden, 1879), 36–43, ch. 27; Joseph I. C. Clarke, *Japan at First Hand* (New York: Dodd, Mead, 1918), ch. 20; Parsons, *Engineer in China*, 273–285; Arthur Maclay, *A Budget of Letters from Japan* (New York: Armstrong, 1886), 144–148; Akira Iriye, *Across the Pacific: An Inner History of American–East Asian Relations* (New York: Harcourt, Brace, 1967), 10–12, 25–28; Martha Clevenger, "Through Western Eyes: Americans Encounter Asians at the Fair," *Gateway Heritage* 17/2 (1996), 46–51.

71. Griffis, *Townsend Harris*, 338; Gulick, *Evolution of the Japanese*, 191, 194, 207–208, 218–219, 425–428; Maclay, *Letters from Japan*, 341; T. R. Jernigan, "Japan's Entry into the Family of Nations," *North American Review* 169 (Aug. 1898), 226; Wiley, *China and Japan*, 380–381; Howard K. Beale, *Theodore Roosevelt and the Rise of America to World Power* (New York: Collier, 1956), 42–43 (quotations).

72. Martin, *The Chinese* (New York: Revell, 1898), 94, 145–148, 228; Martin, *The Lore of Cathay or the Intellect of China* (New York: Revell, 1901), 8–9 (quotations), 23–29; Nevius, *China and the Chinese* (New York: Harper, 1869), 275–281; Gustavus, "The Future Language of China," *CRMJ* (Nov.–Dec. 1877), 474–476; James H. Wilson, *China: Travels and Investigations in the Middle Kingdom* (New York: Appleton, 1887), 82–84 (quotation), 308–309; Chester Holcombe, *The Real Chinaman* (London: B. F. Stevens, 1895), 248; Smith, *Uplift of China*, 47–48; Millard, *New Far East*, 255–266, 268; Coleman, "Missionary Attitudes," 190.

73. Steven Mintz, *Moralists and Modernizers: America's Pre–Civil War Reformers* (Baltimore: Johns Hopkins University Press, 1995), 63; Walter LaFeber, *The Clash: A History of U.S.-Japan Relations* (New York: Norton, 1997), 66.

74. Gavan Daws, *Shoal of Time: A History of the Hawaiian Islands* (New York: Macmillan, 1968), ch. 3 and 158–163.

75. A. D. Hall, *China: Land of Contradictions* (New York: Street and Smith, 1900), 141; and, interpolating from the population estimates provided by Ping-ti Ho, "Salient Aspects of China's Heritage," in Ping-ti Ho and Tang Tsou, eds., *China in Crisis: China's Heritage and the Communist Political System* (Chicago: University of Chicago Press, 1968), 1:8.

76. See *CRMJ* (Feb. 1889), 49–56, (Nov. 1887), 409–413; B. C. Henry, *The Cross and the Dragon* (New York: Randolph, 1885), 428–432; Dana L. Robert, "The Methodist Struggle over Higher Education in Fuzhou, China, 1877–1833," *Methodist History* 34/3 (1996), 183–188.

77. Elliott Osgood, *Breaking Down Chinese Walls* (New York: Revell, 1908), 12–14, 18, 21, 67–76; Hunter Corbett, "The Work of the Protestant Missions in Shantung," *CRMJ* (Mar.–Apr. 1881), 88–89; Jenny Willing, "The Intellectual Uses of the Woman's Foreign Missionary Work," *CRMJ* (May–June 1878), 215–221; Jane Hunter, *The Gospel of Gentility: American Women Missionaries in Turn-of-the-Century China* (New Haven: Yale University Press, 1984), 16–26; Robert, "Education in Fuzhou," 178–183; Graves, *Years in China*, 28; Coleman, "Missionary Attitudes," 187, 194–195.

78. Mintz, *Moralists and Modernizers*, 19–21, 35–37 (quotations); *CRMJ* (May–June 1874), 150–152, (Aug. 1889), 380; Kristin Gleeson, "The Stethoscope and the Gospel: Presbyterian Foreign Medical Missions, 1840–1900," *American Presbyterians* 71/1 (1993), 129–137; Jessie Lutz, *China and the Christian College, 1850–1950* (Ithaca: Cornell University Press, 1971), 21, 29.

79. *CRMJ* (May–June 1874), 144–150, (Mar.–Apr. 1878), 114–119, (Nov.–Dec. 1881), 455–460, (Jan.–Feb. 1884), 8–31, (Oct. 1886),

398–400, (Feb. 1887), 57; Peter Parker, *The Life, Letters, and Journals* (Boston: Congregational Pub. Soc., 1896), 82, 134–135; Graves, *Years in China*, 228–253; Osgood, *Chinese Walls*, ch. 1; Hall, *Land of Contradictions*, 109, 141–142; Peter Buck, *American Science and Modern China, 1876–1936* (Cambridge: Cambridge University Press, 1980), 3–4, 9–16, 82–83; Gleeson, "Stethoscope and Gospel," 130–131.

80. Osgood, *Chinese Walls*, 16–18, 21, 56, 71; Buck, *Science and China*, 9, 13, 21, 32, 68–70; Hunter, *Gospel of Gentility*, 28–30, 39; and the essays by Joel McKay and Harold Platt in Joel Tarr and Gabriel Dupuy, eds., *Technology and the Rise of the Networked City in Europe and America* (Philadelphia: University of Pennsylvania Press, 1988).

81. *CRMJ*—on women and Chinese doctors and assistants—(June 1866), 236–237, (July–Aug. 1880), 302, (Jan.–Feb. 1884), 12, (Apr. 1885), 158, (Jan. 1886), 16–23; on science education and translation activities—(July–Aug. 1879), 292–297, (Sept.–Oct. 1879), 397–398, (Mar.–Apr. 1880), 139–141, (July–Aug. 1880), 242–250, (Mar.–Apr. 1881), 92–95, (May–June 1881), 225–236; Parker, *Life and Letters*, 245; Osgood, *Chinese Walls*, 18–19, 21; Smith, *Uplift of China*, 165–167; Graves, *Years in China*, 248–249, 269–273; Buck, *Science and China*, 31–36; Hunter, *Gospel of Gentility*, 15–16; Lutz, *China and Christian Colleges*, 63–64, 68, 111–112.

82. Marilyn Blatt Young, *American Policy in China, 1895–1901* (Cambridge, MA: Harvard University Press, 1968); Hunt, *Special Relationship*, ch. 6.

83. Iriye, *Across the Pacific*, 71–75; Clarke, *Japan at First Hand*, ch. 5; Millard, *New Far East*, 286–294. On the government-directed transformation of Japan, see Peter Duus, *The Rise of Modern Japan* (Boston: Houghton Mifflin, 1976), chs. 7–9; Thomas C. Smith, *Political Change and Industrial Development in Japan: Government Enterprise, 1868–1880* (Stanford: Stanford University Press, 1955). And for Japanese responses to western learning, see Masao Watanabe, *The Japanese and Western Science*, trans. Otto

Theodor Benfey (Philadelphia: University of Pennsylvania Press, 1976).

84. Donald Roden, "Baseball and the Quest for National Dignity in Meiji Japan," *American Historical Review* 85/3 (1980): 514–534; Gael Graham, "Exercising Control: Sports and Physical Culture in American Protestant Mission Schools in China, 1880–1930," *Signs* (Autumn 1994): 23–48.

85. Gulick, *Evolution of the Japanese*, 20, 27–29, 32–34, 45, 48.

86. Lafeber, *The Clash*, ch. 2. And see the documents included in Peter Duus, ed., *The Japanese Discovery of America* (Boston: Bedford, 1997), 179–183, 187–200.

87. For concern about the Yellow Peril, see Adas, *Machines*, 357–365; Willard Straight, "Chinese Students in Japan," *Nation* (Aug. 1905), 181; Millard, *New Far East*, 17, 262–267, 270–282, ch. 19; Smith, *Chinese Characteristics*, 41; Hall, *Land of Contradictions*, 56–58; Kenneth S. Latourette, *The Development of Japan* (New York: Macmillan, 1920), 90–91. And for an empathetic American attempt to see the issue from an Asian perspective, Sidney L. Gulick, *The White Peril in the Far East* (New York: Revell, 1905), ch. 12.

3. Engineers' Imperialism

1. For Taft's early career, I have relied on Henry F. Pringle, *The Life and Times of William Howard Taft* (New York: Farrar, Reinhart, 1939), and Herbert S. Duffy, *William Howard Taft* (New York: Minton, Balch, 1930).

2. Jacob Schurman, "The Immediate Duty of the United States: An Authorized Interview," *Outlook* 63 (1899), 534.

3. Lewis E. Gleek Jr., *American Institutions in the Philippines (1898–1941)* (Manila: Historical Conservation Society, 1976), 19.

4. Rodney J. Sullivan, *Exemplar of Americanism: The Philippine Career of Dean C. Worcester* (Ann Arbor: Center for South and Southeast Asian Studies, 1991), 13–14, 31–39; Peter Stanley, "'The

Voice of Worcester Is the Voice of God': How One American Found Fulfillment in the Philippines," in Stanley, ed., *Reappraising an Empire* (Cambridge, MA: Harvard University Press, 1984), 120–133.

5. For the text of McKinley's proclamation see W. Cameron Forbes, *The Philippine Islands* (Boston: Houghton Mifflin, 1928), 2:437–438. Otis quoted in Stuart Creighton Miller, *"Benevolent Assimilation": The American Conquest of the Philippines, 1899–1903* (New Haven; Yale University Press, 1982), 66. On population estimates see A. W. Thomas, *The Philippines and the Purpose* (Washington: Jefferson Democrat Pub., 1900), 155.

6. On the roles of Filipino leaders in the maneuvers associated with the annexation and the formulation of policies for the new colonies, see Bonifacio Salamanca, *The Filipino Reaction to American Rule, 1901–1913* (New Haven: Shoe String Press, 1968); and Peter Stanley, *A Nation in the Making: The Philippines and the United States, 1899–1921* (Cambridge, MA: Harvard University Press, 1974).

7. For such arguments, see, e.g., N. B. Scott, *Government of the Philippines* (Washington: GPO, 1900); and Benjamin R. Tillman, "Causes of Southern Opposition to Imperialism," *North American Review* 71 (1900): 439–446.

8. Though many historians have accepted a total of civilian casualties of approximately 200,000, I have relied on the estimate calculated by Ken de Bevoise; see his *Agents of Apocalypse: Epidemic Disease in the Colonial Philippines* (Princeton: Princeton University Press, 1995), esp. 10–13.

9. Philip C. Jessup, *Elihu Root* (New York: Dodd, Mead, 1939), 1:300. "The Situation in the Philippines," *Outlook* (Sept. 12, 1903), 111. See also Michael Adas, "Improving on the Civilising Mission: Assumptions of United States Exceptionalism in the Colonisation of the Philippines," *Itinerario* 22/4 (1998), 44–66.

10. Schurman, "Immediate Duty," 536. It is sobering to realize that this line of reasoning was offered by an official who was *opposed* to the annexation.

11. Roosevelt, "Expansion and Peace," in *The Strenuous Life* (New York: Review of Reviews, 1910), 28–35.

12. Taft to Judge William R. Day, Aug. 16, 1900, Taft Papers, LCMD, ser. 3, reel 31. Roosevelt, *Strenuous Life*, 11, 19 (quotation), 33, 240–241; Schurman, "Immediate Duty," 535; Miller, *Benevolent Assimilation*, 57–58. See also Kristin Hoganson, *Fighting for American Manhood: How Gender Politics Provoked the Spanish-American and Philippine Wars* (New Haven: Yale University Press, 1998), chs. 6–7.

13. Miller, *Benevolent Assimilation*, chs. 7, 8, 10; and essays by Miller, Kenton Clymer, and Glenn Antony May in Stanley, ed., *Reappraising an Empire*.

14. Thomas, *Philippines and Purpose*, 155.

15. See Pringle, *Life of Taft*, 1: chs. 1–2; Peter W. Stanley, "William Cameron Forbes: Proconsul in the Philippines," *Pacific Historical Review* 35 (1966): 285–302; Sullivan, *Exemplar of Imperialism*, esp. 6–9; Mary Fee, *A Woman's Impressions of the Philippines* (Chicago: A. C. McClurg, 1910); Paul T. Gilbert, *The Great White Tribe in Filipinia* (Cincinnati: Jennings and Pye, 1903).

16. Bryn, *The Progress of Invention in the Nineteenth Century* (New York: Muan, 1900), ch. 1 and concl.; Morison, *The New Epoch as Developed by the Manufacture of Power* (New York: Arno, 1903), 1–12, 19–20, 31–33, 128–134; Strong, *Our Country: Its Possible Future and Its Present Crisis* (New York: American Missionary Society, 1885).

17. On the economic downturns and social tensions, see, e.g., Rayford Logan, *The Negro in American Life and Thought: The Nadir, 1877–1901* (New York: Dial, 1954); Joel Williamson, *A Rage for Order* (New York: Oxford University Press, 1986); Alan Kraut, *The Huddled Masses: The Immigrant in American Society, 1880–1921* (Arlington Heights, IL: Harlan-Davidson, 1982); Michael Nash, *Conflict and Accommodation: Coal Miners, Steel Workers and Socialism, 1890–1920* (Westport, CT: Greenwood, 1982).

18. The phrase is Robert Wiebe's, from *The Search for Order, 1877–*

1920 (New York: Hill and Wang, 1967). On the shift to consumerism see William Leach, *Land of Desire: Merchants, Power, and
the Rise of a New American Culture* (New York: Vintage, 1993);
T. J. Jackson Lears, *Fables of Abundance: A Cultural History of
Advertising in America* (New York: Basic Books, 1994), chs. 5–7.

19. Elaine Showalter, *Sexual Anarchy: Gender and Culture at the Fin
de Siècle* (New York: Viking, 1990); Gail Bederman, *Manliness
and Civilization: A Cultural History of Gender and Race in the
United States, 1880–1917* (Chicago: University of Chicago Press,
1995), esp. 10–31 and chs. 3 and 5; T. J. Jackson Lears, *No Place of
Grace: Antimodernism and the Transformation of American Culture, 1880–1920* (New York: Pantheon, 1981).

20. F. G. Gosling, *Before Freud: Neurasthenia and the American
Medical Community, 1870–1910* (Urbana: University of Illinois
Press, 1987), chs. 1–4; Harvey Green, *Fit For America: Health, Fitness, Sport, and American Society* (New York: Pantheon, 1986),
chs. 5–9; Hoganson, *Fighting for American Manhood*, chs. 6 and
8; Peter Stearns, *Be a Man! Males in Modern Society* (New York:
Holmes and Meier, 1990), ch. 5.

21. Cecelia Tichi, *Shifting Gears: Technology, Literature, Culture in
Modernist America* (Chapel Hill: University of North Carolina
Press, 1987), 42–55, 99, 105.

22. Edwin Layton, *The Revolt of the Engineers: Social Responsibility
and the American Engineering Profession* (Baltimore: Johns
Hopkins University Press, 1971), ch. 1; Ruth Oldenziel, *Making
Technology Masculine: Men, Women and Modern Machines in
America, 1870–1945* (Amsterdam: University of Amsterdam Press,
1999), chs. 2–3; Pursell, *Machine in America*, ch. 6; and the essays
by Joel McKay and Harold Platt in Joel Tarr and Gabriel Dupuy,
eds., *Technology and the Rise of the Networked City in Europe
and America* (Philadelphia: Temple University Press, 1988).

23. Layton, *Revolt of the Engineers*, ch. 2; Monte A. Calvert, *The Mechanical Engineer in America, 1830–1910* (Baltimore: Johns
Hopkins University Press, 1967), chs. 4–6; David F. Noble, *Amer-

ica by Design: Science, Technology, and the Rise of Corporate Capitalism (New York: Knopf, 1977), chs. 2–3.

24. Morison quoted in Layton, *Revolt of the Engineers,* 58–59. See Ruth Oldenziel's work, esp. her "Gender and the Meanings of Technology: Engineering in the U.S., 1880–1945" (Ph.D. diss., Yale University, 1992), 23–27, 91–94. See also Tichi, *Shifting Gears,* 20–26, 97–112.

25. Carnegie, *Triumphant Democracy* (New York: Scribner, 1886), 232. In my discussion of gender issues in engineering I draw mainly on Oldenziel's *Making Technology Masculine.* For more on these patterns, see the special issue of *Technology and Culture* 31/1 (1997) devoted to "Gender Analysis and the History of Technology," esp. the essays by Nina Lerman, Oldenziel, and Judith McGaw.

26. See, for examples, Morison, *The New Epoch,* esp. 96–97, 128–134, and ch. 5; Fred Colvin, *60 Years with Men and Machines* (New York: McGraw-Hill, 1947); Layton, *Revolt of the Engineers,* ch. 3.

27. Martha Banta, *Taylored Lives: Narrative Productions in the Age of Taylor, Veblen, and Ford* (Chicago: University of Chicago Press, 1993), 68–74 and chs. 3–5; Layton, *Revolt of the Engineers,* ch. 6; Daniel Nelson, *Managers and Workers: Origins of the New Factory System in the United States 1880–1920* (Madison: University of Wisconsin Press, 1975), esp. chs. 3–4; Judith A. Merkle, *Management and Ideology* (Berkeley: University of California Press, 1980), chs. 1–3.

28. "Publicola," *Duty of the American People to the Philippines* (1898), 16–17, LC.

29. PCR (1901), pt. 1, 122; pt. 2, 282; (1903), pt. 1, 255, 262; pt. 2, 57; pt. 3, 15–16, 27; William Howard Taft, *Present Day Problems* (New York: Dodd, Mead, 1908), 3; J. W. Beardsley, *Bureau of Engineering Report,* Aug. 15, 1904, BIA, 350, 7152; Taft to Samuel Felton, Aug. 10, 1900, Taft Papers, ser. 3, reel 31.

30. Forbes, *Philippine Islands,* 1:408–409.

31. See Senator Scott's June 1900 speech in Scott, "Government of the Philippines," esp. 10–13.

32. Crow, *America and the Philippines* (Garden City, NY: Doubleday, Page, 1914), 99. See also Jacob Schurman's attempts to reconcile his early opposition to annexation with his leadership of the First Commission; Schurman, *Philippine Affairs* (New York: Scribner, 1902), 84–85.

33. Publicola, *Duty*, 17. See also John P. Finley, "Race Development by Industrial Means among the Moros and Pagans of the Southern Philippines," *Journal of Race Development* 3/3 (1913), 346–347.

34. Taft to William P. Frye, Jan. 7, 1901; Taft's Annual Report to Elihu Root, secretary of war, Jan. 24, 1901, Taft Papers, ser. 3, reel 31. Roosevelt, "Our Policy and Our Work in the Philippines," preface to Mrs. Campbell Dauncy, *The Philippines: An Account of their People, Progress, and Condition* (Boston: J. B. Millet, 1910), 18.

35. Smith, "Message of the Governor-General to the Philippine Commission and the Philippine Assembly, Constituting the Philippine Legislature," in Francis Burton Harrison Papers, LCMD, container 42, 1 (quotations), 2–7. "Speech Made by Governor-General W. Cameron Forbes at the Popular Banquet in his Honor," Manila, Aug. 30, 1913, ibid., container 51. Forbes, *Philippine Islands*, 2:455–456; see also Stanley, *Nation in the Making*, 139–140, 194–195.

36. See Michael Adas, *Machines as the Measure of Men: Science, Technology and Ideologies of Western Dominance* (Ithaca: Cornell University Press, 1989), esp. ch. 4.

37. Charles Elliott, *The Philippines to the End of the Commission Government* (Indianapolis: Bobbs-Merrill, 1917), 279–280; Forbes, *Philippine Islands*, 2:456; PCR (1901), pt. 2, 71. See also Fred Atkinson, *The Philippine Islands* (Boston: Ginn, 1905), 6; Hugo Miller, *Economic Conditions in the Philippines* (New York: Ginn, 1913), 20; Dudley McGovney, *Civil Government in the Philippines* (Chicago: Scott, Foresman, 1903), 48–49.

38. Robert W. Rydell, *All the World's a Fair* (Chicago: University of Chicago Press, 1984); Bederman, *Manliness and Civilization*, esp. 31–41.

39. Rydell, *World's a Fair*, ch. 6; Martha R. Clevenger, "Through Western Eyes: Americans Encounter Asians at the Fair," *Gateway Heritage* (Fall 1996), 43–46; Paul Kramer, "Making Cessions: Race and Empire Revisited at the Philippine Exposition, St. Louis, 1901–1905," *Radical History Review* 73/1 (1999): 74–114.

40. Many of Taft's letters during his years as head of the commission and later as governor express these views, as do accounts of the islands by ex-officials, journalists, teachers, missionaries, and travelers.

41. See, e.g., *PCR* (1901), pt. 2, 49, 61–62; Crow, *America and Philippines*, ch. 7; Sullivan, *Exemplar of Imperialism*, 41.

42. Griffis, "America in the Far East II: The Anglo-Saxon in the Tropics," *Outlook* 60/15 (Dec. 10, 1898), 902–907; Roosevelt, *Strenuous Life*, 243–244; Howard K. Beale, *Theodore Roosevelt and the Rise of America to World Power* (New York: Collier, 1956), 41–48; Charles Burke Elliot, *The Philippines to the End of the Military Regime* (Indianapolis: Bobbs-Merrill, 1917), 84–85 (quotations). Kidd, *The Control of the Tropics* (London: Macmillan, 1898). These social evolutionist presuppositions suffused the social theorizing of engineers in this era; see Layton, *Revolt of the Engineers*, 55–58, 141.

43. Publicola, *Duty*, 16–19; Charles Morris, *Our Island Empire* (Philadelphia: Lippincott, 1899), 381–382; Taft to Edward Colston, Apr. 24, 1901, Taft Papers, ser. 3, reel 32, 8; Elliott, *End of Commission*, 268–270; Crow, *America and Philippines*, 43–44, 48, 212; L. Donald Warren, *Isles of Opportunity* (Washington: Review and Herald, 1928), 63–64, 70, 119. Quoted phrase from Frank L. Strong, *The United States and the Philippines* (Manila: Staples-Howe, 1913), 4.

44. Taft to William Worthington, July 20, 1900, ser. 3, reel 31, 4; Taft to John Spooner, Sept. 3, 1900, ser. 3, reel 31, 6; and esp. Taft to Edward Colston, Apr. 24, 1901, ser. 3, reel 32, 4–5, Taft Papers. See also Schurman, "Immediate Duty," 535; Morris, *Island Empire*, 420; Elliott, *End of Commission*, 187, 189; Gilbert, *White Tribe*, 293–295; Fee, *Woman's Impressions*, 90–91.

45. See Victor G. Heiser, "Sanitation in the Philippines and Its Possible Effect on Other Oriental Countries," *Philippine Review* 2 (1917), 121–122, Harrison Papers, container 51; Crow, *America and Philippines*, 106–112; Elliott, *End of Commission*, 188–190, 194–197, 215–216; PCR (1903), pt. 1, 271–272.

46. William H. Taft, *Special Report of the Secretary of War to the President on the Philippines* (Washington: GPO, 1908), 25–26; PCR (1901), pt. 1, 15; Elliott, *End of Military Regime*, 111, and *End of Commission*, 269; Edith Moses, *Unofficial Letters of an Official's Wife* (New York: Appleton, 1908), 16–17; Atkinson, *Philippine Islands*, 6, 266–267; Crow, *America and Philippines*, 52, 57; Gilbert, *White Tribe*, 299–300, 302; Nicholas Roosevelt, *The Philippines: A Treasure and a Problem* (New York: J. H. Sears, 1926), 5; Arthur Judson Brown, *The New Era in the Philippines* (New York: Revell, 1903), 76–77.

47. Gilbert, *White Tribe*, 294–296, 300; Atkinson, *Philippine Islands*, 267–268, 271, 274–275; G. A., "Filipino Characteristics," *Harper's Weekly* (Mar. 4, 1899), 226; Taft quoted in Oscar M. Alfonso, *Roosevelt and the Philippines, 1897–1909* (Quezon City: University of Philippines Press, 1970), 45–46; Ralph Minger, "Taft, MacArthur, and the Establishment of Civil Government in the Philippines," *Ohio Historical Quarterly* 70 (1961), 316, 320.

48. See Syed Hussein Alatas, *The Myth of the Lazy Native* (London: Frank Cass, 1977). For American expressions of these themes, see Jacob Schurman, "Our Duty to the Philippines," *Independent* 51 (Dec. 1899), 3466–3467; Moses, *Unofficial Letters*, 354; Morris, *Island Empire*, 417–419, 470–471; Gilbert, *White Tribe*, 296–297; Brown, *New Era*, 55; *Report of the Special Mission to the Philippines* (Manila: Bureau of Printing, 1921), 17–18.

49. Crow, *America and Philippines*, 57, 104–106, 119; Forbes, *Philippine Islands*, 1:330, 517–519; Elliot, *End of Commission*, 268, quotation; Moses, *Unofficial Letters*, 350–351; Fee, *Woman's Impressions*, 101, 103, 138–139; Brown, *New Era*, 75, 81–87.

50. The vast literature on American racial patterns ranges from the seminal accounts of C. Vann Woodward and Gunnar Myrdal to

Joel Williamson's *The Crucible of Race: Black-White Relations in the American South* (New York: Oxford University Press, 1984) and David Roediger's *The Wages of Whiteness* (New York: Verso, 1991).

51. For contemporary accounts, see, e.g., Crow, *America and Philippines*, esp. 230–233; Moses, *Unofficial Letters*, 355; Mrs. Campbell Dauncy, *An Englishwoman in the Philippines* (New York: Dutton, 1906). And see Stanley, *Nation in the Making*, 164–165; Stanley Karnow, *In Our Image: America's Empire in the Philippines* (New York: Ballantine, 1989), esp. 211–215; Kenton J. Clymer, *Protestant Missionaries in the Philippines, 1898–1916* (Urbana: University of Illinois Press, 1986), 74–75, 90–92; Virginia Licuanan, *Filipinos and Americans* (Manila: Baguio Country Club, 1982).

52. Schurman, *Philippine Affairs*, 63; John Bancroft Devins, *An Observer in the Philippines* (Boston: American Tract Soc., 1905), 193, 198, 200; Gilbert, *White Tribe*, 302; Warren, *Isles of Opportunity*, 65–66, 73–74; Morris, *Island Empire*, 397, 417, 420; David P. Barrows, *A History of the Philippines* (New York: American Book Co., 1905), 319; Elliot, *End of Military Regime*, 80–83; Taft, *Special Report*, 26; Norbert Lyons, "Some Observations on Race Contact," *Report of the 33rd Annual Lake Mohonk Conference* (Oct. 19, 1916), 148–150; Moses, *Unofficial Letters*, 345–346; *Report of the Special Mission*, 11; Brown, *New Era*, 55, 59.

53. Dauncy, *Englishwoman*, 258–259. See also Adas, *Machines as the Measure*, 302–304.

54. Harrison, *The Corner-Stone of Philippine Independence: A Narrative of Seven Years* (New York: Century, 1922), 55–59.

55. Government of the Philippines, *Annual Report of the Director of Public Works* (Manila, 1912) in BIA, 350, 7152–70; *Report of the Special Mission*, 21, Harrison Papers, container 40.

56. Taft to Charles T. Baker, July 13, 1901, Taft Papers, ser. 3, reel 33.

57. For samples of these characterizations, see Publicola, *Duty*, 9, 15; G. A., "Filipino Characteristics," 226; Daniel G. Brinton, "The Races of the Philippine Archipelago," *Scientific American Supple-*

ment (Dec. 17, 1898), 19212; Taft to William R. Day, Aug. 16, 1900, 2, and Taft to Edward Colston, Apr. 24, 1901, 4–5, Taft Papers, ser. 3, reels 31 and 32; Atkinson, *Philippine Islands*, 271–272; Elliott, *End of Military Regime*, 60.

58. Stearns, *Be a Man*, 109, 111, 113–116, 126–288, 136, 139. And see Cynthia E. Russett, *Sexual Science: The Victorian Construction of Womanhood* (Cambridge, MA: Harvard University Press, 1989), esp. 54–63 and ch. 2; Gosset, *Before Freud*, 97–100; Carol Z. Stearns and Peter N. Stearns, *Anger: The Struggle for Emotional Control in American History* (Chicago: University of Chicago Press, 1986), ch. 4; Sarah Stage, *Female Complaints: Lydia Pinkham and the Business of Women's Medicine* (New York: Norton, 1979), 64–88; Antony Rotondo, "Body and Soul: Changing Ideals of American Middle-Class Manhood," *Journal of Social History* 16 (1983): 22–38; Hoganson, *Fighting for American Manhood*, chs. 1 and 5; Bederman, *Manliness and Civilization*.

59. Morris, *Island Empire*, 419. See also Gilbert, *White Tribe*, 295.

60. Elizabeth Ammons, "The Engineer as Cultural Hero and Willa Cather's First Novel, *Alexander's Bridge*," *American Quarterly* 38 (1986), esp. 750–751; Tichi, *Shifting Gears*, 117–132. The term "heroic" was gendered masculine as recently as the 1983 edition of *Webster's New Universal Unabridged Dictionary*.

61. Quotation from *PCR* (1901), pt. 1, 61. See also G. A., "Filipino Characteristics," 226; Gilbert, *White Tribe*, 161, 297–299; Fee, *Woman's Impressions*, 101, 104–106; Elliott, *End of Military Regime*, 84–85.

62. See Barrows, *History of the Philippines*, 10; *Report of the Special Mission*, 11; Atkinson, *Philippine Islands*, 275; Moses, *Unofficial Letters*, 350–351; Elliot, *End of Military Regime*, 110; Clymer, *Protestant Missionaries*, 80–81.

63. De Bevoise, *Agents of Apocalypse*, ch. 3.

64. Baker to Taft, May 16, 1901, 3. Taft would reiterate these sentiments in numerous letters soliciting candidates for posts in the colonial bureaucracy. Taft Papers, ser. 3, reel 32.

65. E. J. Westerhouse, "Provincial Public Works Organization of the Philippine Islands and Its Engineering Personnel," *Engineering News* 70/20 (Nov. 13, 1913), 958–959, BIA, 350, 7152; Miller, *Benevolent Assimilation*, ch. 10; John Morgan Gates, *Schoolbooks and Krags: The United States Army in the Philippines* (Westport, CT: Greenwood, 1973); Clymer, *Protestant Missionaries*, 22.

66. Renato Rosaldo, *Ilongot Headhunting, 1883–1974* (Stanford: Stanford University Press, 1980), 2–14; Clymer, *Protestant Missionaries*, 67–74; and contemporary accounts, such as Worcester's "Field Sports among the Wild Men of Northern Luzon," *National Geographic* 22/3 (1911), 215–267.

67. For examples see Roy MacLeod and Deepak Kumar, eds., *Technology and the Raj* (New Delhi: Sage, 1995); J. A. A. van Doorn, *De Laatste Eeuw van Indië: Ontwikkeling en Ondergang van een Koloniaal Project* (Amsterdam: Bert Bakker, 1994), chs. 4–7; David del Testa, "Imperial Corridor: Progress, Politics and Disappointment along the Hanoi-Saigon Railroad, 1898–1945" (Ph.D. diss., University of California at Davis, 2000).

68. Adas, *Machines as the Measure*, esp. 318–342.

69. For the text of McKinley's instructions, which were drafted and transmitted by Root, see Forbes, *Philippine Islands*, 2:439–445; quoted portions, 442–443.

70. Roosevelt, *The Philippines*, 15.

71. In McKinley's "benevolent assimilation" proclamation, in Forbes, *Philippine Islands*, 2:437–438; Edward C. Pierce, *The "Single Tribe" Fiction* (Chicago: American Anti-Imperialist League, 1900), 14–16; Atkinson, *Philippine Islands*, 9, 264–265; and Elihu Root, quoted in Glenn Antony May, *Social Engineering in the Philippines, 1900–1913* (Westport, CT: Greenwood, 1980), 7.

72. Taft quoted in Alfonso, *Roosevelt and the Philippines*, 16. For colonizers' intentions see Gleek, *American Institutions*, 5, 8–9, 212, 207; May, *Social Engineering*, 6–7, 11, 17, 88, 104, 140–141; Devins,

Observer in the Philippines, 108; Warren, *Isles of Opportunity*, 116; Roosevelt, *The Philippines*, 16–18, 237–239.

73. Elliott, *End of Military Regime*, 54–60; Harrison, *Corner-Stone*, ch. 21.

74. On the Philippines see Norman Owen, ed., *Compradore Colonialism* (Ann Arbor: University of Michigan Press, 1971); and essays by Michael Cullinane and Reynaldo Ileto in Stanley, *Reappraising an Empire*. For a contemporary denunciation of British policies, see Valentine Chirol, *Indian Unrest* (New York: Scribner, 1910).

75. Schurman, "Our Duty to the Philippines," 3467; Taft to John M. Harlan, Sept. 22, 1900, 3, Taft Papers, ser. 3, reel 31; PCR (1st, 1901), 71; (2nd, 1901), 60, 283; Elliot, *End of Commission*, 272; David Barrows, *A Decade of American Government in the Philippines* (New York: World Book, 1914), 46.

76. Strong, *United States and Philippines*, 2–3.

77. The photographs appeared in *World Outlook* 1 (Apr. 1915), 1–5, Harrison Papers, container 57. See also Westerhouse, "Provincial Public Works," 960–961; Forbes, *Philippine Islands*, 1: opp. p. 376; Crow, *America and Philippines*, opp. pp. 68, 69, 192, 218; *Report of the Special Mission*, 31; Elliott, *End of Commission*, 280, 288; Dean C. Worcester, *The Philippines Past and Present* (New York: Macmillan, 1921), 880.

78. See PCR (1901), pt. 1, 71, pt. 2, 283; Taft, *Present Day Problems*, 26–27; Taft to Elihu Root, Aug. 21, 1900, Taft Papers, ser. 3, reel 31; Stanley, *Nation in the Making*, 87–88, 104, and "Forbes Proconsul," 293; Warren, *Isles of Opportunity*, 70–71; Elliott, *End of Commission*, 271, 287–288, 318; Crow, *America and Philippines*, 32, 218–219, 227.

79. Warren, *Isles of Opportunity*, 68; De Bevoise, *Agents of Apocalypse*, 175–184; Reynaldo Ileto, "Cholera and the Origins of the American Sanitary Order in the Philippines," in David Arnold, ed., *Imperial Medicine and Indigenous Societies* (Manchester:

Manchester University Press, 1988), 125–148; Rodney Sullivan, "Cholera and Colonialism in the Philippines," in Roy MacLeod and Milton Lewis, eds., *Disease, Medicine and Empire* (London: Routledge, 1988).

80. Ileto, "Cholera," 132. On the American propensity to employ military metaphors in dealing with insect pests and disease, see Edmund Russell III, "'Speaking of Annihilation': Mobilizing for War against Human and Insect Enemies, 1914–1945," *Journal of American History* 82/4 (Mar. 1996): 1505–29.

81. Heiser, "Sanitation," 117–123; Elliot, *End of Commission*, ch. 9; essays by Joel Tarr and Letty Anderson in Tarr and Dupuy, *Rise of the Networked City*; Tarr, *The Search for the Ultimate Sink: Urban Pollution in Historical Perspective* (Akron: University of Akron Press, 1996), esp. pt. 2 and ch. 14.

82. My overview of health and sanitation programs is based primarily on De Bevoise, *Agents of Apocalypse*, 109–117, 135–141, 153–163, 185–190; Victor Heiser, *An American Doctor's Odyssey* (New York: Norton, 1936); Elliot, *End of Commission*, 186–187, 194–195, 205–218; Forbes, *Philippine Islands*, 1:ch. 8; Worcester, *Past and Present*, ch. 16.

83. Clymer, *Protestant Missionaries*, 76–77; Crow, *America and Philippines*, 118–119; *Report of the Special Mission*, 11.

84. Elliott, *End of Commission*, 185, 196; Forbes, *Philippine Islands*, 1:332–333; Heiser, "Sanitation," 120, 122–124; Taft, *Present Day Problems*, 23.

85. De Bevoise, *Agents of Apocalypse*, 158–163; PCR (1903), vol. 1, 4–5, 297–298.

86. Henry Jackson Waters, *The Development of the Philippines* (Manila: Bureau of Printing, 1915); Finley, "Race Development"; Crow, *America and Philippines*, 192–205; and the sections on agriculture in the successive reports of the Philippine commissions.

87. David Barrows, *Decade of American Government*, 50; Waters, *Development*, 10; Harrison, *Corner-Stone*, 241–242.

88. Taft, *Present Day Problems*, 29–31; Francis B. Harrison, "Agricul-

tural Co-operation in the Philippines," *Economic World* (Nov. 1915), Harrison Papers, container 51; May, *Social Engineering*, ch. 8.

89. For samples of these views, see Stanley, *Nation in the Making*, 55–56, 187–188, 198; Alfonso, *Roosevelt and the Philippines*, 44–45, 191–192.

90. For explicit skepticism, see *PRC* (1903), pt. 2, 59–60. For the larger debate, see May, *Social Engineering*, ch. 6. May's analysis (chs. 6–7) is a key source for my discussion here.

91. Taft, *Present Day Problems*, 22–23; *PCR* (1903), pt. 2, 881–883; Worcester, *Past and Present*, 508–509; Warren, *Isles of Opportunity*, 28–30, 72–73, 76–77; May, *Social Engineering*, 119–120.

92. On sports as character building, see Donald J. Mrozek, *Sport and American Mentality, 1880–1910* (Knoxville: University of Tennessee Press, 1983), 28–37 and chs. 3–5 and 7; Green, *Fit for America*, pts. 2–3. On these ideas in the Philippines, see Crow, *America and Philippines*, 118–124; Elliott, *End of Commission*, 241; Forbes, *Philippine Islands*, 454–457, 460; Warren, *Isles of Opportunity*, 69; Worcester, "Field Sports," 215–267, esp. 215–217.

93. Elliott, *End of Commission*, 274–275, 288; Roosevelt, *The Philippines*, 176–178; Stanley, *Nation in the Making*, 104, 140, 176; Barrows, *History of Philippines*, 46; *PCR* (1901), pt. 2, 282; Moses, *Unofficial Letters*, 355; Crow, *America and Philippines*, 171; Atkinson, *Philippine Islands*, 52 (quoting Taft). The islands' technological dependence on the United States is illustrated by the resolution of the first session of the Philippine legislature in March 1908, which called for reduced tariffs on machinery. See *Journal of the Philippine Commission* (Manila: Bureau of Printing, 1908), 90–91.

94. May, *Social Engineering*, 140–141.

95. See, e.g., the typescript copy of portions of the Westerhouse article that later appeared in *Engineering News*, BIA, 350, 7152. Westerhouse's critical commentary regarding lack of congressional funding has been removed from the published version. For

more on these issues, see Jones O. Garfield, *The Unhappy Conditions in the Philippine Islands* (Oakland, CA: Oakland Tribune, 1915), and the retrospectives published by Forbes, Elliot, and Nicholas Roosevelt. See also Stanley, *Nation in the Making*, 191, 272.

96. Westerhouse typescript, BIA; Elliott, *End of Commission*, 300–316; PCR (1901), pt. 2, 60–62; Gleek, *American Institutions*, 206–207; May, *Social Engineering*, 160–167.

97. My analysis of the colonial-*compradore* alliances is much indebted to the works by Peter Stanley, Bonifacio Salamanca, Glenn May, and Norman Owen cited above. On Filipino political patterns see the writings of Carl Landé, esp. *Leaders, Factions and Parties: The Structure of Philippine Politics* (New Haven: Yale University Press, 1965); and Remigio E. Agpalo, *The Political Elite and the People* (Manila: College of Public Administration, 1972).

98. Stanley, *Nation in the Making*, 166, 180–181.

99. Elliot, *End of Commission*, 196–198; Gleek, *American Institutions*, 210, 232–234; Stanley, *Nation in the Making*, 70; "Translation of a conversation overheard on a train between Tokio and Yokohama," Apr. 24, 1915?, Harrison Papers, container 38; Warren, *Isles of Opportunity*, 100–101; Roosevelt, *The Philippines*, 14; Fee, *Woman's Impressions*, 96–98.

100. See Para Villanueva de Kalaw, "The Filipino Woman in the Past and Present," 32–36; Conrado Benítez, "The Physical Upbuilding of the Filipinos," 28–31; and Theodoro R. Yangco, "Wealth and Prosperity by the Saving Habit," 57–59, all in *Philippine Review* 2 (1917); as well as Vincente L. Rafael, *Discrepant Histories: Translocal Essays in Philippine History* (Philadelphia: Temple University Press, 1995). On the "new Filipino woman," see Gleek, *American Institutions*, 46–49.

101. A typescript of the pageant was presented to former governor Francis Burton Harrison, who had done more than any other to

transfer power and economic influence to the *compradore* elites. See Harrison Papers, container 40.

102. On resistance in the post-conquest decades, see Reynaldo Clemena Ileto, *Payson and Revolution* (Quezon City: Ateneo de Manila University Press, 1979), esp. chs. 5–6; David R. Sturtevant, *Popular Uprisings in the Philippines* (Ithaca: Cornell University Press, 1976); on the post–World War II rebellions, Benedict J. Kerkvliet, *The Huk Rebellion* (Berkeley: University of California Press, 1977), and Eduardo Lachica, *Huk: Philippine Agrarian Society in Revolt* (Manila: Solidaridad, 1971); on the post-Huk decades, see essays by Kerkvliet and Brian Fegan in James C. Scott and Ben Kerkvliet, eds., *Everyday Forms of Protest in South-East Asia* (London: Frank Cass, 1986), and Kerkvliet, "Patterns of Philippine Resistance and Rebellion, 1970–1986," *Philipinas* 6 (Spring 1986), 35–52. For an overview of the Marcos era see Karnow, *In Our Image*.

4. Foundations of an American Century

1. Alan B. Cole, ed., *With Perry in Japan: The Diary of Edward Yorke McCauley* (Princeton: Princeton University Press, 1942), 27; quotation from Joseph Bucklin Bishop, *The Panama Gateway* (New York: Scribner, 1913), 197.

2. David McCullough, *The Path between the Seas: The Creation of the Panama Canal, 1870–1914* (New York: Simon and Schuster, 1977), 473, 249–259.

3. Quotations from James Bryce, *South America: Observations and Impressions* (London: Macmillan, 1912), 36; Frederic Haskin, *The Panama Canal* (New York: Doubleday, 1914), 20, 23–26, 29–30; Roosevelt's speech in W. Leon Pepperman, *Who Built the Panama Canal?* (London: J. M. Dent, 1915), 400; Ira E. Bennett, *History of the Panama Canal: Its Construction and Builders* (Washington: Historical Pub. Co., 1915), 114, 116, 118. And see J. Michael

Hogan, *The Panama Canal in American Politics* (Carbondale; Southern Illinois University Press, 1986), 45–54.

4. William R. Scott, *The Americans in Panama* (New York: Statler, 1912), 1–2.

5. The aspersions were frequently revived in works such as Ralph Avery, *America's Triumph at Panama* (Chicago: L. W. Walter, 1913), 55.

6. Quotations from Joseph Bucklin Bishop, *Uncle Sam's Panama Canal and World History* (New York: John Wanamaker, 1913), 1; Willis J. Abbot, *Panama and the Canal: The Story of Its Achievement, Its Problems and Prospects* (New York: Dodd, Mead, 1914), 459; John Foster Fraser, *Panama and What It Means* (London: Cassel, 1913), 1. For dismissive accounts of the French project, see Avery, *America's Triumph*, 45–68; Hugh C. Weir, *The Conquest of the Isthmus* (New York: Putnam, 1909), 8, 15–16, 27–28.

7. Gorgas, *Sanitation in Panama* (New York: Appleton, 1915); Goethals, *The Panama Canal: An Engineering Treatise* (New York: McGraw-Hill, 1916), 2 vols.

8. Sibert and Stevens, *The Construction of the Panama Canal* (New York: Appleton, 1915), 11–14; 60–61, 86, 93–94. See also Ian Cameron, *The Impossible Dream: The Building of the Panama Canal* (New York: Morrow, 1972), chs. 11–14.

9. On the obstacles of topography and climate, see Lindon W. Bates, *The Panama Canal: System and Projects* (Published by the author, 1905). Kidd, *The Control of the Tropics* (London: Macmillan, 1898).

10. Avery, *America's Triumph*, 56–58; Bryce, *South America*, 31; Bennett, *Panama Canal*, 119–120; McCullough, *Path between the Seas*, 131–147, 405–426, 465–468. My thinking on these issues has been shaped by the as yet unpublished work by Paul Sutter on the sanitary regime in the Canal Zone.

11. Huntington, *Civilization and Climate* (New Haven: Yale University Press, 1915), esp. chs. 16, 18.

12. Bishop, *Panama Gateway*, 194, 197.

13. Akira Iriye, *Pacific Estrangement: Japanese and American Expansionism, 1897–1911* (Cambridge, MA: Harvard University Press, 1972); Walter LaFeber, *The Clash: A History of U.S.-Japan Relations* (New York: Norton, 1997), ch. 3; William McElwee, *The Art of War: Waterloo to Mons* (Bloomington: Indiana University Press, 1974), 241–255.

14. Walter LaFeber, *The Panama Canal: The Crisis in Historical Perspective* (Oxford: Oxford University Press, 1979), 52–57. During World War I, anxieties over German aggression against the canal did play a significant role in U.S. policy in the Caribbean, especially the occupation of Haiti in 1915. See Peter H. Smith, *Talons of the Eagle: Dynamics of U.S.-Latin American Relations* (New York: Oxford University Press, 1996), 52, 59–60.

15. McCullough, *Path between the Seas*, 161–181, 279–294, chs. 11–13; Walter LaFeber, *Inevitable Revolutions: The United States in Central America* (New York: Norton, 1993), 31–39; Louis Perez, *Cuba and the Platt Amendment, 1902–1934* (Pittsburgh: University of Pittsburgh Press, 1986); Merze Tate, *The United States and the Hawaiian Kingdom: A Political History* (New Haven: Yale University Press, 1965), ch. 8.

16. J. F. Stevens, *An Engineer's Recollections* (New York: McGraw-Hill, 1936), 52 (quotations).

17. Weir, *Conquest of the Isthmus*, chs. 1, 9, 12; Fraser, *Panama and What It Means*, 3, 4, 10; Stevens quoted in Bishop, *Panama Gateway*, 300–305.

18. David McBride, *Missions for Science: U.S. Technology and Medicine in America's African World* (New Brunswick, NJ: Rutgers University Press, 2002), ch. 2; Michael L. Conniff, *Black Labor on a White Canal: Panama, 1904–1981* (Pittsburgh: University of Pittsburgh Press, 1985), ch. 3; McCullough, *Path between the Seas*, 576–585.

19. Smith, *Talons of the Eagle*, 52.

20. Hans Schmidt, *The United States Occupation of Haiti, 1915–1934* (New Brunswick, NJ: Rutgers University Press, 1995), 135–137, 158,

167–175, 182–183; Perez, *Cuba and the Platt Amendment*, esp. ch. 3; Howard Gillette Jr., "The Military Occupation of Cuba, 1899–1902: Workshop for American Progressivism," *American Quarterly* 25/4 (1973): 410–425; Bruce Calder, *The Impact of Intervention: The Dominican Republic during the U.S. Occupation of 1916–1924* (Austin: University of Texas Press, 1984), 34–37, 49–54, 81–90.

21. Frederic Haskin, *The Panama Canal* (New York: Doubleday, Page, 1914), 368–376 (quotations 20, 375); Robert Rydell, *All the World's a Fair: Visions of Empire at American International Expositions, 1876–1916* (Chicago: University of Chicago Press, 1984), ch. 8; Bill Brown, "Science Fiction, the World's Fair, and the Prosthetics of Empire, 1910–1915," in Amy Kaplan and Donald E. Pease, eds., *Cultures of United States Imperialism* (Durham: Duke University Press, 1993), 140–151.

22. See Gerhard Ritter, *The Sword and the Scepter: The Problem of Militarism in Germany* (Coral Gables: University of Miami Press, 1972), 3:134–50, 165–176, 288–343.

23. Daniel M. Smith, *War and Depression: America 1914–1939* (St. Louis: Forum Press, 1972), 25–26; William Leuchtenburg, *The Perils of Prosperity, 1914–1932* (Chicago: University of Chicago Press), 14–15.

24. Smith, *War and Depression*, 25–31, 35–38; Alfred F. Hurley, *Billy Mitchell: Crusader for Air Power* (Bloomington: Indiana University Press, 1975), 23–24, 32–33; Lindy Biggs, *The Rational Factory: Architecture, Technology, and Work in America's Age of Mass Production* (Baltimore: Johns Hopkins University Press, 1996), chs. 5–6; John T. Schlebecker, *Whereby We Thrive: A History of American Farming, 1607–1972* (Ames: Iowa State University Press, 1975), 206–211.

25. American deaths totaled over 112,000, but more than half of these were from the 1918–1919 influenza epidemic. On U.S. contributions to the convoy system, see Edward M. Coffman, *The War to End All Wars: The American Military Experience in World War I* (Madison: University of Wisconsin Press, 1986), ch. 4.

26. David M. Kennedy, *Over Here: The First World War and American Society* (New York: Oxford University Press, 1980), 212–221, 225.

27. Michael Adas, *Machines as the Measure of Men: Science, Technology, and Ideologies of Western Dominance* (Ithaca: Cornell University Press, 1989), 364–401.

28. In addition to works by the authors mentioned, see Michael S. Reynolds, *Hemingway's First War: The Making of "A Farewell to Arms"* (Princeton: Princeton University Press, 1976), esp. chs. 4–5; and Kennedy, *Over Here*, 222–225.

29. John Terraine, *To Win a War: 1918, the Year of Victory* (Garden City, NY: Doubleday, 1981); John Toland, *No Man's Land: 1918, the Last Year of the Great War* (Garden City, NY: Doubleday, 1980).

30. Mitchell, *Winged Defense: The Development and Possibilities of Modern Air Power* (New York: Putnam, 1925), 29–30, 34, 126–127, 214. Michael Sherry, *The Rise of American Air Power* (New Haven: Yale University Press, 1984), esp. 21–31, 51–58, 101, 116–117.

31. Hurley, *Crusader for Air Power*, chs. 3–7; Mitchell, *Winged Defense*, 13–14, 24–26, 86–89; Alexander P. De Seversky, *Victory through Air Power* (New York: Simon and Schuster, 1942), 24–27; Donald Mrozek, *Air Power and the Ground War in Vietnam* (Washington: Pergamon-Brassey, 1989), 11–13.

32. Officer quoted in Kennedy, *Over Here*, 203. On the supply problem see Larry L. Bland, ed., *The Papers of George Catlett Marshall* (Baltimore: Johns Hopkins University Press, 1981), 1:113–114, 119, 128.

33. On the mobilization see Grosvenor Clarkson, *Industrial America in the World War: The Strategy behind the Line, 1917–1918* (Boston: Houghton Mifflin, 1923). On links to social reform or engineering, see Kennedy, *Over Here*, ch. 2; Ronald Schaffer, *America in the Great War: The Rise of the Welfare State* (New York: Oxford University Press, 1991); Nancy K. Bristow, *Making Men Moral: Social Engineering During the Great War* (New York: New York University Press, 1996).

34. Dorothy Ross, *The Origins of American Social Science* (Cambridge: Cambridge University Press, 1991), ch. 10; John M. Jordan, *Machine-Age Ideology: Social Engineering and American Liberalism, 1911–1939* (Chapel Hill: University of North Carolina Press, 1994), 2–4, 7–8, 22–23, 112–115, 120–122, and ch. 4; Emily S. Rosenberg, *Spreading the American Dream: American Economic and Cultural Expansion, 1890–1945* (New York: Hill and Wang, 1982), 40–42; Donald Fisher, *Fundamental Development of the Social Sciences: Rockefeller, Philanthropy, and the United States Social Research Council* (Ann Arbor: University of Michigan Press, 1993).

35. Veblen, "The Place of Science in Modern Civilization," *American Journal of Sociology* 11/5 (1906), 585–588, 597–599; Jordan, *Machine-Age Ideology*, 14–19.

36. Jordan, *Machine-Age Ideology*, 15 (Veblen quotation), 74, 78 (Dewey quotation), 93; Sidney Kaplan, "Social Engineers as Saviors: Effects of World War I on Some American Liberals," *Journal of the History of Ideas* 17/4 (1956): 347–369; James A. Neuchterlein, "The Dream of Scientific Liberalism: *The New Republic* and American Progressive Thought, 1914–1920," *Review of Politics* 42/1 (1980): 167–190.

37. Michael Adas, "Contested Hegemony: The Great War and the Afro-Asian Assault on the Civilizing Mission Ideology," *Journal of World History* 15/1 (2004): 31–63.

38. Veblen, "Dementia Praecox" (1922), in Loren Baritz, ed., *The Culture of the Twenties* (Indianapolis: Bobbs-Merrill, 1970), 28–40.

39. John M. Levy, *Urban America: Processes and Problems* (Saddle River, NJ: Prentice-Hall, 2000), 27.

40. Richard Guy Wilson, Dianne H. Pilgrim, and Dickran Tashijian, *The Machine Age in America, 1918–1941* (New York: Harry N. Abrams, 1986).

41. Aaron Norman, *The Great Air War* (New York: Macmillan, 1968), chs. 7, 9, 11; Coffman, *To End All Wars*, ch. 7.

42. See Duhamel, *America the Menace: Scenes from the Life of the Future* (London: Allen and Unwin, 1931). Duhamel's skepticism was countered by European thinkers such as Philip Gibbs, who argued that "the fate of industrial European civilisation hangs on American decisions." See Gibbs, *The Hope of Europe* (London: Heinemann, 1921), 137–138 (quotation), 158.

43. On Lindbergh's ambivalence about technology see Leonard S. Reich, "From the *Spirit of St. Louis* to the SST: Charles Lindbergh, Technology, and Environment," *Technology and Culture* 36/2 (1995): 351–393. On the flight and its technological ramifications, see Dominick A. Pisano and F. Robert van der Linden, *Charles Lindbergh and the Spirit of S. Louis* (New York: Harry N. Abrams, 2002); Scott A. Berg, *Lindbergh* (New York: Putnam, 1998), chs. 5–7.

44. Wilson, *Machine-Age America*, 69, 91, 103.

45. Lilienthal, *TVA: Democracy on the March* (New York: Harper, 1944), esp. chs. 8–10.

46. Ibid., chs. 19–20.

47. David Ekbladh, "Mr. TVA: 'Grass Roots' Development, David Lilienthal, and the Rise and Fall of the Tennessee Valley Authority as a Symbol for U.S. Overseas Development, 1933–1973," *Diplomatic History* 26 (Summer 2002): 335–374; William E. Leuchtenburg, *The FDR Years: On Roosevelt and His Legacy* (New York: Columbia University Press, 1995), 159–161; Carroll Pursell, *The Machine in America: A Social History of Technology* (Baltimore: Johns Hopkins University Press, 1995), 263–269.

48. Donald Worster, *Dust Bowl: The Southern Plains in the 1930s* (Oxford: Oxford University Press, 1979), pts. 2–3, and *Nature's Economy: A History of Ecological Ideas* (Cambridge: Cambridge University Press, 1977), ch. 12; R. Douglas Hurt, *The Dust Bowl: An Agricultural and Social History* (Chicago: Nelson-Hall, 1981), 17–47; Ernest L. Schusky, *Culture and Agriculture: An Ecological Introduction to Traditional and Modern Farming Systems* (New York: Bergin and Garvey, 1989), 99–123.

49. Mary Brown Bullock, *An American Transplant: The Rockefeller Foundation and Peking Union Medical College* (Berkeley: University of California Press, 1980), 1–16.

50. Rosenberg, *Spreading the American Dream*, 28–33.

51. Edwin Clausen, "The Eagle's Shadow: Chinese Nationalism and American Educational Influence, 1900–1927," *Asian Profile* 16/5 (1988), 422–425; James C. Thompson Jr., *While China Faced West: American Reformers in Nationalist China, 1928–1937* (Cambridge, MA: Harvard University Press, 1969), 35–38.

52. McBride, *Missions for Science*, esp. 95–102; Schmidt, *Occupation of Haiti*, 156–159, 178–181, 186–187.

53. Bullock, *American Transplant*, 7–8, 14–15, 22, 38, 40–43; Thompson, *American Reformers*, 38–39; Marilyn Bailey Ogilvie and Clifford J. Choquette, "Western Biology and Medicine in Modern China: The Career and Legacy of Alice M. Boring (1883–1955)," *Journal of the History of Medicine and Allied Sciences* 48/2 (1993): 198–215.

54. For Latin America, see Richard E. Brown, "Public Health and Imperialism," *Monthly Review* 29/4 (1977): 21–35; Marcos Cueto, "The Cycles of Eradication: The Rockefeller Foundation and Latin American Public Health, 1918–1940," in Paul Weindling, ed., *International Health Organizations and Movements* (Cambridge: Cambridge University Press, 1995), 222–223, 226–229. For China, Frank Ninkovich, "The Rockefeller Foundation, China, and Cultural Change," *Journal of American History* 70/4 (1984): 799–820; Brown, *American Transplant*, 21–22, 42–46; Thompson, *American Reformers*, 131–134.

55. Cueto, "Cycles of Eradication," 222, 225, 226–229; Deborah Fitzgerald, "Exporting American Agriculture: The Rockefeller Foundation in Mexico, 1943–53," *Social Studies of Science* 16/3 (1986): 457–483; Thompson, *American Reformers*, 27–29, 75, 116–123; Randall E. Stross, *The Stubborn Earth: American Agriculturalists on Chinese Soil, 1898–1937* (Berkeley: University of California Press, 1986).

56. John Farley, "The International Division of the Rockefeller Foundation: The Russell Years, 1920–1934," in Weindling, *International Health Organizations*, 203–221; Cueto, "Cycles of Eradication," 225, 227–228, 232–233, 237–238; Christian Brannstrom, "Polluted Soil, Polluted Souls: The Rockefeller Hookworm Eradication Campaign in São Paulo, Brazil, 1917–1926," *Historical Geography* 25/1 (1997): 25–45; Ka-che Yip, "Health and Nationalist Reconstruction: Rural Health in Nationalist China, 1928–1937," *Modern Asian Studies* 26/2 (1992): 395–415; Ninkovich, "Rockefeller in China," 801–803; Bullock, *American Transplant*, 17–20.

57. See L. G. Morgan, *The Teaching of Science to the Chinese* (Hong Kong: Kelly and Walsh, 1933), x–xii, 59–63; George R. Twiss, *Science and Education in China* (Shanghai: Commercial Press, 1925), 39–40, 48–60.

58. Thompson, *American Reformers*, 22–27; Ninkovich, "Rockefeller in China," 818–820.

5. Imposing Modernity

1. Richard Polenberg, *War and Society: The United States, 1941–1945* (Westport, CT: Greenwood, 1972), 244. Clive Pointing, *Armageddon: The Second World War* (London: Sinclair-Stevenson, 1995), 294; Gerhard Weinberg, *A World at Arms: A Global History of World War II* (Cambridge: Cambridge University Press, 1994), 894–897. Theodore H. White and Annalee Jacoby, *Thunder Out of China* (New York: William Sloan, 1946), ch. 11; Mark Selden, *The Yenan Way in Revolutionary China* (Cambridge, MA: Harvard University Press, 1971), 121–122, 155–156; Huýnh Kim Khánh, *Vietnamese Communism 1925–1945* (Ithaca: Cornell University Press, 1982), 299–302.

2. David Noble, *The Religion of Technology* (New York: Knopf, 1998), ch. 8; Omer Bartov, *Murder in Our Midst: The Holocaust, Industrial Killing, and Representation* (New York: Oxford University Press, 1996).

3. Estimates of American combat deaths range from a high of 400,000, offered by James Stokesbury, *A Short History of World War II* (New York: Morrow, 1980), to "about" 300,000 to 320,000, as in, e.g., Weinberg, *World At Arms*, 894; Pointing, *Armageddon*, 294, 296; Lawrence S. Wittner, *Rebels Against War: The American Peace Movement, 1941–1960* (New York: Columbia University Press, 1969), 109–111.

4. Henri Michel, *The Second World War* (New York: Praeger, 1975), 82.

5. Alan S. Milward, *War, Economy and Society, 1939–1945* (Berkeley: University of California Press, 1977), esp. 63–74.

6. The apt characterization is from Russell Weigley, *The American Way of War* (New York: Columbia University Press, 1973), 146.

7. Michel, *Second World War*, 428.

8. Quoted in Edward H. Judge and John W. Langdon, eds., *The Cold War: A History through Documents* (Upper Saddle River, NJ: Prentice Hall, 1999), 11–12.

9. Robert H. Connery, *The Navy and the Industrial Mobilization in World War II* (Princeton: Princeton University Press, 1951), 5; Wittner, *Rebels against War*, 111–115; and Milward, *War and Society*, 63. Mitchell, *Second World War*, 428, 435–436, 819; Richard Overy, *Why the Allies Won* (New York: Norton, 1995), 190–192, 225; Pointing, *Armageddon*, 135–36, 295–296; Polenberg, *War and Society*, 241–242.

10. As early as mid-1943, 55 percent of Americans polled were convinced that the United States was most responsible for what they assumed would soon be an allied victory in the war. See Wittner, *Rebels against War*, 103.

11. Overy, *Why the Allies Won*, 214.

12. Stalin quoted in Donald M. Nelson, *Arsenal of Democracy: The Story of American War Production* (New York: Harcourt, Brace, 1946), ix. On lend-lease totals see Henri, *Second World War*, 428, 450; Milward, *War and Society*, 50–51, 71–74; Pointing, *Armageddon*, 78–79; Overy, *Why the Allies Won*, 192; and George C. Her-

ring, "Experiment in Foreign Aid, 1941–1945" (Ph.D. diss., University of Virginia, 1965), esp. chs. 6–7 and 284–314. For more on U.S. wartime assistance to China and the Soviet Union, see A. N. Young, *China and the Helping Hand* (Cambridge, MA: Harvard University Press, 1963); and Herring, *Aid to Russia 1941–1946: Strategy, Diplomacy, and the Origins of the Cold War* (New York: Columbia University Press, 1973).

13. William L. O'Neill, *A Democracy at War* (Cambridge, MA: Harvard University Press, 1993), 119–125.

14. Nelson, *Arsenal of Democracy*, 31.

15. Overy, *Why the Allies Won*, 210, 221–222, 224–225.

16. Ibid., 190–192, 210, 224–227.

17. Michael Kammen, "The Problem of American Exceptionalism: A Reconsideration," *American Quarterly* 45/1 (1993), 1–43; Michael Adas, "From Settler Colony to Global Hegemon: Integrating the Exceptionalist Narrative of the American Experience into Global History," *American Historical Review* 106/5 (Dec. 2001), 1692–1720. Morgenthau, "The Pathology of Power," *American Perspective* 4 (Winter 1950), 9.

18. See. e.g., Wittner, *Rebels Against War*, 99–100, 122; Weinberg, *World at Arms*, 899.

19. The retention of the Japanese emperor was an obvious departure from the determination to extirpate the old order. But the allies repeatedly insisted on Hirohito's abdication as a condition of surrender almost to the end of the war, and that insistence may have contributed to prolonging the conflict. See Herbert Feis, *The Atomic Bomb and the End of World War II* (Princeton: Princeton University Press, 1961). On postwar reconstruction see John Dower, *Embracing Defeat: Japan in the Wake of World War II* (New York: Norton, 1999); David W. Ellwood, *Rebuilding Europe: Western Europe, America, and Postwar Reconstruction* (London: Longman, 1992).

20. Marc Trachtenberg, *A Constructed Peace: The Making of the European Settlement, 1945–1963* (Princeton: Princeton University Press,

1999), chs. 1–4 and 8; Melvin Lefler, *A Preponderance of Power: National Security, the Truman Administration, and the Cold War* (Stanford: Stanford University Press, 1992), esp. chs. 1–5.

21. On the often messianic quality of these rivalries, see François Furet, *The Passing of an Illusion: The Idea of Communism in the Twentieth Century* (Chicago: University of Chicago Press, 1999), esp. viii, ix, 2; Jonathan Becker, *Soviet and Russian Press Coverage of the United States* (New York: St. Martin's, 1999), 68, 70, 112; Reinhold Niebuhr, "Pagan Goddess," *Time* (June 26, 1950), 61–62. On the American obsession with the threat of communist subversion, see Stephen Whitfield, *The Culture of the Cold War* (Baltimore: Johns Hopkins University Press, 1991).

22. See, e.g., Edmund Stevens, "Eyewitness Account from Soviet Union," *Life* (May 22, 1950); Leo Rosten, "How the Politburo Thinks," *Look* (Mar. 13, 1951); John Turkevich, "The Scientist in the U.S.S.R," *Atlantic* (Jan. 1953), 45–49; Vice Admiral Leslie C. Stevens, "The Russian People," *Atlantic* (May 1952); L. C. Hart, "The Biggest Swindle in History," *Look* (Apr. 11, 1950); Richard M. Nixon, "Russia as I Saw It," *National Geographic* 116/6 (Dec. 1959).

23. John Scott, "Russia through Russian Eyes," *Life* (Sept. 26, 1949), 114–130; Edward Crankshaw, "A Communist Strikes at the Heart of Communism," *Life* (July 29, 1957), 43–54. The most unrelenting and widely available of these "defector" or apostate accounts were Arthur Koestler's *Darkness at Noon* (New York: Macmillan, 1949) and the essays in R. H. S. Crossman, ed., *The God That Failed: Six Studies in Communism* (New York: Hayer, 1949).

24. Paul Milyukov, "Eurasianism and Europeanism in Russian History," *Festschrift Th. G. Masaryk zum 80. Geburtstage* (Bonn: Verlag von Friedrich Cohen, 1930), 1, 225–236; B. H. Sumner, *A Short History of Russia* (New York: Harcourt, Brace and World, 1949), 171–172; Karl Wittfogel, *Oriental Despotism: A Comparative Study of Total Power* (New Haven: Yale University Press, 1957); Adam B. Ulam, *The Bolsheviks: The Intellectual and Political His-*

tory of the Triumph of Communism in Russia (New York: Macmillan, 1965), 2; Bertram D. Wolfe, *Three Who Made a Revolution: A Biographical History* (New York: Dell, 1964), 15–16, 39–41, 399–405.

25. David C. Engerman, "Modernization from the Other Shore: American Observers and the Costs of Soviet Economic Development," *American Historical Review* 105/2 (2000), 383–416.

26. David Foglesong, "Roots of 'Liberation': American Images of the Future of Russia in the Early Cold War, 1948–1953," *International History Review* 21/1 (1999), 72–74.

27. Quotations from editorials on "Voices of People's Capitalism," *Life* (Feb. 18, 1957), 36; "Foreign Aid and Our Moral Credo," *Life* (Apr. 22, 1957), 42.

28. "Why Russia Doesn't Want to Go to War Now," *Look* (Apr. 19, 1955), 43, 45.

29. Stevens, "The Russian People," 29–30 (quoted portions); Admiral G. Kirk, "The People in Russia," *Look* (Apr. 22, 1952), 36; Emmet John Hughes, "Changing Russia," *Life* (Feb. 8, 1954), 115–131; "The Kremlin's Huckster," *Time* (Apr. 4, 1950), 29; *Life* (Apr. 29, 1957), 42; "A Night at the Movies," *Time* (Feb. 20, 1950), 42–43.

30. Thayer, "Why Russia Doesn't Want War," 43–45 (quoted phrases); William O. Douglas, "Soviet Colonialism—Product of Terror," *Look* (Dec. 12, 1955), 37–38; Nixon, "Russia as I Saw It," 725, 738–740.

31. Crankshaw, "Is Stalin in Trouble?" *Look* (Jan. 29, 1952), 26–31.

32. "Advertising American Style," *Life* (May 2, 1949), 59–62; *Time* (Apr. 10, 1950), 29; *Look* (Apr. 22, 1952), 34–35; *Life* (Feb. 15, 1954), 109; *Time* (Apr. 24, 1950), 19; *Look* (Apr. 19, 1955), 45–46; *Life* (May 22, 1950), 125–138.

33. "Recovery in the West," *Life* (May 16, 1949); "This Is Germany: The Soviet Zone," *Look* (May 9, 1950); *Time* (May 5, 1950), 22–23; "Iron Curtain Countries," *Life* (Dec. 5, 1949).

34. *Newsweek* (Aug. 3, 1959), 16–17; Nixon, "Russia as I Saw It," 717–723. And see Emily S. Rosenberg, "Consuming Women: Images

of Americanization in the American Century," in Michael Hogan, ed., *The American Legacy: U.S. Foreign Relations in the American Century* (New York: Cambridge University Press, 1999), 445–454.

35. See, e.g., Douglas, "Soviet Colonialism," 41–43; Nixon, "Russia as I Saw It," 718–734, 743–746; "Women: Russia's Second-Class Citizens," *Look* (Nov. 30, 1954), 114–119; "This I Like So Much!" *Life* (Aug. 6, 1951), 19.

36. Morton Schwartz, *Soviet Perceptions of the United States* (Berkeley: University of California Press, 1978), 7–8, 18–20, 24–27, 30; Becker, *Soviet Press Coverage*, 69–71, 110–116.

37. Stephen P. Gilbert, *Soviet Images of America* (New York: Crane, Russak, 1975), 55–58; Schwartz, *Soviet Perceptions*, 6–8, 10–14, 17, 30.

38. Becker, *Soviet Press Coverage*, 69.

39. Quoted in David Callahan and Fred I. Greenstein, "The Reluctant Racer: Eisenhower and U.S. Space Policy," in Roger D. Launius and Howard E. McCurdy, eds., *Spaceflight and the Myth of Presidential Leadership* (Urbana: University of Illinois Press, 1997), 20.

40. Consistent with cold war sensibilities, I use the term "command communism" as a collective designation for authoritarian, Marxist-inspired regimes that aspire to totalitarian control of their subjects. Under these regimes, the state channels national resources primarily into strengthening the military, highly centralized development projects, and government-controlled social support programs.

41. *Look* (May 22, 1951), 1, 133 (quoted headline); "Defense and the Danger Line," *Life* (May 27, 1957), 35; "The Red Army Can Be Beaten," *Look* (Apr. 10, 1951), 27–31; Michael Beschloss, "Kennedy and the Decision to Go to the Moon," in Launius and McCurdy, *Spaceflight and Leadership*, 51–52.

42. "Common Sense or Sputnik," *Life* (Oct. 21, 1957), 35; "Arguing the Case for Being Panicky," *Life* (Nov. 18, 1957), 126–127. "The Feat That Shook the Earth," *Look* (Oct. 21, 1957), 25.

43. See the introduction and essays by Paul Koistinen, Mark Pilisuk and Thomas Hayden, and Walter Adams in Carroll W. Pursell Jr., ed., *The Military-Industrial Complex* (New York: Harper and Row, 1972); James Alden Barber Jr., "The Military-Industrial Complex," in Stephen E. Ambrose and Barber, eds., *The Military and American Society* (New York: Free Press, 1972); Thomas J. McCormick, *America's Half Century: The United States Foreign Policy in the Cold War and After* (Baltimore: Johns Hopkins University Press, 1989), 92–98 and ch. 5.

44. "A Birthday Flexing of Red Biceps," *Life* (Nov. 18, 1957), 35–37; Rip Bulkeley, *The Sputnik Crisis and Early United States Space Policy* (Bloomington: Indiana University Press, 1991), 3–7 (Jackson quotation); *Foreign Relations of the United States* 24 (1957), 162–163 (Dulles quotation); Callahan and Greenstein, "Eisenhower and Space," 25–28, 30–33; Beschloss, "Kennedy and the Moon," 59; Walter McDougall, *The Heavens and the Earth: A Political History of the Space Age* (New York: Basic Books), 62, 144.

45. Morris Watnick, "The Appeal of Communism to the Peoples of Underdeveloped Areas," *Economic Development and Cultural Change* 1/1 (1952–1953), 22–36.

46. Nils Gilman, *Mandarins of the Future: Modernization Theory in Cold War America* (Baltimore: Johns Hopkins University Press, 2003), chs. 3–4; Robert Latham, *The Liberal Moment: Modernity, Security, and the Making of the Postwar International Order* (New York: Columbia University Press, 1997); Howard Brick, *Age of Contradiction: American Thought and Culture in the 1960s* (New York: Twayne, 1998), chs. 2–3.

47. See esp. Dean C. Tipps, "Modernization Theory and the Comparative Study of Societies: A Critical Perspective," *Comparative Studies in Society and History* 15/2 (1973), 200, 208–209, 211.

48. Rostow, *The Stages of Economic Growth* (Cambridge: Cambridge University Press, 1964 ed.), 8–9, 22. See also "Technology and the Economic Theorist: Past, Present and Future," in Rostow, *History, Policy and Economic Theory* (Boulder: Westview, 1990), 317–354.

49. Tipps, "Modernization Theory," 199, 203–204; Reinhard Bendix, "Tradition and Modernity Reconsidered," *Comparative Studies in Society and History* 9/3 (1967), 318, 320–322, 329–331; Walter Huggins, "American History in Comparative Perspective," *Journal of American Studies* 2/1 (1977), 33, 35, 37.

50. *Stages of Growth*, 4, 6–7, and ch. 3; Rostow, *Politics and the Stages of Growth* (Cambridge: Cambridge University Press, 1971), 26, 29–30.

51. Rostow, *Stages of Growth*, 4–5; *Politics and Growth*, chs. 2–3.

52. See, e.g., Joseph Kahl, *The Measurement of Modernism* (Austin: University of Texas Press, 1968); Alex Inkeles, "Making Men Modern: On the Causes and Consequences of Individual Change in Six Developing Countries," *American Journal of Sociology* 75/2 (1969); Daniel Lerner, *The Passing of Traditional Society* (New York: Free Press, 1958).

53. See, e.g., Max Millikan and Rostow, *A Proposal: Key to Effective Foreign Policy* (New York: Harper, 1957), 39–40, 131, 149–151.

54. See the transcripts of Rostow's speeches in the following issues of the *DSB*, "Shaping the Future" (Feb. 3, 1964), 182–183; "The United States and the New Asia" (Dec. 19, 1966), 913–914; "The Present Stage of the Cold War" (Nov. 5, 1962), 679–680; "Development: Some Lessons" (Sept. 16, 1963), 427, 429–430.

55. "American Strategy," *DSB* (Apr. 16, 1962), 630, and "Economic Development in Asia," *DSB* (May 31, 1965), 853.

56. See Robert Lewis, *Science and Industrialization in the USSR* (New York: Holmes and Meier, 1979); Richard Stites, *Revolutionary Dreams: Utopian Vision and Experimental Life in the Russian Revolution* (New York: Oxford University Press, 1989), 145–164.

57. Alexander Geshenkron, *Economic Backwardness in Historical Perspective* (Cambridge, MA: Harvard University Press, 1962); Robert Heilbroner, *The Great Ascent: The Struggle for Economic Development in Our Time* (New York: Harper, 1963), esp. chs. 5, 7.

58. David Schmitz, *Thank God They're on Our Side: The United States and Right-Wing Dictatorships* (Chapel Hill: University of

North Carolina Press, 1999); Michael Hunt, *Ideology and U.S. Foreign Policy* (New Haven: Yale University Press, 1987).

59. Many of these suppositions were shared by British and French colonial officials who from the 1930s through the 1950s sought to implement paternalistic reform strategies, even though they never seriously promoted full industrialization. See Frederick Cooper, *Decolonization and African Society: The Labor Question in French and British Africa* (Cambridge: Cambridge University Press, 1996).

60. Kate Brown, "Gridded Lives: Why Kazakhstan and Montana Are Nearly the Same Place," *American Historical Review* 106/1 (2001), 28.

61. Quotation from Bob Sutcliffe, "Development after Ecology," in V. Bhaskar and A. Glyn, eds., *The North, The South and the Environment: Ecological Constraints and the Global Economy* (New York: Earthscan, 1995), 237; Rostow, "Is the American Style Viable?" *Educational Forum* 40/4 (1976), 462.

62. Carolyn Merchant, *The Death of Nature: Women, Ecology and the Scientific Revolution* (San Francisco: Harper and Row, 1980), esp. chs. 7–10.

63. Eugene Skolnikoff, *Science, Technology, and American Foreign Policy* (Cambridge, MA: MIT Press, 1967), 99–100.

64. Pinchot, *The Fight for Conservation* (Garden City, NY: Harcourt Brace, 1910), 43. This overview of American environmentalism is informed mainly by Roderick Frazier Nash, *American Environmentalism* (New York: McGraw-Hill, 1990) and *Wilderness in the American Mind* (New Haven: Yale University Press, 1974); Samuel P. Hays, *Conservation and the Gospel of Efficiency: The Progressive Conservation Movement, 1890–1920* (Cambridge, MA: Harvard University Press, 1959) and *Beauty, Health and Permanence: Environmental Politics in the United States, 1955–1985* (Cambridge: Cambridge University Press, 1987).

65. Ramachandra Guha, "Radical American Environmentalism and Wilderness Preservation: A Third World Critique," *Environmental*

Ethics 2/2 (1989): 71–83; Madhav Gadgil and Ramachandra Guha, *This Fissured Land: An Ecological History of India* (Berkeley: University of California Press, 1992), 232–236, 243–245.

66. Nash, *Wilderness in the American Mind*, ch. 12.

67. Ramachandra Guha, *Environmentalism: A Global History* (New York: Longman, 1999), ch. 5; Philip Shabecoff, *A Fierce Green Fire: The American Environmental Movement* (New York: Hill and Wang, 1993), 107–110, chs. 5–6.

68. Galbraith, "How Much Should a Country Consume?" in Henry Jarrett, ed., *Perspectives on Conservation* (Baltimore: Johns Hopkins University Press, 1958), 92.

69. This overview is based largely on Murray Feshbach and Alfred Friendly Jr., *Ecocide in The U.S.S.R.* (New York: Basic Books, 1992); and Douglas Weiner, *Models of Nature: Ecology, Conservation, and Cultural Revolution in Soviet Russia* (Bloomington: Indiana University Press, 1988).

70. Trotsky quoted in Guha, *Environmentalism*, 126. As Thomas Hughes has reminded us, Lenin and other Bolshevik leaders were even more enthusiastic about a future shaped by Fordism and Taylorism than many American acolytes of these supreme system builders. Hughes, *American Genesis* (New York: Penguin, 1989), 8.

71. Stalin personally oversaw the sharp upward revision in the 1937 census returns, in part to cover staggering losses due to forced collectivization and the early purges. See Feshbach and Friendly, *Ecocide in the U.S.S.R.*, 31, 33.

72. Judith Shapiro, *Mao's War against Nature: Politics and the Environment in Revolutionary China* (Cambridge: Cambridge University Press, 2001); Vaclav Smil, *The Bad Earth: Environmental Degradation in China* (Armonk, NY: M. E. Sharpe, 1984); Smil, *China's Environmental Crisis: An Inquiry into the Limits of National Development* (Armonk, NY: M. E. Sharpe, 1993). And see Jasper Becker, *Hungry Ghosts: Mao's Secret Famine* (New York: Free Press, 1996).

73. On the American predilection for large-scale projects, see William Lederer and Eugene Burdick, *The Ugly American* (New York: Fawcett, 1958) esp. 119–120, 237–238. Contrary to the impression given by Lederer and Burdick, the Soviets followed much the same approach.

74. Rostow, "Development: Some Lessons," 426–429; "Unsolved Problems of International Development," *International Development Review* (Dec. 1965), 15–18; "Economic Development in Asia," *DSB* (May 31, 1965), 847–849; "The Nationalization of Takeoff," *DSB* (May 27, 1967), 824–829.

75. See, e.g., James C. Scott, *Seeing like a State: How Certain Schemes to Improve the Human Condition Have Failed* (New Haven: Yale University Press, 1998), ch. 7; Suzanne Moon, "Takeoff or Self-Sufficiency? Ideologies of Development in Indonesia, 1957–1961," *Technology and Culture* 39/2 (1998): 187–212; Nick Cullather, "Damming Afghanistan: Modernization in a Buffer State," *Journal of American History* 89/2 (2003): 535; Deborah Fitzgerald, "Exporting American Agriculture: The Rockefeller Foundation in Mexico, 1943–53," *Social Studies of Science* 16 (1986): 447–483; Henry C. Hart, *New India's Rivers* (Bombay: Orient Longmans, 1956). See also Michael Lipton, "Urban Bias and Food Policy in Poor Countries," *Food Policy* 1/1 (1975): 41–52.

76. Marx, *The First Indian War of Independence, 1857–1859* (Moscow: Foreign Languages Publishing House, n.d.), esp. 17–21.

77. See, e.g., Kahl, *Measurement of Modernism*, 18–19.

78. For varying perspectives on these issues see John L. H. Keep, *The Russian Revolution: A Study in Mass Mobilization* (New York: Norton, 1976); Sheila Fitzpatrick, *The Russian Revolution, 1917–1932* (Oxford: Oxford University Press, 1982); Robert Conquest, *The Harvest of Sorrow: Soviet Collectivization and the Terror-Famine* (New York: Oxford University Press, 1986); Scott, *Seeing like a State*, chs. 5–6; Eric Strauss, *Soviet Agriculture in Perspective: A Study of Its Successes and Failures* (New York: Praeger, 1969), chs. 4–6.

79. Watnick, "Appeal of Communism," 27–32.

80. Roger Kanet and Boris Ipatov estimated in 1980 that over 75 percent of Soviet aid to Africa went to energy and heavy industry and less than 10 percent to agriculture. See their "Soviet Aid and Trade in Africa," in Warren Weinstein and Thomas H. Henriksen, eds., *Soviet and Chinese Aid to African Nations* (New York: Praeger, 1980), 16–23; Christopher Stevens, *The Soviet Union and Black Africa* (New York: Holmes and Meier, 1976), 74–87. For Soviet influence in China, see Mark Selden, *The Yenan Way in Revolutionary China* (Cambridge, MA: Harvard University Press, 1971); Maurice Meisner, *Mao's China: A History of the People's Republic* (New York: Free Press, 1977), esp. chs. 9–10, 13–14; Thomas Rawski, *China's Transition to Industrialism* (Ann Arbor: University of Michigan, 1980).

81. George T. Yu, "The Tanzania-Zambia Railway: A Case Study of Chinese Economic Aid to Africa," in Weinstein and Henriksen, *Soviet and Chinese Aid*, 117–119, 140; Bruce D. Larkin, *China and Africa 1949–1970: The Foreign Policy of the People's Republic of China* (Berkeley: University of California Press, 1971), ch. 4, esp. 97–98.

82. Philip Snow, *The Star Raft: China's Encounter with Africa* (New York: Weidenfeld and Nicolson, 1988), chs. 4–5; Wolfgang Bartke, *The Economic Aid of the PR China to Developing Socialist Countries* (Munich: K. G. Saur, 1989), 29; and esp. Scott, *Seeing like a State*, ch. 7.

83. Kerr, John T. Dunlop, Frederick H. Harbison, and Charles A. Myers, *Industrialism and Industrial Man: The Problems of Labor and Management in Economic Growth* (Cambridge, MA: Harvard University Press, 1960); Lerner, *Passing of Traditional Society*; Inkeles, "Making Men Modern." As Katherine Scott has argued, this tendency to omit women from development theory is also prevalent in Marxist-inspired dependency analysis. See Scott, *Rethinking Modernization and Dependency Theory* (Boulder: Lynne Rienner, 1995), ch. 5.

84. Jane J. Jaquette, "Women and Modernization Theory: A Decade of Feminist Criticism," *World Politics* 34/2 (1982), 268–269.

85. See, e.g., Kahl, *Measurement of Modernism*, 4–6, 18–20; Katherine Scott, *Gender and Development* (Boulder: Rienner, 1955), 5–10, 23–41, 125.

86. See Ester Boserup's pioneering *Women's Role in Economic Development* (New York: St. Martin's, 1970). But her arguments should be considered in light of critiques such as those of Lourdes Beneria, Gita Singh, and others in *Signs* 7/2 (Winter 1981).

87. Rostow, "Development: Some Lessons," 426–427; "Democracy in Developing Nations," *DSB* (Feb. 17, 1964), 255.

88. This overview relies heavily on Barbara Rogers, *The Domestication of Women: Discrimination in Developing Countries* (New York: St. Martin's, 1979); Caroline O. N. Moser, *Gender, Planning and Development: Theory, Practice and Training* (London: Routledge, 1993), esp. chs. 3–4; Scott, *Gender and Development*, chs. 3–4; articles such as Doranne Jacobson's "Indian Women in the Process of Development," *Journal of International Affairs* 30/2 (1976–77): 211–242; and collections such as Mayra Buvinić, Margaret A. Lycette, and William Paul McGreevey, eds., *Women and Poverty in the Third World* (Baltimore: Johns Hopkins University Press, 1983); and Beverly Lindsay, ed., *Comparative Perspectives of Third World Women: The Impact of Race, Sex, and Class* (New York: Praeger, 1980).

89. Lederer and Burdick, *Ugly American*, ch. 19.

90. Rogers, *Domestication of Women*, 10.

91. Sheila Rowbotham, *Women, Resistance and Revolution: A History of Women and Revolution in the Modern World* (New York: Vintage, 1974); Elisabeth Croll, *Feminism and Socialism in China* (London: Routledge, 1978).

92. See Maxine Molyneux, "Women in Socialist Societies: Problems of Theory and Practice," in Kate Young, Carol Wolkowitz, and Roslyn McCullagh, *Of Marriage and the Market: Women's Subordination Internationally and Its Lessons* (London: Routledge and

Kegan Paul, 1984), 167–202; Elisabeth Croll, "Women in Rural Production and Reproduction in the Soviet Union, China, Cuba and Tanzania: Socialist Development Experiences," *Signs* 7/3 (Winter 1981): 360–398; Scott, *Gender and Development*, ch. 6.

93. See, e.g., Moon, "Takeoff or Self-Sufficiency?" 187–212.

94. For European precedents for these attitudes, see Adas, *Machines as the Measure of Men*, esp. chs. 2, 5. On contempt for local knowledge, see Scott, *Seeing like a State*, chs. 7–9. Also see Bruno Latour, *Science in Action: How to Follow Scientists and Engineers through Society* (Cambridge, MA: Harvard University Press, 1987); Michael E. Latham, *Modernization as Ideology: American Social Science and "Nation Building" in the Kennedy Era* (Chapel Hill: University of North Carolina Press, 2000), 46–59.

95. See Richard Critchfield, "Grain Man," *World Monitor* (Oct. 1990), 45–51; Lester Brown, *Seeds of Change: The Green Revolution and Development in the 1970s* (New York: Praeger, 1970). Amartya Sen and others later stressed maldistribution, social inequities, and differential access to food supplies, rather than climate-related shortages. See Sen, *Poverty and Famines: An Essay on Entitlement and Deprivation* (New York: Oxford University Press, 1981).

96. Nick Cullather, "Miracles of Modernization: The Green Revolution and the Apotheosis of Technology," *Diplomatic History* 28/2 (2004), 227–228, 232–234, 240–241 (quotation); Richard E. Bissell, "After Foreign Aid—What?" *Washington Quarterly* (Summer 1991), 26.

97. Ernest L. Schusky, *Culture and Agriculture* (New York: Bergin and Harvey, 1981), ch. 7; Andrew Pearse, *Seeds of Plenty, Seeds of Want: Social and Economic Implications of the Green Revolution* (Oxford: Clarendon Press, 1980); Francine Frankel, *India's Green Revolution: Economic Gains and Political Costs* (Princeton: Princeton University Press, 1971).

98. Schusky, *Culture and Agriculture*, 129–133.

99. Ibid., 126, 131, 133–134, 136–137.

100. Cullather, "Miracles of Modernization," 246–252.

101. These patterns have been explored extensively by the dependency school, perhaps most cogently by F. H. Cardoso and Albert Hirschman. For overviews see Heilbroner, *Great Ascent*; and Gunnar Myrdal, *Rich Lands and Poor: The Road to World Prosperity* (New York: Harper, 1957).

102. Adas, *Machines as the Measure*, esp. chs. 4–5; S. Irfan Habib, "Science, Technical Education and Industrialisation: Contours of a *Bhadralok* Debate," in Roy MacLeod and Deepak Kumar, eds., *Technology and the Raj: Western Technology and Technical Transfers to India 1700–1947* (New Delhi: Sage, 1995), 235–249.

103. Rostow, "Economic Development, 427; and *Politics and the Stages of Growth*, 61–62; Skolnikoff, *Technology and Foreign Policy*, 198–202; Alaba Ogunsanwo, *China's Policy in Africa 1858–1971* (Cambridge: Cambridge University Press, 1974), 87 (quotation); Guha, *Environmentalism*, 109–110; Goran Hyden, *No Shortcuts to Progress: African Development Management in Perspective* (Berkeley: University of California Press, 1983), 1–6; Timothy Mitchell, *Rule of Experts: Egypt, Techno-Politics, Modernity* (Berkeley: University of California Press, 2002), esp. sec. 3.

104. *Toward Freedom: The Autobiography of Jawaharlal Nehru* (New York: John Day, 1942), 228–230.

105. Scott, *Seeing like a State*, chs. 7–9; Gadgil and Guha, *Ecology and Equity: The Use and Abuse of Nature in Contemporary India* (London: Routledge, 1995), chs. 5–6; Mitchell, *Rule of Experts*, ch. 7; and essays by Jagdish Sinha, Dinesh Abrol, and V. V. Krishna in McLeod and Kumar, *Technology and the Raj*.

106. For compilations that include extensive excerpts from Gandhi's writings on these issues, see *Young India, 1924–1926* (Madras, 1927); *Hind Swaraj or Indian Home Rule* (1909; Ahmedabad: Navajivan, 1938); and *Socialism of My Conception* (Bombay, 1957). Gandhi's disciple J. C. Kumarappa published important works on the village movement and localized economic organiza-

NOTES TO PAGES 273-277

tion. See also Shiva Nand Jha, *A Critical Study of Gandhian Economic Thought* (Agra, 1955); Romesh Diwan and Mark Lutz, eds., *Essays in Gandhian Economics* (New York: Intermediate Technology Group, 1987).

107. See Stephen Cotgrove, "Technology, Rationality and Domination," *Social Studies of Science* 5 (1975): 55–68.

108. Schumacher, *Small Is Beautiful: Economics as if People Mattered* (New York: Harper and Row, 1973).

109. Lederer and Burdick, *Ugly American*, chs. 17–19, and their "Foreign Aid and Our Moral Credo," *Life* (Apr. 22, 1957), 42.

110. Francine R. Frankel, *India's Political Economy, 1947–1977: The Gradual Revolution* (Princeton: Princeton University Press, 1978); Michael Brecher, *Nehru: A Political Biography* (London: Oxford, 1959), ch. 12; Guha and Gadgil, *Fissured Land*, 181–185.

111. See, e.g., Michael Watts, *Silent Violence: Food, Famine and Peasantry in Northern Nigeria* (Los Angeles: University of California Press, 1983); Richard Franke and Barbara Chasin, *Seeds of Famine: Ecological Destruction and the Development Dilemma in the West African Sahel* (Montclair, NJ: Allenheld, Osmun, 1980); Guha and Gadgil, *Fissured Land*, chs. 6–7, and *Ecology and Equity*, chs. 1–3; Moon, "Takeoff or Self-Sufficiency"; Scott, *Seeing like a State*, esp. chs. 7–8; Goran Hyden, *Beyond Ujamaa in Tanzania* (Berkeley: University of California Press, 1980); John Waterbury, *The Egypt of Nasser and Sadat: The Political Economy of Two Regimes* (Princeton: Princeton University Press, 1983), pt. 3.

112. For works that explore related ways in which development specialists and indigenous bureaucrats finessed the need to address underlying socioeconomic problems and substantial reform, and the ways in which social scientific theorizing can reinforce these proclivities, see James Ferguson, *The Anti-Politics Machine: "Development," Depoliticization, and Bureaucratic Power in Lesotho* (Cambridge: Cambridge University Press, 1990); John Harriss, *Politicizing Development: The World Bank and Social Capital* (London: Anthem Press, 2001).

113. See, e.g., John W. Sewell and the Staff of the Overseas Development Council, eds., *The United States and World Development Agenda 1980* (New York: Praeger, 1980), 2–5; David Morawetz, *Twenty-Five Years of Economic Development* (Washington: World Bank, 1977), 12–14.

114. F. E. Trainer, "Reconstructing Radical Development Theory," *Alternatives* 14/4 (1989), 481–493; Amulya Kumar N. Reddy, "Alternative Technology: A Viewpoint from India," *Social Studies of Science* 5 (1975), 331–333.

6. Machines in the Vietnam Quagmire

1. My account of the clash at Ap Bac and the ways it exemplified U.S. interventions in Vietnam is based on Neil Sheehan, *A Bright Shining Lie: John Paul Vann and America in Vietnam* (New York: Vintage, 1988), bk. 3; Malcolm Browne, *The New Face of War* (Indianapolis: Bobbs-Merrill, 1965), 9–16; and David M. Toczek, *The Battle of Ap Bac, Vietnam* (Westfield, CT: Greenwood, 2001).

2. David Halberstam, *The Best and the Brightest* (New York: Random House, 1969), 205–209, 256, 263; *Reporting Vietnam: American Journalism, 1959–1975* (New York: Library of America, 1998), 18–78; Herbert Y. Schandler, "U.S. Military Victory in Vietnam: A Dangerous Illusion," in Robert McNamara et al., *Argument without End: In Search of Answers to the Vietnam Tragedy* (New York: Public Affairs Press, 1999), 327–328; George Herring, *America's Longest War: The United States and Vietnam, 1950–1975* (New York: McGraw-Hill, 1996), 112 (quotation).

3. *PP*, 2:18, 31–32, 171; Sheehan, *Bright Shining Lie*, bk. 4; David Halberstam, *The Making of a Quagmire* (New York: Random House, 1964), ch. 11. Herring, *Longest War*, 110–113.

4. Lloyd C. Gardner, *Approaching Vietnam: From World War II through Dienbienphu* (New York: Norton, 1988), 30–39, 300–302; Warren Kimble, *Forged in War: Roosevelt, Churchill, and the Second World War* (New York, 1997), 138–140, 298–305; Stein

Tønnesson, *The Vietnamese Revolution of 1945: Roosevelt, Ho Chi Minh and de Gaulle in a World at War* (London: Sage, 1991), 13–19, 62–66, ch. 7.

5. *PP*, 1:97–107; Ellen Hammer, *The Struggle for Indochina, 1940–1955* (Stanford: Stanford University Press, 1954), 313–329. For a general account see Alain Ruscio, *La guerre française d'Indochine* (Brussels: Editions Complexe, 1992).

6. Mark Bradley, "Imagining Vietnam: The United States in Radical Vietnamese Anti-Colonial Discourse," *Journal of American–East Asian Relations* 4/2 (1995): 299–329.

7. Hammer, *Struggle for Indochina*, 330–337; George McTurnan Kahin and John W. Lewis, *The United States in Vietnam* (New York: Dell, 1969), ch. 3.

8. Halberstam, *Best and Brightest*, 87–93, 137–138; *PP*, 1:100–101. On the crisis from a Laotian perspective, see Hugh Toyé, *Laos: Buffer State or Battleground* (London: Oxford University Press, 1968).

9. See, e.g., Brian Van De Mark, *Into the Quagmire: Lyndon Johnson and the Escalation of the Vietnam War* (New York: Oxford, 1995); Larry Berman, *Planning a Tragedy: The Americanization of the War in Vietnam* (New York: Norton, 1982); Lloyd Gardner, *Pay Any Price: Lyndon Johnson and the Wars for Vietnam* (Chicago: Ivan Dee, 1995); David Kaiser, *American Tragedy: Kennedy, Johnson and the Origins of the Vietnam War* (Cambridge, MA: Harvard University Press, 2000); Fredrik Logevall, *Choosing War: The Lost Chance for Peace and the Escalation of War in Vietnam* (Berkeley: University of California Press, 1999).

10. Harry G. Summers Jr., *On Strategy: A Critical Analysis of the Vietnam War* (Novato, CA: Presidio, 1982), 120.

11. Van De Mark, *Into the Quagmire*, 207; William Shawcross, *Sideshow: Kissinger, Nixon and the Destruction of Cambodia* (New York: Simon and Schuster, 1979), 87, 89; Herring, *Longest War*, 246.

12. Logevall, *Choosing War*, 196, 230, 248.

13. Van De Mark, *Into the Quagmire*, 186–192.

14. Paul Edwards, *The Closed World: Computers and the Politics of Discourse in War America* (Cambridge, MA: MIT Press, 1996), ch. 4; Michael Clare, *War without End: American Planning for the Next Vietnams* (New York: Random House, 1972), pt. 1; Deborah Shapley, *Promise and Power: The Life and Times of Robert McNamara* (Boston: Little, Brown, 1993), chs. 6–7. On McNamara's disillusionment see his *In Retrospect: The Tragedy and Lessons of Vietnam* (New York: Random House, 1995), chs. 7–9; and Shapley, *Promise and Power*, pt. 4.

15. Halberstam, *Best and Brightest*, 348–349.

16. Shapley, *Promise and Power*, 328.

17. Edwards, *Closed World*, 127.

18. See, e.g., Rostow quoted in Herring, *Longest War*, 88; and Maxwell Taylor's testimony in House Committee on Appropriations, *Department of Defense Appropriations for 1964*, 88th Congress, 1st sess., pt. 1, 483–484. On the laboratory analogy see James William Gibson, *The Perfect War: Technowar in Vietnam* (Boston: Atlantic Monthly Press, 1986), 78–82.

19. Shapley, *Promise and Power*, 146–151; Halberstam, *Best and Brightest*, 248 (quotation), 247–257.

20. See, e.g., *PP*, 2:18, 19–23, 32–33; Logevall, *Choosing War*, 62–63, 228–229, 234, 242–243; Loren Baritz, *Backfire* (New York: Morrow, 1985), 259–271.

21. McNamara, *Argument without End*, 304.

22. Ellsberg, *Papers on the War* (New York: Simon and Schuster, 1972), 28n16.

23. See Frances FitzGerald, *Fire in the Lake: The Vietnamese and the Americans in Vietnam* (Boston: Little, Brown, 1972).

24. Pham Van Dong quoted in the PBS documentary "Vietnam: A Television History," based on Stanley Karnow's *Vietnam: A History* (New York: Viking, 1983).

25. See, e.g., McNamara, *Argument without End*, 40–42; McNamara, *In Retrospect*, 116–117, 214–215, 218–220, 321–322.

26. See Lê Thanh Khoi, *Le Viêt-Nam: histoire et civilisation* (Paris:

Editions de minuit, 1955); Keith W. Taylor, *The Birth of Vietnam* (Berkeley: University of California Press, 1983); Troung Buu Lam, *Patterns of Vietnamese Response to Foreign Intervention, 1858–1900* (New Haven: Yale University Press, 1967); Alexander B. Woodside, *Community and Revolution in Modern Vietnam* (Boston: Houghton Mifflin, 1976). On cultural misperceptions on both sides of the conflict see Mark Bradley, *Imagining Vietnam and America: The Making of Post-Colonial Vietnam, 1919–1950* (Chapel Hill: University of North Carolina Press, 2000).

27. Though the possibility was raised in policy deliberations; see, e.g., Maxwell Taylor quoted in Van De Mark, *Into the Quagmire*, 98–99. See also Halberstam, *Making of a Quagmire*, 113; Logevall, *Choosing War*, 369.

28. See Halberstam, *Making of a Quagmire*, chs. 13–14.

29. McNamara, *In Retrospect*, 32–33, 39–40, 322.

30. George Ball, "The Lessons of Vietnam: Have We Learned or Only Failed?" *New York Times Magazine* (Apr. 1, 1972), 13 (quoted phrase); Van De Mark, *Into the Quagmire*, 219; Shapley, *Promise and Power*, 315.

31. See, e.g., Ellen J. Hammer, "The Bao Dai Experiment," *Pacific Affairs* 23/1 (1960); Thomas A. Ennis, "Indo-China: The Aftermath of Colonialism," *Current History* 23/132 (1952). On dissenting voices in the state department, see Stanley Karnow, *Vietnam: A History* (New York: Viking, 1983), 177–178; and among public intellectuals, Logevall, *Choosing War*, 56–57.

32. Robert J. Lifton, *Home from the War: Vietnam Veterans: Neither Victims nor Executioners* (New York: Simon and Schuster, 1973), 59, 64; Caputo, *A Rumor of War* (New York: Ballantine, 1978), 193.

33. Roger Hilsman, *To Move a Nation: The Politics of Foreign Policy in the Administration of John F. Kennedy* (Garden City, NY: Doubleday, 1967), 523; see also 437–438, 443–450, 453–454, 467, 509.

34. Sheehan, *Bright Shining Lie*, 269–292; Hilsman, *To Move a Na-*

tion, 441–442, 453, 457–458, 479–481; McNamara, *In Retrospect*, 47, 49, 58–59; Halberstam, *Making of a Quagmire*, 73–75, 147–148, 154–162.

35. Herring, *Longest War*, 101–102, 124, 171, 176, 181, 255; Shapley, *Promise and Power*, 150–151, 250–251, 351, 359, 414, 431, 443.

36. Edwards, *Closed World*, 3–8; Gibson, *Perfect War*, 396–399; Paul Dickson, *The Electronic Battlefield* (Bloomington: Indiana University Press, 1976).

37. Baritz, *Backfire*, 289–291; Gibson, *Perfect War*, 107–129, 189–192; Eric M. Bergerud, *Red Thunder, Tropic Lightning: The World of a Combat Division in Vietnam* (New York: Penguin, 1993), 99–105; Caputo, *Rumor of War*, 155–171, 190–197.

38. *PP*, 2:4, 14.

39. Rostow, "Guerrilla Warfare in the Underdeveloped Areas," *DSB* (Aug. 7, 1961), 234, 237; Max F. Millikan and W. W. Rostow, *A Proposal: Key to an Effective Foreign Policy* (New York: Harper, 1957), 37, 39–40, 55, 121–122, 128, 131, 149–151.

40. Zachary Karabell, *Architects of Intervention: The United States, the Third World and the War, 1946–1962* (Baton Rouge: Louisiana State University Press, 1999), 25–33.

41. Inaugural Address, Jan. 20, 1949, *Public Papers of the Presidents of the United States, Harry S Truman, 1949* (Washington: GPO, 1949), 116; Remarks to the American Society of Civil Engineers, Nov. 2, 1949, ibid., 546–547.

42. On the NSC's recommendations, see John Gaddis, *Strategies of Containment: A Critical Appraisal of Postwar American National Security Policy* (New York: Oxford University Press, 1982), ch. 4.

43. On policymakers' touting of the Korean example, see David Ekbladh, "How to Build a Nation," *Wilson Quarterly* (Winter 2004): 12–20; for more pessimistic assessments of South Korea's economic progress, Bruce Cumings, *Korea's Place in the Sun: A Modern History* (New York: Norton, 1997), ch. 6. Diem's rule has been chronicled in many accounts, including Denis Warner, *The Last Confucian* (Baltimore: Penguin, 1963) and the perceptive

writings of journalists such as David Halberstam, Neil Sheehan, and Robert Shaplen.

44. Hilsman, *To Move a Nation*, 431–435.

45. Michael Latham, *Modernization as Ideology: American Social Sciences and "Nation Building" in the Kennedy Era* (Chapel Hill: University of North Carolina Press, 2000), 174, 191–197. See also Bradley, *Imagining Vietnam*, 46–51.

46. See, e.g., Milton Osborne, *Strategic Hamlets in South Viet-nam; A Survey and Comparison* (Ithaca: Cornell University Press, 1965).

47. Pierre Brocheux, *The Mekong Delta: Ecology, Economy, and Revolution, 1860–1960* (Madison: University of Wisconsin Press, 1995), ch. 1; Gerald Hickey, *Village in Vietnam* (New Haven: Yale University Press, 1964), chs. 1–2.

48. Eric Bergerud, *The Dynamics of Defeat: The Vietnam War in Hau Nghia Province* (Boulder: Westview, 1991), 35–37, 51–52; Hilsman, *To Move a Nation*, 431, 441; Douglas S. Blaufarb, *The Counterinsurgency Era: U.S. Doctrine and Performance* (New York: Free Press, 1977), 114–115.

49. Blaufarb, *Counterinsurgency*, 111, 122–124; Latham, *Modernization as Ideology*, 183–185; Hilsman, *To Move a Nation*, 424–425, 441 (concentration camp reference), 452–453; Bergerud, *Dynamics of Defeat*, 52; Orville Schell and Barry Weisberg, "Ecocide in Indochina," in Weisberg, ed., *Ecocide in Indochina: The Ecology of War* (San Francisco: Canfield, 1970), 26.

50. Halberstam, *Making of a Quagmire*, 95, 175, 179–183; Hilsman, *To Move a Nation*, 446–447, 455; PP, 2:455–457.

51. Bradley, *Imagining Vietnam*, 49.

52. Quotations from Samuel Zaffiri, *Westmoreland: A Biography of William C. Westmoreland* (New York: Morrow, 1994), 106; Hilsman, *To Move a Nation*, 442; Halberstam, *Making of a Quagmire*, 192–193. And see Bergerud, *Dynamics of Defeat*, 95–97.

53. William C. Westmoreland, *A Soldier Reports* (New York: Doubleday, 1976), 268, 178.

54. Works often cited in official exchanges were Douglas Pike, *Vietcong: The Organization and Techniques of the National Liberation Party of South Vietnam* (Cambridge, MA: MIT Press, 1966); and George K. Tanham, *Communist Revolutionary Warfare: From the Vietminh to the Vietcong* (New York: Praeger, 1961).

55. Donald Mrozek, *Air Power and the Ground War in Vietnam* (Washington: Pergamon-Brassey, 1989), 62, 124–125, 157–160; Bergerud, *Dynamics of Defeat*, 83–93, and *Red Thunder*, 94–100; Ronald H. Spector, *After Tet: The Bloodiest Year in Vietnam* (New York: Vintage, 1993), 42–44, 56–59.

56. Westmoreland, *A Soldier Reports*, 152–153; Zaffiri, *Westmoreland*, 161–162, 195, 211; Bergerud, *Dynamics of Defeat*, ch. 4; Blaufarb, *Counterinsurgency Era*, 100–103, 118–119, 126; Schandler, "Dangerous Illusion," 319, 321–323, 335–336, 351–362.

57. Bergerud, *Red Thunder*, 99–106.

58. William Shawcross, *Side-Show: Kissinger, Nixon and the Destruction of Cambodia* (New York: Simon and Schuster, 1979).

59. Bergerud, *Red Thunder*, 109–110, 125, 135–139; Al Santori, *Everything We Had: An Oral History of the Vietnam War* (New York: Random House, 1981), 26–27, 42–43, 49–50.

60. Bergerud, *Red Thunder*, 147.

61. For some of the best accounts of these conditions, see Caputo, *Rumor of War*, pt. 1; Bergerud, *Red Thunder*, ch. 3 and 149–163; Michael Herr, *Dispatches* (New York: Avon, 1968); Mark Baker, *Nam: The Vietnam War in the Words of the Soldiers Who Fought There* (New York: Berkley, 1981).

62. Herr, *Dispatches*, 13, 42; Bergerud, *Red Thunder*, 93, 100, 119.

63. Tom Manold and John Penycate, *The Tunnels of Cu Chi* (New York: Berkley, 1985); Beth Levine, "Headfirst into Underground Battle," *Military History* (Feb. 1987), 43–48.

64. Browne, *New Face of War*, 17–29, 267–268; Spector, *After Tet*, 12, 13, 78–83, 87–88; Sheehan, *Bright Shining Lie*, 207–208, 216–227, 243–244; Halberstam, *Making of a Quagmire*, 189–190; Bergerud, *Red Thunder*, 109–111, 117–122, 124–125.

65. Quoted phrases from Lifton, *Home from the War*, 65.

66. Alain Jaubert, "Zapping the Vietcong by Computer," *New Scientist* (Mar. 30, 1972), 685–688; and essays by William Pepper, Thomas Bodenheimer, and George Roth in Weisberg, *Ecocide in Indochina*.

67. Hue-Tam Ho Tai, *Radicalism and the Origins of the Vietnamese Revolution* (Cambridge, MA: Harvard University Press, 1992), esp. chs. 3, 7; William Duiker, *Sacred War: Nationalism and Revolution in a Divided Vietnam* (New York: McGraw-Hill, 1995), 66–67, 87, 144–145, 191; Pike, *Viet Cong*, 174–178, 242, 264; Brown, *New Face of War*, 161–168, 171. Kerrey quoted in Gregory L. Vistica, "What Happened in Thanh Pong?" *NYT Magazine* (Apr. 29, 2001), 68; see also Santoli, *Everything They Had*, 68–69.

68. Zaffiri, *Westmoreland*, 162.

69. Theodore White, *Thunder out of China* (New York: William Sloan, 1946), 274–276.

70. Bergerud, *Dynamics of Defeat*, 61–64, 71, 73–84; Spector, *After Tet*, ch. 4.

71. See Le Ly Hayslip, *When Heaven and Earth Changed Places* (New York: Doubleday, 1989); Halberstam, *Making of a Quagmire*, 188–189; Bergerud, *Dynamics of Defeat*, 55, 63–67; Blaufarb, *Counterinsurgency Era*, 98–102.

72. John Lewallen, *Ecology of Devastation: Indochina* (Baltimore: Penguin, 1971), esp. chs. 2–5; and the essays in J. B. Nielands et al., *Not Since the Romans Salted the Land* (Ithaca: Glad Day, 1970), some of which are excerpted in Weisberg, *Ecocide in Indochina*.

73. Seth Mydans, "Vietnam Sees War's Legacy in Its Young," *New York Times* (May 16, 1999), 12.

74. Pepper, "Children of Vietnam," in Weisberg, *Ecocide in Indochina*, 100–101.

75. Quoted in Peter Braestrup, *Big Story: How the American Press and Television Reported and Interpreted the Crisis of Tet 1968 in Vietnam and Washington* (New Haven: Yale University Press, 1983), 193.

76. Robert K. Brigham, *Guerrilla Diplomacy: The NLF's Foreign Relations and the Viet Nam War* (Ithaca: Cornell University Press, 1999), ch. 1; Mark Clodfelter, *The Limits of Air Power: The American Bombing of North Vietnam* (New York: Free Press, 1989), 48, 62–63.

77. M. M. Postan, "Walt Rostow: A Personal Appreciation," in Charles Kindleberger and Guido di Tella, eds., *Economics in the Long View: Essays in Honor of W. W. Rostow* (New York: New York University Press, 1982), 1:4–5; Clodfelter, *Limits of Air Power*, 50.

78. *PP*, 3:107–111; Halberstam, *Best and Brightest*, 161–162; Clodfelter, *Limits of Air Power*, 44–51, 57; Mrozek, *Air Power*, 22–23.

79. Gerhard Weinberg, *A World at Arms: A Global History of World War II* (Cambridge: Cambridge University Press, 1994), 574–581; William L. O'Neill, *A Democracy at War: America's Fight at Home and Abroad in World War II* (Cambridge, MA: Harvard University Press, 1993), chs. 11, 13; Michael Sherry, *The Rise of American Air Power: The Creation of Armageddon* (New Haven: Yale University Press, 1987), esp. chs. 8–9.

80. *PP*, 3:205–206, 212–215; 4:62–67, 75–77, 107–112, 129–130, 180–185, 222–228, 231–232; Mrozek, *Air Power*, 179, 181; Clodfelter, *Limits of Air Power*, 66–67, 93–94, 99; Logevall, *Choosing War*, 228–229.

81. Clodfelter, *Limits of Air Power*, chs. 2–4; Jon M. Van Dyke, *North Vietnam's Strategy for Survival* (Palo Alto: Pacific Books, 1972).

82. Van Dyke, *North Vietnam's Strategy*, ch. 11; Clodfelter, *Limits of Air Power*, 84, 90, 131, 134–135; Herring, *Longest War*, 163–165; Mrozek, *Air Power*, 92; *PP*, 4:136.

83. Clodfelter, *Limits of Air Power*, 107, 111–112, 140; Van Dyke, *North Vietnam's Strategy*, chs. 3, 5; Schandler, "Dangerous Illusion," 336–348; McNamara, *Argument without End*, ch. 6. On the government in the North, see Ngoc-Luu Nguyen, "Peasants, Party and Revolution: The Politics of Agrarian Transformation in Northern Vietnam, 1930–1975" (Ph.D. diss., University of Amsterdam, 1987).

84. For an appeal to consider the potential of air power to wrest victory from defeat, see Summers, *On Strategy*, 77, 85–89, 117–124, 153–157; for evidence that renders this claim dubious at best, see Clodfelter, *Limits of Air Power*, chs. 5–7; Gibson, *Perfect War*, chs. 12–13; Baritz, *Backfire*, 221, 224.

85. Johnson quoted in Van De Mark, *Into the Quagmire*, 191. See Thompson, *Defeating Communist Insurgency: The Lessons of Malaya and Vietnam* (New York: Praeger, 1966), 112–113, 124–125; Rostow, "Problems and Constructive Trends on the World Scene," *DSB* 54 (July 18, 1966), 81.

86. Nguyen Thi Dieu, *The Mekong River and the Struggle for Indochina* (Westport, CT: Greenwood, 1999), pp. 101–109, 149, 153–161; Gardner, *Pay Any Price*, pp. 6–7, 52–53, 99–111, 123–124, 187–195; David Ekbladh, "'Mr. TVA': Grass-Roots Development, David Lilienthal, and the Rise and Fall of the Tennessee Valley Authority as a Symbol for U.S. Overseas Development, 1933–1973," *Diplomatic History* 26/3 (2002), 336–337, 340–346, 348–353, 363–366, 369, 370–371, 373–374.

87. Sheehan, *Bright Shining Lie*, bk. 6.

88. This overview of the CORDS initiative is based largely on Sheehan, *Bright Shining Lie*; *PP*, 2:487–498; and George Herring, *LBJ and Vietnam: A Different Kind of War* (Austin: University of Texas Press, 1994), 63–88.

89. See esp. Spector, *After Tet*; Bergerud, *Dynamics of Defeat*, chs. 6–10.

90. For differing assessments of the Phoenix program, see Zaffiri, *Westmoreland*, 211–214; Dale Andradé, *Ashes to Ashes: The Phoenix Program and the Vietnam War* (Lexington, MA: Heath, 1990); Douglas Valentine, *The Phoenix Program* (New York: Morrow, 1990); Santori, *Everything We Had*, 199–202.

91. For the strongest statement of this thesis, see Summers, *On Strategy*, 77–78, 87–89, 101–102, 168–172.

92. George Ball stressed these factors in "The Lessons of Vietnam," 13, 40, 43. See also Jeffrey Race, *War Comes to Long An: Revolu-*

tionary Conflict in a Vietnamese Province (Berkeley: University of California Press, 1972), esp. xiv-xviii, ch. 1, 50–57, 159–179, 245–260.

93. Herr, *Dispatches*, 183.

7. Technowar in the Persian Gulf

1. According to a *USA Today* poll, Mar. 2, 1991. A *New York Times/ CBS News* poll the week before gave him an 87% approval rating. See *NYT* (Mar. 2, 1991), 1, 7. For Bush's declaration of the end of the war, see *NYT* (Feb. 28, 1991), 1.

2. I will refer to the U.S.-led coalition's expulsion of Iraqi forces from Kuwait as the 1991 Gulf War or Operation Desert Storm rather than as the First Gulf War. From a global perspective, the Iran-Iraq War in the 1980s was the First Gulf War.

3. *NYT* (Mar. 2, 1991), 1.

4. John Mueller, *Policy and Opinion in the Gulf War* (Chicago: University of Chicago Press, 1994), 304; Richard P. Hallion, *Storm over Iraq: Air Power and the Gulf War* (Washington: Smithsonian Institution Press, 1992), 17–21, 157–161; John Kifner, "New G.I.'s Shake off Vietnam Ghost," *NYT* (Feb. 25, 1991).

5. Yuen Foong Khong, *Analogies at War: Korea, Munich, Dien Bien Phu, and the Vietnam Decisions of 1965* (Princeton: Princeton University Press, 1991), 3–6, 77, 100, 176–190.

6. Joe Stork, "Iraq and the War in the Gulf," *Merip Reports* (June 1981), 12–16; Marion Farouk-Sluggett and Peter Sluggett, *Iraq since 1958: From Revolution to Dictatorship* (London: Tauris, 1990), ch. 7.

7. See Samir al-Khalil, *Republic of Fear: The Politics of Modern Iraq* (Berkeley: University of California Press, 1989); Efraim Karsh and Inari Rautsi, *Saddam Hussein: A Political Biography* (New York: Free Press, 1991); Alan Bullock, *Hitler: A Study in Tyranny* (New York: Harper and Row, 1964).

8. Evidence suggests that Bush took the threat seriously in the early

days of the crisis. See Bob Woodward, *The Commanders* (New York: Simon and Schuster, 1991), 226–227, 247–253, ch. 19.

9. The photos were purchased by ABC but were never televised, according to Marie Gottschalk in "Operation Desert Cloud: The Media and the Gulf War," *World Policy Journal* 9/3 (1992), 460.

10. On the Halabja gassings see a report by Stephen Pelletiere, Douglas Johnson, and Leif Rosenberger of the U.S. Army War College: *Iraqi Power and U.S. Security in the Middle East* (Washington: GPO, 1990).

11. John Tierney, "Baffled Occupiers, or the Missed Understandings," *NYT* (Oct. 22, 2003), 4.

12. Elizabeth Drew, "Letter from Washington," *New Yorker* (Feb. 4, 1991), 82–86; Douglas Kellner, *The Persian Gulf TV War* (Boulder: Westview, 1992), 65–71. On western images of Arabs see Edward Said, *Orientalism* (London: Routledge and Kegan Paul, 1978); Douglass Little, *American Orientalism: The United States and the Middle East since 1945* (Chapel Hill: University of North Carolina Press, 1992), esp. ch. 1.

13. Between August 1, 1990, and February 28, 1991, references to Vietnam in the print media and on television far exceeded those to any other topic mentioned in connection with the Gulf crisis. Everett E. Dennis et al., *The Media at War: The Press and the Persian Gulf Conflict* (New York: Gannett Foundation Media Center, 1991), 42.

14. On Vietnam-inspired arguments against the buildup see Joshua Muravchik, "The End of the Vietnam Paradigm?" *Commentary* (May 1991), 17–23.

15. Yelena Melkumyan, "Soviet Policy and the Gulf Crisis," in Ibrahim Ibrahim, ed., *The Gulf Crisis: Background and Consequences* (Washington: Center for Contemporary Arab Studies, 1992), 76–91; Lawrence Freedman and Ephraim Karsh, *The Gulf Conflict, 1990–1991: Diplomacy and War in the New World Order* (Princeton: Princeton University Press, 1993), 76–80.

16. Hallion, *Storm over Iraq*, 177–188; Jeffrey Record, *Hollow Victory:*

A *Contrary View of the Gulf War* (Washington: Brassey, 1993), 66–68.

17. Jean Smith, *George Bush's War* (New York: Henry Holt, 1992), 77, 80–81, 95, 99.

18. For different perspectives on Iraqi-Kuwaiti relations, see H. Rahman, *The Making of the Gulf War: Origins of Kuwait's Long-Standing Territorial Dispute with Iraq* (Berkshire, Eng.: Ithaca Press, 1997); Walid Khalidi, "The Gulf Crisis: Origins and Consequences," *Journal of Palestine Studies* 20/2 (1991): 5–28.

19. For opposing views on the rationales see Christopher Layne, "Why the Gulf War Was Not in the National Interest," and Joseph S. Nye Jr., "Why the Gulf War Served the National Interest," *Atlantic Monthly* (July 1991), 54–81. My analysis here is based on reporting in the print media during the crisis, esp. *NYT*, the *Guardian* (London), the *Washington Report on Middle East Affairs, Middle East International* (London), as well as the 1996 PBS *Frontline* documentary on the war, written and produced by Eamonn Matthews.

20. Freedman and Karsh, *Gulf Conflict*, ch. 12; Adel Safty, "Dateline Iraq: Confrontation, War, and the Great Game of Balance of Power," *International Studies* 29/4 (1992), 430–432; Michael Tanzer, "Oil and the Gulf Crisis," in Phyllis Bennis and Michel Moushabeck, eds., *Beyond the Storm: A Gulf Crisis Reader* (New York: Olive Branch, 1991), 263–267.

21. See Majid Khadduri, *The Gulf War: The Origins and Implications of the Iraq-Iran Conflict* (New York: Oxford University Press, 1988).

22. On Iraq's WMD programs see Pelletiere, Johnson, and Rosenberger, *Iraqi Power*. On the impact of WMD charges on support for military action, see Steve Niva, "The Battle Is Joined," and Nasser Aruri, "Human Rights and the Gulf Crisis: The Verbal Strategy of George Bush," in Bennis and Moushabeck, *Beyond the Storm*, 63–64, 311–312.

23. Westmoreland, "Address at the Annual Luncheon of the Associa-

tion of the United States Army," Washington, Oct. 14, 1969, entered into *Congressional Record*, Oct. 16, 1969, 30348–30349.

24. On preparations to fight technowars, see hearings of the Committee on Armed Services, U.S. Senate, *Investigation into Electronic Battlefield Program*, 91st Cong., 2nd sess., Nov. 18, 19, and 24, 1970 (Washington: GPO, 1971), 3–39; on post-Vietnam projects to develop high-tech military systems, see Michael T. Klare, *War without End: American Planning for the Next Vietnams* (New York: Vintage, 1972), ch. 7.

25. "Regime Thought War Unlikely, Iraqis Tell U.S.," *NYT* (Feb. 12, 2004), 1, 16; Lawrence Freedman and Efraim Karsh, "How Kuwait Was Won: Strategy in the Gulf War," *International Security* 16/2 (1991), 6, 12, 15–19.

26. On the post-Vietnam transformations, see Hallion, *Storm over Iraq*, chs. 2–3; and Robert Coram, *Boyd: The Fighter Pilot Who Changed the Art of War* (Boston: Little, Brown, 2002). For the military hardware used in the 1991 war, see James F. Dunnigan and Austin Bay, *From Shield to Storm: High-Tech Weapons, Military Strategy, and Coalition Warfare in the Persian Gulf* (New York: Morrow, 1992).

27. The influential critiques included Harry G. Summers Jr., *On Strategy: A Critical Analysis of the Vietnam War* (Novato, CA: Presidio, 1982); William W. Momyer, *Air Power in Three Wars* (Washington: GPO, 1978); and U. S. Grant Sharp, *Strategy for Defeat: Vietnam in Retrospect* (San Rafael, CA: Presidio, 1978).

28. Hallion, *Storm over Iraq*, ch. 4; Andrew Bacevich, *American Empire: The Realities and Consequences of U.S. Diplomacy* (Cambridge, MA: Harvard University Press, 2002), 46–47, 164–166, 226–227.

29. Record, *Hollow Victory*, 57–60. When the air force chief of staff, General Michael Dugan, defied this cautious posture in September 1990, he was relieved of his command. See Woodward, *The Commanders*, ch. 20.

30. Woodward, *The Commanders*, 326–330.

31. Michael A. Palmer, *Guardians of the Gulf: A History of America's Expanding Role in the Persian Gulf, 1833–1992* (New York: Free Press, 1992), 164.

32. Record, *Hollow Victory*, ch. 4; Stephen Biddle, "Victory Misunderstood: What the Gulf War Tells Us about the Future of Conflict," *International Security* 21/2 (1996), 139–141, 154–173; Dunnigan and Bay, *From Shield to Storm*, 86–91; Michael J. Mazarr, Don M. Snider, and James A. Blackwell Jr., *Desert Storm: The Gulf War and What We Learned* (Boulder: Westview, 1993), 113–117.

33. One who did note it at the time was R. W. Apple Jr., who covered both wars. See "Hueys and Scuds: Vietnam and Gulf Are Wars Apart," *NYT* (Jan. 23, 1991), 1.

34. This account is largely based upon William J. Perry, "Desert Storm and Deterrence," *Foreign Affairs* 70/4 (1991), 68–73; Hallion, *Storm over Iraq*, ch. 6; Norman Friedman, *Desert Victory: The War for Kuwait* (Annapolis: Naval Institute Press, 1991), chs. 8–9; Mazarr, Snider, and Blackwell, *Desert Storm*, ch. 5; and David Halberstam, *War in a Time of Peace: Bush, Clinton, and the Generals* (New York: Scribner, 2001), ch. 5.

35. Mazarr, Snider, and Blackwell, *Desert Storm*, 93; Gregg Easterbrook, "'High Tech' Isn't Everything," *Newsweek* (Feb. 18, 1991), 49.

36. General Merrill McPeak quoted in Hallion, *Storm over Iraq*, 201.

37. Quoted ibid., 203.

38. On the "AirLand" phase of the war see Michael R. Gordon and Bernard E. Trainor, *The General's War* (Boston: Little, Brown, 1995), chs. 13–19; Mazarr, Snider, and Blackwell, *Desert Storm*, ch. 6; Freedman and Karsh, *The Gulf Conflict*, chs. 28–29; Peter Tsouras and Elmo C. Wright, "The Ground War," in Bruce W. Watson et al., *Military Lessons of the Gulf War* (London: Greenhill, 1991), 81–120.

39. These were the rationales stressed by Bush, Scowcroft, and the military at the time and subsequently invoked to rebut criticisms

of postwar diplomacy. See George Bush and Brent Scowcroft, *A World Transformed* (New York: Knopf, 1998), 488–492.

40. For Bush administration responses see the *Frontline* "War in the Gulf" documentary. On Schwarzkopf's lack of instructions see Bush and Scowcroft, *World Transformed*, 490.

41. A view Richard Hallion argues in *Storm over Iraq*. See also Record, *Hollow Victory*, 114; Palmer, *Guardians of the Gulf*, 204–205.

42. Douglas Kinnard, *The War Managers: American Generals Reflect on Vietnam* (New York: Da Capo, 1977).

43. See Daniel C. Hallin, *The "Uncensored War": The Media and Vietnam* (New York: Oxford University Press, 1986).

44. Jan Servaes, "Was Grenada a Testcase for the 'Disinformation War'?" *Media Development* 3 (1991): 41–44.

45. Swofford, *Jarhead: A Marine's Chronicle of the Gulf War and Other Battles* (New York: Scribner, 2003), 13–14. On media access to troops in Vietnam, see Hallin, *Uncensored War*, 6–9, 126–140.

46. Peter Hammar, "Broadcast Technology in the Midst of War," *Broadcast Engineering* (June 1991): 54–76; Gottschalk, "Operation Desert Cloud," 453–459, 472–473.

47. Hallion, *Storm over Iraq*, 221–223, 237, 247–248; Mazarr, *Desert Storm*, 150.

48. Gottschalk, "Operation Desert Cloud," 455, 465–473; Tom Engelhardt, *The End of Victory Culture* (New York: Basic Books, 1994), 292–295; Kellner, *Persian Gulf TV War*, 157–164, 176–182; Daniel C. Hallin, "Images of Vietnam and the Persian Gulf in U.S. Television," in Susan Jeffords and Lauren Rabinovitz, eds., *Seeing through the Media: The Persian Gulf War* (New Brunswick: Rutgers University Press, 1994), 54–55.

49. Quoted in Kellner, *Persian Gulf TV War*, 159.

50. See, e.g., "SLAMs Hit Iraqi Target in First Combat Firing," *Aviation Week and Space Technology* (Jan. 28, 1991); "Night Vision Systems Yield Payoff in Persian Gulf War," ibid. (Feb. 3, 1992); William J. Broad, "As Antimissile Era Dawns, Planners Eye Panoply of Weapons," *NYT* (Feb. 5, 1991); "The War: Deadly Science,"

Newsweek (Feb. 18, 1991); Malcolm W. Browne, "Invention That Shaped the Gulf War: The Laser-Guided Bomb," *NYT* (Feb. 26, 1991).

51. H. Bruce Franklin, "From Realism to Virtual Reality: Images of America's Wars," in Jeffords and Rabinovitz, *Seeing through the Media*, 41.

52. Michelle Kendrick, "Kicking the Vietnam Syndrome: CNN's and CBS's Video Narratives of the Persian Gulf War," in Jeffords and Rabinovitz, *Seeing through the Media*, 69.

53. Three decades later Hubert Van Es, the photographer, revealed that the building was not the embassy. See "Thirty Years at 300 Millimeters," *NYT* (Apr. 29, 2005), 25.

54. Kendrick, "Video Narratives," 71–73; Hallin, "Images of Vietnam and the Persian Gulf," 53–56 (quotation).

55. *Operation Desert Shield* (New York: Harris, 1991). Quoted caption, 33. My thanks to the students in my 2004 undergraduate history seminar on American interventions overseas for drawing my attention to the Desert Storm trading cards, and especially Orin Puniello for lending me his collection.

56. "Our Women in the Desert," *Newsweek* (Sept. 10, 1990), 22–25. Dana L. Cloud, "Operation Desert Comfort," in Jeffords and Rabinovitz, *Seeing through the Media*, 155–170; Lauren Rabinovitz, "Soap Opera Woes: Genre, Gender, and The Persian Gulf War," ibid., 196–201; Kellner, *Persian Gulf TV War*, 74–76.

57. Venise T. Berry and Kim E. Karloff, "Perspectives on the Persian Gulf War in Popular Black Magazines," in Jeffords and Rabinovitz, *Seeing through the Media*, 257–259; Kellner, *Persian Gulf TV War*, 73–74.

58. See Paul Mann, "Mammoth Air/Ground Assault Defeats Iraq in Gulf War," *ASWT* (Mar. 4, 1991), 22; Swofford, *Jarhead*, 220–228.

59. Quotations from John Balzar of the *Los Angeles Times*, who viewed the videos at briefings for air force officers, cited in Engelhardt, "War as Total Television," 92–93. On similar censored tapes of Iraqi pilots, see Tom Matthews and Douglas

Waller, "The Secret History of the War," *Newsweek* (Mar. 18, 1991), 32; on the "turkey shoot," see Hallion, *Storm over Iraq*, 234–236; and Michael Kelly, *Martyrs' Day: Chronicle of a Small War* (New York: Vintage, 1994), 227–232.

60. Gottschalk, "Operation Desert Cloud, 463–464, 480; Kellner, *Persian Gulf TV War*, 375, 404; Mazarr, Snider, and Blackwell, *Desert Storm*, 156–157.

61. Kelly, *Martyrs' Day*, 175.

62. Michael Sherry, *The Rise of American Air Power: The Creation of Armageddon* (New Haven: Yale University Press, 1987), 51–58, 144, 226–229.

63. Kellner, *Persian Gulf TV War*, 397–210; Hallion, *Storm over Iraq*, 198–200.

64. On bombing damage see "Defeat of Iraq Sparks Debate on Which Air Role Was Crucial," *ASWT* 136/4 (1992), 62–63; Record, *Hollow Victory*, 109–115; Gottschalk, "Operation Desert Cloud," 452–453, 462–465; Kellner, *Persian Gulf TV War*, 161, ch. 7; for lower estimates of destruction, Hallion, *Storm over Iraq*, 198–200. On the embargo see Geoff Simons, *Targeting Iraq: Sanctions and Bombing in US Policy* (London: Saqi, 2002); David Rieff, "Were the Sanctions Right?" *NYT Magazine* (July 27, 2003), 41–46.

65. Farouk-Sluggett and Sluggett, *Iraq since 1958*, ch. 7, and "Iraq since 1986: The Strengthening of Saddam," *Middle East Report* (Nov.–Dec. 1990), 19–24; Stork, "Iraq and the War in the Gulf," 9–17; Adeed Dawisha, "Iraq: The West's Opportunity," *Foreign Policy* (1980–1981), 142–144.

66. Marion Farouk-Sluggett, "Liberation or Repression? Pan-Arab Nationalism and the Women's Movement in Iraq," in Derek Hopwood, Habib Ishow, and Thomas Koszinowski, eds., *Iraq: Power and Society* (Oxford: St. Antony's College, 1993), 64–73; Raja Habib Khuzai and Songul Chapouk, "Iraq's Hidden Treasure," *NYT* (Dec. 3, 2003), 31; Cynthia Enloe, "The Gendered Gulf," in Jeffords and Rabinovitz, *Seeing through the Media*, 215–218.

67. For modernizers' rationalizations for supporting dictatorial regimes, see Nils Gilman, *Mandarins of the Future: Modernization Theory in Cold War America* (Baltimore: Johns Hopkins University Press, 2003), 185–190.

68. Richard Sale, "Saddam Key in Early CIA Plot," UPI Release, Apr. 20, 2003; Roger Morris, "A Tyrant 40 Years in the Making," *NYT* (Mar. 14, 2003), 29; Bruce W. Jentleson, *With Friends like These: Reagan, Bush, and Saddam Hussein, 1982–1990* (New York: Norton, 1994), chs. 1–2; Douglas Little, "Mission Impossible: The CIA and the Cult of Covert Action in the Middle East," *Diplomatic History* 28/5 (2004), 696–697.

69. This characterization is deployed to good effect by Nicholas Kristof and Sheryl WuDunn in *China Wakes: Struggle for the Soul of a Rising Power* (New York: Times Books, 1994). For a critical analysis of the main contours and effects of the transmutation, see Richard Smith, "The Chinese Road to Capitalism," *New Left Review* 199 (1993): 55–99.

70. Nick Eberstadt, "The Perversion of Foreign Aid," *Commentary* (June 1985), 19, 21, 23–25, 27–28; Richard Bissell, "After Foreign Aid What?" *Washington Quarterly* (Summer 1991), 23, 26–27, 32; Ann Crittenden, "Foreign Aid," *NYT Magazine* (June 6, 1982), 67; American Foreign Policy Council, *Modernizing Foreign Assistance* (Westport, CT: Praeger, 1992), 17–18.

71. "The Good News," *NYT* (Nov. 28, 2003), 43.

72. Crittenden, "Aid under Fire," 66–76; Eberstadt, "Perversion of Aid," 24–26. On the Alliance for Progress and the Peace Corps, see Brent Ashabranner, *A Moment in History: The First Ten Years of the Peace Corps* (Garden City, NY: Doubleday, 1971); Michael Latham, *Modernization as Ideology: American Social Science and "Nation Building" in the Kennedy Era* (Chapel Hill: University of North Carolina Press, 2000), 99–101. David M. Barrett, ed., *Lyndon Johnson's Vietnam Papers: A Documentary Collection* (College Station: Texas A&M University Press, 1997), 8.

73. The key studies, published by the New American Library (New

York) were *The Limits to Growth* (1972); *Mankind at the Turning Point* (1974); and *Rio: Reshaping the International Order* (1977).

74. See Andreas De Rhoda's interview with Kahn in the *Christian Science Monitor* (July 28, 1976), 14–15; and Kahn, *The Next 200 Years: A Scenario for America and the World* (New York: Morrow, 1976).

75. Heilbroner, *Inquiry into the Human Prospect* (New York: Harper, 1974), 50–52.

76. Eberstadt, "Perversion of Aid," 24–27.

77. Deborah Shapley, *Promise and Power: The Life and Times of Robert McNamara* (Boston: Little, Brown, 1993), 480–481, 502–524; Eberstadt, "Perversion of Aid," 22–27; Elliott R. and Victoria A. Morss, *Foreign Aid: An Assessment of New and Traditional Strategies* (Boulder: Westview, 1982), 26–27.

78. Robert P. Morgan, *Science and Technology for International Development* (Boulder: Westview, 1984), 17, 22–23, 82–98, 120–121; Committee on International Relations, *Proposal for a Program in Appropriate Technology* (Washington: GPO, 1977); Edward C. Wolf, *Beyond the Green Revolution: New Approaches for Third World Agriculture* (Washington: Worldwatch Institute, 1986); Morss, *Foreign Aid*, 27, 32–37, 96–103; Eugene B. Skolnikoff, *Science, Technology, and American Foreign Policy* (Cambridge, MA: MIT Press, 1967), 196–199; Shapley, *Promise and Power*, 531–537; Crittenden, "Aid under Fire," 76; and for the persistence of these trends in India, Saritha Rai, "Tiny Loans Have Big Impact on Poor," *NYT* (Apr. 12, 2004), C3. For a developing country's perspective see "Philippine Strategy for Sustainable Development," *Philippine Development* 18/4 (1991): 15–25.

79. Amartya Sen, "Population: Delusion and Reality," *New York Review of Books* (Sept. 22, 1994), 62–71; Rekha Mehra, "Women, Empowerment, and Economic Development," *Annals of the American Academy of Political and Social Science* 554 (1997): 136–149; Clair Apodaca, "The Effects of Foreign Aid on Women's Attainment of Their Economic and Social Human Rights," *Journal*

of *Third World Studies* 18/2 (2000), 205–206; Caroline O. Moser, "Gender Planning in the Third World: Meeting Practical and Strategic Gender Needs," in Tina Wallace and Candida March, eds., *Changing Perceptions: Writings on Gender and Development* (Oxford: Oxfam, 1991), 158–171; Ramachandra Guha, *The Unquiet Woods: Ecological Change and Peasant Resistance in the Himalaya* (Delhi: Oxford University Press, 1989), ch. 7; Paul Harrison, *The Third World Tomorrow* (London: Penguin, 1980), ch. 16.

80. *Sub-Saharan Africa: From Crisis to Sustainable Growth* (Washington: World Bank, 1989), 3.

81. Shapley, *Promise and Power*, 514–519; 538–553; Eberstadt, "Perversion of Aid," 28–31; Morss, *U.S. Foreign Aid*, 103–106; Morgan, *Science and Technology*, 18, 143.

82. Morgan, *Science and Technology*, 81–82; J. S. Sarma, *Growth and Equity: Policies and Implementation in Indian Agriculture* (Washington: International Food Policy Research Institute, 1981), Research Report no. 28.

83. United Nations Development Program, *Human Development Report 1996* (New York: Oxford University Press, 1996); Apodaca, "Effects of Foreign Aid," 207–211, 215–216; Mehra, "Women and Development," 139; Emma Zapata Martelo, "Modernization, Adjustment, and Peasant Production: A Gender Analysis," *Latin American Perspectives* 23/1 (1996): 118–130; Diane Elson, "Structural Adjustment: Its Effect on Women," Marivi Arregui and Clara Baez, "Free Trade Zones and Women Workers," and Peggy Antrobus, "Women in Development," in Wallace and March, eds., *Changing Perceptions*; Celia Dugger, "World Bank Challenged: Are the Poor Really Helped?" *NYT* (July 28, 2004), 4.

Epilogue: The Paradox of Technological Supremacy

1. Well-documented estimates are surprisingly difficult to locate. I have derived these from statistics in a number of books on each

war as well as in Noble Frankland, *Encyclopedia of 20th Century Warfare* (New York: Orion, 1989); Thomas S. Arms, *Encyclopedia of the Cold War* (New York: Facts on File, 1994); Patrick Brogar, *World Conflicts* (Lanham, MD: Scarecrow Press, 1998); and Stanley L. Kutler, ed., *The Encyclopedia of the Vietnam War* (New York: Scribner, 1996).

2. James Atlas, "Among the Lost: Illusions of Immortality," *NYT* (Oct. 7, 2001), 5.

3. Council of Economic Advisors, *Economic Report of the President* (Washington, 2000), 346, 419; John M. Levy, *Urban America: Processes and Problems* (Saddle River, NJ: Prentice-Hall, 2000), 30; Kenneth Jackson, *Crabgrass Frontier: The Suburbanization of the United States* (New York: Oxford University Press, 1985). On the development of systems to provide such amenities, see William Cronon, *Nature's Metropolis: Chicago and the Great West* (New York: Norton, 1991).

4. On the history and design of the trade center, I have relied primarily on James Glanz and Eric Lipton, "The Height of Ambition," *NYT Magazine* (Sept. 8, 2002), and their *City in the Sky: The Rise and Fall of the World Trade Center* (New York: Times Books, 2003); Angus Gillespie, *Twin Towers: The Life of New York City's World Trade Center* (New Brunswick, NJ: Rutgers University Press, 1999); and Eric Darton, *Divided We Stand: A Biography of New York's World Trade Center* (New York: Basic Books, 1999).

5. Michael Lewis, "The 'Look at Me' Strut of a Swagger Building," *NYT* (Jan. 6, 2002), 5.

6. Prior to the construction of the twin towers, the maximum percentage of floor space left for rentals was 62 percent. Gillespie, *Twin Towers*, 78.

7. See the PBS documentary "Why the Towers Fell," produced for *Nova* (Boston: WGBH Educational Foundation, 2002); Richard Bernstein, *Out of the Blue: The Story of September 11, 2001, from Jihad to Ground Zero* (New York: Times Books, 2002); William

Langewiesche, *American Ground: Unbuilding the World Trade Center* (New York: North Point, 2002). Later investigations have called into question some of the conclusions in these early analyses, but definitive answers have yet to emerge; see "Study Suggests Design Flaws Didn't Doom Towers," *NYT* (Oct. 20, 2004), 1 and B3.

8. Paul Boyer, *By the Bomb's Early Light: American Thought and Culture at the Dawn of the Atomic Age* (Chapel Hill: University of North Carolina Press, 1985), chs. 25–26.

9. To date the fullest information we have on intelligence lapses linked to the attacks was produced by the National Commission on Terrorist Attacks Upon the United States and published in the commission's *9/11 Report* (Washington: GPO, 2004).

10. A misconception that was apparently shared by Arabs who had internalized western stereotypes. See As'ad Abu Khalil, *Bin Laden, Islam and America's New "War on Terrorism"* (New York: Seven Stories, 2002), 78.

11. Richard E. Bissell, "After Foreign Aid—What?" *Washington Quarterly* (Summer 1991), 24; American Foreign Policy Council, *Modernizing Foreign Assistance* (Westport, CT: Praeger, 1992), 63–66.

12. See, e.g., Lucien Pye, "Political Science and the Crisis of Authoritarianism," *American Political Science Review* 84/1 (1990), 7.

13. Bush's address, on this and other key issues, followed the guidelines previewed in Condoleezza Rice, "Promoting the National Interest," *Foreign Affairs* 79/1 (2000): 45–62.

14. David Ekbladh, "Globalisation and the Language of Modernisation," *World Affairs* 6/2 (2002): 52–64.

15. For a globalist's recognition of the need for social safety nets, see Thomas Friedman, *The Lexus and the Olive Tree* (New York: Archon, 2002). On the unequal distribution of the benefits of globalization in the 1990s, see, e.g., Amy Chua, *World on Fire: How Exporting Free Market Democracy Breeds Ethnic Hatred and Global Instability* (New York: Anchor, 2003); Belinda Coote, *The*

Trade Trap (Oxford: Oxfam, 1992); Chakravarthi Raghavan, *Recolonization: GATT, the Uruguay Round and the Third World* (Penang: Third World Network, 1990); Cornelia Aldana, *Contract for Underdevelopment: Subcontracting for Multinationals in the Philippine Semiconductor and Garment Industries* (Manila: IBON Databank, 1989).

16. See David Halberstam, *War in a Time of Peace: Bush, Clinton, and the Generals* (New York: Scribner, 2001).

17. According to Fareed Zakaria the U.S. military budget is greater than those of the next fifteen countries combined. See "Our Way: The Challenge for the World's Only Superpower," *New Yorker* (Oct. 14 and 21, 2002), 74.

18. See, e.g., bin Laden's proclamations "Jihad against Jews and Crusaders" (Feb. 23, 1998) and "The Sword Fell" (Oct. 7, 2001) as well as his State Department Profile for 1998, in John Prados, ed., *America Confronts Terrorism: A Documentary Record* (Chicago: Ivan Dee, 2002), 12–13, 173, 176–178.

19. Mark Bowden, *Black Hawk Down: A Story of Modern War* (New York: Penguin, 1999), 20–30, 90–97; Seymour M. Hersh, "The Missiles of August," *New Yorker* (Oct. 12, 1998), 34–41.

20. Bernstein, *Out of the Blue*, 119–124, 132.

21. Harries, "Madeleine Albright's 'Munich Mindset,'" *NYT* (Dec. 19, 1996). See also Eric Schmitt, "Gulf Syndrome: Americans Decide War May Not Be Quite So Scary," *NYT* (Nov. 30, 1997), Week in Review, 1, 5.

22. Bowden, *Black Hawk Down*; Dexter Filkins, "Flaws in U.S. Air War Left Hundreds of Civilians Dead," *NYT* (July 21, 2002), 1, 10; Peter L. Bergen, *Holy War, Inc.: Inside the Secret World of Osama bin Laden* (New York: Free Press, 2001), 21–22, 26, 95, 99.

23. Matthew Brzezinski, "Autopilot: Can the Next War Be Fought with No Soldiers at All?" *NYT Magazine* (Apr. 20, 2003), 80.

24. Ibid.

25. Berger, *Holy War, Inc.*, ch. 10; Scott Atran, "The Upgraded Networks of Global Terrorism," *International Herald Tribune* (Mar. 17, 2004), 7.

26. Prados, *America Confronts Terrorism*, 8.

27. Gregg Easterbook, "American Power Moves beyond the Mere Super," *NYT* (Apr. 27, 2003), sec. 4, 1; Bill Keller, "The Thinkable," *NYT Magazine* (May 4, 2003), 51–53, 101–102.

28. For two of the more influential expressions of these views, see Samuel P. Huntington, "The Clash of Civilizations," *Foreign Affairs* 72 (1993): 22–49 (and the subsequent book with the same title); Bernard Lewis, "The Roots of Muslim Rage," *Atlantic Monthly* (Sept. 1990), 47–60.

29. See Marshall Hodgson, *The Venture of Islam* (Chicago: University of Chicago Press, 1974), 1:305–308; William C. Atkinson, *A History of Spain and Portugal* (Harmondsworth: Penguin, 1963), 56–61; Albert Hourani, *A History of the Arab Peoples* (Cambridge, MA: Harvard University Press, 1991), 42–43, 118–119, 235–238; Marion Woolfson, *Prophets in Babylon: Jews in the Arab World* (London: Faber and Faber, 1980).

30. See, e.g., Tariq Ali, *The Clash of Fundamentalisms: Crusades, Jihads and Modernity* (London: Verso, 2002).

31. Halberstam, *War in a Time of Peace*, 468–480, 485–487; R. W. Apple Jr., "On Killing from beyond Harm's Way," *NYT* (Apr. 18, 1999), 4; Elizabeth Becker, "Military Leaders Tell Congress of NATO Errors in Kosovo," *NYT* (Oct. 15, 1999), 8; Michael Hirsh, "At War with Ourselves: In Kosovo, America Confronts Its Own Ideals," *Harper's* (July 1999), 60–70. As Hirsh points out, Clinton's reluctance to deploy ground troops was bolstered by NATO's opposition to such a commitment.

32. John W. Dower, "Occupation: A Warning from History," in Lloyd C. Gardner and Marilyn B. Young, eds., *The New American Empire* (New York: New Press, 2005), 182–197; Dower, *Embracing Defeat: Japan in the Wake of World War II* (New York: Norton, 1999).

33. James Dao, "The New Air War: Fewer Pilots, More Hits and Scarcer Targets," *NYT* (Nov. 29, 2001), B1, B4.

34. Christian Parenti, *The Freedom: Shadows and Hallucinations in Occupied Iraq* (New York: New Press, 2004); Sarah Chayes, "Af-

ghanistan's Future, Lost in the Shuffle," *NYT* (July 1, 2003); Jay Bookman, "U.S. High Horse Now Riderless," *Atlanta Journal Constitution* (July 17, 2003); Amy Waldman and Dexter Filkins, "2 Fronts: Quick Wars, but Bloody Peace," *NYT* (Sept. 19, 2003); Steven R. Weisman, "Holding Iraq Together, by Tending to Its Parts," *NYT* (May 30, 2004), Week in Review, 5.

35. Jeff Madrick, "The Iraqi Time Bomb," *NYT Magazine* (Apr. 6, 2003), 48–51.

36. Chas W. Freeman Jr., "Even a Superpower Needs Help," *NYT* (Feb. 26, 2003), 25; Michael Ignatieff, "The Burden," *NYT Magazine* (Jan. 2, 2003), 50–51.

37. A neglect most apparent in Paul Kennedy's *The Rise and Fall of the Great Powers* (New York: Vintage, 1987).

38. Adam Clymer, "In the Fight for Privacy, States Set Off Sparks," *NYT* (July 6, 2003), sec. 4, 1, 3. On the economic and human rights costs of homeland security, see Matthew Brzezinski, "Fortress America: Is This Where You Want to Live?" *NYT Magazine* (Feb. 23, 2003).

ACKNOWLEDGMENTS

When this book became the main focus of my research and writing in the late 1990s, I reckoned that I could complete it in three or four years. After all, for decades I had read widely on American imperialism and foreign relations, extensive work in foreign languages would not be required, and most of the primary sources I would need were available in the Boston–Washington corridor. But as I plunged more deeply into the voluminous secondary literature on the colonization of North America and U.S. expansion across the Pacific, the magnitude of my task became clear. Its completion owes much to the high quality of the history written on the diverse case examples that I explore, which cover nearly four centuries of American and global history. My access to these works, and to the primary documents I have consulted, was made possible by the superb assistance of the staffs of the Library of Congress and the National Archives, the British Library, the New York Public Library, and the libraries of the University of California at Berkeley, Princeton University, the University of Pennsylvania, Cornell University, and Rutgers University. Funding in the early stages of the project was provided by the John Simon Guggenheim Foundation, and I am especially grateful for the leave and research support I received from Rutgers University.

Over the years when I was working through the case studies and formulating my arguments, presentations in both academic and public venues gave me wonderful opportunities to test my ideas and learn from the expertise of specialists working in disparate time periods and subfields. My thinking has been shaped by the questions and suggestions of colleagues, both faculty and graduate students, in research seminars at the University of California at Berkeley, Cornell University, the University of Delaware, the University of Illinois, Johns Hopkins University, Leiden University, Northwestern University, Pennsylvania State University, the University of Pennsylvania, Rutgers University, Seikei University (Tokyo), Stanford University, and the Madison and Milwaukee branches of the University of Wisconsin. The input of participants in NEH summer seminars at the University of California at Santa Cruz and Pennsylvania State University and workshops at Leiden University, Yale University, and the Massachusetts Institute of Technology shaped my thinking in important ways. I also benefited from audience responses to public lectures delivered at the University of Florida, Hampshire College, McMaster University, the University of Toronto, the University of Virginia, Whitman College, and the World Conference of Historians (Oslo). These sessions often led to fruitful one-on-one exchanges. Though the list of colleagues involved in these is long, I would especially like to thank Rosalind Williams, Laura Hein, Fumiko Nishizaki, Peter Fritzsche, Paul Sutter, Ron Kline, Dan Beaver, John Weaver, Lindy Biggs, Peter Kolchin, Arwan Mohun, David Schmitz, David Ludden, Lynn Lees, Anders Stephanson, Sharon Kingsland, and Wim van den Doel. And I owe a special debt to Ram Guha, Jim Scott, and Phil Curtin for the inspiration and understandings they have contributed both to *Dominance by Design* and to so much of my previous scholarship.

Without the research assistance of a succession of able graduate students at Rutgers—Steve Adams, Karen Balcolm, Richard Keller, Brady Brower, Kevin Deuschle, Ray Ashare, Margaret Smith, Lindsay Braun, Carmen Khair, Louisa Rice, and above all Steve Jankiewicz—this book might never have been completed. A number of colleagues read substantial parts of various versions of the manuscript and offered useful suggestions for revision. I am particularly grateful in this regard to Ram Guha, Matt Guterl, Michael Hunt, and Nick Cullather. I have also gained a good deal from expert readings of specific chapters by Paul Clemens, Gary Darden, Jasper van den Kerk, Robert Shaffer, David Ekbladh, Michael Mahoney, David Fogelson, Fumiko Nishizaki, Justin Hart, Richard Keller, Denise Quirk, Rudy Bell, Bonnie Smith, Mark Wasserman, John Chambers, David Engerman, Kirstin Hoganson, Antoinette Burton, Liam O'Brien, Jane Landes, Greg Roeber, Jack Levy, Ed Rhodes, David Wilson, Christine Skwiot, Zachary Matusheski, and Carl Langbert. As he has for decades, Lloyd Gardner shared his understandings regarding the workings of American diplomatic history, and Bill Mooney taught me a great deal about how to tell a story. Jamal Farhat applied his photographic skills to the images that I hope capture the mood and content of the chapters. My wife, Jane, and my daughter, Claire, contributed both perceptive responses and stylistic suggestions, and my son, Joel, provided insights regarding the production and reception of frontier art. The remarkable editorial skills of Camille Smith at Harvard University Press proved invaluable to whatever success I have had in working analysis and argument into an accessible and compelling narrative. Above all, I am indebted to Joyce Seltzer, who convinced me this was the book I needed to write, and who has been a vital source of encouragement and intellectual stimulation. Working with her over the past decade has been both challenging and rewarding.

INDEX

Advertising agencies, 110
Afghanistan, 399, 402; Soviet war in, 354, 396, 404; U.S. war in, 28, 403, 405, 406–407, 410, 411, 412–413
Africa, 17, 46, 151, 154, 270; Africans viewed as savages, 56; agriculture in, 210; China as development model for, 259–260; conflicts in, 391; decolonization movements in, 25; European colonialism in, 165; First World War and, 198; missionaries in, 211; as most impoverished continent, 378–379; as "peripheral" region, 304; resistance to colonial rule in, 197; social breakdown and development, 276; "Third World" and, 241; U.S. intervention in, 399
African Americans, 21–22, 140, 144, 157; guards in Perry embassy, 3, 20–21; in Gulf War, 367–368
Agent orange, 321
Agriculture: Amerindians and, 11, 44–45, 49, 51–53, 71, 90, 103; in China, 106; developmentalist ideologies and, 250; energy needs and, 209; in English colonies, 36, 37, 44; English farming techniques, 56, 61; environmental toll of, 86–87; farm machinery, 5, 21, 68, 78, 87; Green Revolution and, 266–270; mechanization of, 7, 110–111; migratory cultures and, 49; overseas expansion and, 106, 108; percentage of U.S. population employed in, 388; in Philippines, 170, 174–175; policy toward peasants, 257–259; productivity of, 79, 253; railroads and, 7; Second World War and, 222; shift from subsistence to market production, 276; in Soviet Union, 232; western frontier expansion and, 78–79
Aguinaldo, Emilio, 128, 137, 180
Aidid, Mohammad Farrah, 401